Magnetic Polymer Composites and Their Emerging Applications

Magnetic composite particles offer much potential for use in a variety of applications, including manufacturing, environmental protection, microfluidics, microelectronics, and biomedicine. *Magnetic Polymer Composites and Their Emerging Applications* explores leading research on the fabrication, characterization, properties, and all reported applications of magnetic polymer composites.

Features:

- Discusses synthesis, properties, and modern fabrication technologies of magnetic polymer composites
- Describes the biocompatibility, suitability, and toxic effects of these materials
- Covers a variety of applications including those in biomedicine, wastewater treatment, soft robotics, 3D/4D printing, and agriculture
- Details opportunities and future directions in magnetic polymer composites and their surface decorations

This unique book serves as a road map for materials engineers, as well as researchers, academics, technologists, and students working in sensor technology.

Magnetic Polymer Composites and Their Emerging Applications

Edited by
Sayan Ganguly, Shlomo Margel, and Poushali Das

CRC Press
Taylor & Francis Group
Boca Raton London New York

CRC Press is an imprint of the
Taylor & Francis Group, an **informa** business

Designed cover image: © Shutterstock, Jurik Peter

First edition published 2025
by CRC Press
2385 NW Executive Center Drive, Suite 320, Boca Raton FL 33431

and by CRC Press
4 Park Square, Milton Park, Abingdon, Oxon, OX14 4RN

CRC Press is an imprint of Taylor & Francis Group, LLC

ISBN: 978-1-032-59331-9 (hbk)
ISBN: 978-1-032-59332-6 (pbk)
ISBN: 978-1-003-45423-6 (ebk)

DOI: 10.1201/9781003454236

Typeset in Times
by KnowledgeWorks Global Ltd.

Contents

Preface

The field of materials science and engineering is always changing, so making and using new materials is now essential for technological growth. This book explores the interesting world of magnetic polymer composites, explaining how they are made, what their properties are, and the wide range of uses that could change many industries. As editors and contributors, we're starting this trip with the goal of making a complete guide that goes into the methods for making magnetic polymer composites, how they are characterized, and their special properties. These materials, which have the good qualities of both polymers and magnetic particles, have gotten a lot of attention because they can be changed, are light, and can be used for more than one thing.

The first part of the book talks about the basic ideas behind magnetic polymer composites. It also talks about different ways to make them, from old-fashioned methods to newer, more advanced ones like in situ polymerization and electrospinning. It is important to understand the details of these synthesis methods so that the mixtures can be made to fit specific needs.

The sections that follow go into more detail about the different ways that magnetic polymer composites can be used in different areas. From biomedical uses like drug transport and hyperthermia therapy to environmental uses like treating wastewater and sensing, we show how flexible these materials are and how they could help solve important problems.

The book also talks about the use of magnetic polymer composites in technology, sensing technologies, and getting energy from the environment. New uses in these areas show how these materials are changing the way next-generation electronics are made.

In each chapter, experts in the field share their ideas, study results, and points of view. This book is a great resource for researchers, engineers, and students of all levels, from those who are just starting out to those who have been doing this for a long time. Because magnetic polymer composites are multi-disciplinary, researchers from many fields need to work together on them. We hope that this book will help to bring these researchers together.

As we start to look into magnetic polymer composites and the new uses they can have, we want to thank the people who have shared their knowledge and experiences. This book is possible because of their hard work and determination to move the field forward.

Magnetic Polymer Composites and Their Emerging Applications is meant to be a useful guide that encourages researchers and engineers to learn more and help materials science and engineering continue to grow.

About the Editors

Dr. Sayan Ganguly is a researcher at University of Waterloo, Canada. Before that he pursued his postdoctoral research at Bar-Ilan University, Israel. He obtained his Ph.D. from the Indian Institute of Technology, Kharagpur. He obtained his B.Sc degree in chemistry (Honours) at Ramakrishna Mission Vidyamandira, Belur Math, University of Calcutta and then his post-graduation B.Tech from University of Calcutta. He then went on to obtain his M.Tech from the University of Calcutta in polymer science and technology. His primary research interests include hydrogels, polymer nanocomposites, EMI shielding, magnetic nanoparticles, 3D printing, and conducting polymer composites. He has published more than 80 papers and chapters in international journals and books. He is currently editing books with CRC Press.

Prof. Shlomo Margel is a professor at Bar-Ilan Institute for Nanotechnology and Advanced Materials (BINA), Department of Chemistry, Bar-Ilan University, Israel. He received his Ph.D. from the Department of Materials Science at the Weizmann Institute, Israel. He completed his postdoctoral studies at the California Institute of Technology (Caltech), Department of Inorganic Chemistry, in 1977 and then served two years as a senior scientist at the Jet Propulsion Laboratory at Caltech. He was a visiting scientist in 1986–87 at DuPont, Central Research and Development; in 1992 at the Polymer Section, University of Ulm, Germany; in 1997 at the Department of Physical Electronics, Tokyo Institute of Technology, Japan; and in 2005 at the Institute for Soldiers' Nanotechnologies, MIT, Cambridge, Massachusetts. He joined the Bar-Ilan Chemistry Department in 1987. Prof. Margel was also the head of the National Committee for Chemistry in High School Education and the president of the Israel Chemical Society between 2006 and 2009. Recently Prof. Margel was nominated by the Israel Academy of Sciences to serve as chair of the National Committee of Chemistry toward International Union of Pure and Applied Chemistry (IUPAC). Prof. Margel is a polymer chemist whose main interest lies in the fields of polymers, biopolymers, colloidal chemistry, surface chemistry, and biotechnology. He has published some 350 publications, has been awarded 45+ patents, and is the author of a few chapters in several books.

Dr. Poushali Das is working as a senior postdoctoral research scientist at McMaster University, Faculty of Engineering, Canada. Prior to this, she held a post-doctoral position in the Department of Chemistry, Bar-Ilan University, Israel. She completed her Ph.D. from the Indian Institute of Technology, Kharagpur, India. She has received prestigious awards including the DST INSPIRE Scholar Award (Government of India), Horizon Europe Marie Skłodowska-Curie Award (European Commission), and H. G. Thode Postdoctoral Award (Canada). She has been serving as a topic editorial board member, guest editor, and reviewer of reputed journals and as an international consultant. She is currently editing books with CRC Press. Her research interests include multi-functional luminescent quantum dots and applications in sensors and the biomedical field, polymer/quantum dot nanocomposites, MXene/polymer nanocomposites, and sonochemical synthesis of nanocomposites.

Contributors

Raquel G. D. Andrade
Physics Centre of Minho and Porto
    Universities and LaPMET Associate
    Laboratory
University of Minho, Campus de
    Gualtar
Braga, Portugal

Venkata Badarinath
Department of Pharmaceutics
Santhiram College of Pharmacy
Andhra Pradesh, India

P. Balaji
Department of Pharmacology
School of Pharmaceutical Sciences
Vels Institute of Science, Technology
    and Advanced Studies
Pallavaram, Chennai, India

Nadia Bhatti
Department of Chemistry
University of Sahiwal
Sahiwal, Pakistan

Prajitha Biju
Department of Pharmaceutics
Yenepoya Pharmacy College &
    Research Centre
Yenepoya (Deemed to be University),
    Naringana
Mangalore, Karnataka, India

Elisabete M. S. Castanheira
Physics Centre of Minho and Porto
    Universities and LaPMET Associate
    Laboratory
University of Minho, Campus de
    Gualtar
Braga, Portugal

Ipsita Chinya
Department of Chemical Engineering
Porbandar, Gujrat, India

Sreeja Nath Chowdhury
Rubber Technology Center
Indian Institute of Technology
Kharagpur, India

Ronald Darwin C.
Department of Pharmacology
School of Pharmaceutical Sciences
Vels Institute of Science, Technology
    and Advanced Studies
Pallavaram, Chennai, India

H. N. Deepakumari
Department of Chemistry
Bharathi College
Bharathinagara, Maddur, Karnataka, India

Mohamed Zerein Fathima
Department of Pharmaceutical
    Chemistry and Analysis
School of Pharmaceutical Sciences
Vels Institute of Science, Technology &
    Advanced Studies
Pallavaram, Chennai, India

Sridevi Ganesan
Department of Pharmaceutics
School of Pharmaceutical Sciences
Vels Institute of Science, Technology
    and Advanced Studies
Pallavaram, Chennai, India

Sayan Ganguly
Department of Chemistry
University of Waterloo
Waterloo, Ontario, Canada

Bina Gidwani
Columbia Institute of Pharmacy
Raipur, Chhattisgarh, India

Christian Chapa González
Grupo de Investigación Nanomedicina
  UACJ
Instituto de Ingeniería y Tecnología
Laboratorio de Integración de Datos y
  Evidencia en Revisiones de Salud y
  Ciencia
Universidad Autónoma de Ciudad Juárez
Ciudad Juárez, México

Sanjay Kumar Gupta
Rungta College of Pharmaceutical
  Sciences and Research
Bhilai, Durg, Chhattisgarh, India

Javeed Iqbal
Department of Chemistry
University of Sahiwal
Sahiwal, Pakistan

S. Jeganath
Department of Pharmaceutics
School of Pharmaceutical Sciences
Vels Institute of Science, Technology
  and Advanced Studies
Pallavaram, Chennai, India

Karthickeyan Krishnan
School of Pharmaceutical Sciences
Vels Institute of Science, Technology
  and Advanced Studies
Pallavaram, Chennai, India

Rahul Kumar
Department of Pharmaceutical
  Chemistry, Columbia Institute of
  Pharmacy, Raipur, India

Rasapelly Ramesh Kumar
Department of Pharmaceutical
  Chemistry
Marri Laxman Reddy Institute of Pharmacy
Dundigal, Hyderabad, Telangana, India

Konatham Tcja Kumar Reddy
University College of Technology
Osmania University
Amberpet, Hyderabad, Telangana, India

Sivuyisiwe Mapukata
DSI/Mintek Nanotechnology
  Innovation Center, Mintek
Advanced Materials Division
Randburg, South Africa

Sonia Maqbool
Department of Chemistry
University of Sahiwal
Sahiwal, Pakistan

Shlomo Margel
Department of Chemistry
Bar-Ilan Institute for Nanotechnology
  and Advanced Materials
Bar-Ilan University
Ramat-Gan, Israel

Bambesiwe M. May
Institute for Nanotechnology and Water
  Sustainability
College of Science, Engineering, and
  Technology
University of South Africa, Florida Campus
Roodepoort, Johannesburg, South Africa

Monika Mishra
Department of Chemistry
School of Applied and Life Sciences
Uttaranchal University
Dehradun, India

Nonkululeko Miya
DSI/Mintek Nanotechnology
  Innovation Center
Advanced Materials Division
Randburg, South Africa

Alima Misiriya
Yenepoya Pharmacy College &
  Research Centre, Yenepoya (Deemed
  to be University), Naringana,
  Mangalore, Karnataka, India

Narayana Raju Padala
Department of Pharmaceutics
Sri Vasavi Institute of Pharmaceutical
    Sciences
Tadepalligudem, Andhra Pradesh, India

Ritu Painuli
Department of Chemistry
School of Applied and Life Sciences
Uttaranchal University
Dehradun, India

Shanmugasundaram Palani
School of Pharmaceutical Sciences
Vels Institute of Science, Technology
    and Advanced Studies
Pallavaram, Chennai, India

Anamika Pandey
Department of Chemistry
School of Applied and Life Sciences
Uttaranchal University
Dehradun, India

Ravindra Kumar Pandey
Columbia Institute of Pharmacy
Raipur, Chhattisgarh, India

Shivam Pandey
Department of Chemistry
School of Applied and Life Sciences
Uttaranchal University
Dehradun, India

Anasuya Patil
Department of Pharmaceutics
KLE College of Pharmacy
Bengaluru, India

Sindhu Priya E S
Department of Pharmaceutics
Yenepoya Pharmacy College &
    Research Centre
Yenepoya (Deemed to be University),
    Naringana
Mangalore, Karnataka, India

Raju A.
Department of Pharmacology
St. Joseph's College of Pharmacy
Cherthala Alappuzha, Kerala, India

Kavati Ramkrishna
Department of Pharmaceutics
Pulipati Prasad College of
    Pharmaceutical Sciences
Khammam, Telangana, India

Ankita Rawat
Department of Chemistry
School of Applied and Life Sciences
Uttaranchal University
Dehradun, India

Aziz-ur-rehman
Department of Chemistry
Government College University
Lahore, Pakistan

Ligia R. Rodrigues
Centre of Biological Engineering
    and LABBELS Associate
    Laboratory
University of Minho, Campus de
    Gualtar
Braga, Portugal

Varsha Sahu
Department of Pharmaceutical
    Sciences
Utkal University
Bhubaneshwar, Odhisa, India

Shiv Shankar Shukla
Columbia Institute of Pharmacy
Raipur, Chhattisgarh, India

Ajay Singh
Department of Chemistry
School of Applied and Life
    Sciences
Uttaranchal University
Dehradun, India

I. Somasundaram
School of Pharmaceutical Sciences
Vels Institute of Science, Technology
    and Advanced Studies
Pallavaram, Chennai, India

Sandhya Vasanth
Department of Pharmaceutics
Yenepoya Pharmacy College &
    Research Centre
Yenepoya (Deemed to be University),
    Naringana
Mangalore, Karnataka, India

Sérgio R. S. Veloso
Physics Centre of Minho and Porto
    Universities and LaPMET Associate
    Laboratory
University of Minho, Campus de Gualtar
Braga, Portugal

Astha Verma
Shri Rawatpura Sarkar Institute of
    Pharmacy
Kumhari, Durg, Chhattisgarh, India

Manisha Verma
Columbia Institute of Pharmacy
Raipur, Chhattisgarh, India

Amber Vyas
University Institute of Pharmacy
Pt. Ravi Shankar Shukla University
Raipur, Chhattisgarh, India

1 Introduction to Magnetic Polymer Composites

Sayan Ganguly

1.1 INTRODUCTION

Magnetic polymer composites (MPCs) are a fascinating and versatile group of materials because of the interesting combination of magnetic and polymeric properties they exhibit. These composites combine the best features of magnetic materials with polymers, opening up a whole new class of functional materials with wide-ranging potential uses [1]. MPCs are composite materials with a polymer matrix and a magnetic nanoparticle or filler scattered throughout. A wide variety of materials, from ferrites and iron oxides to nanoparticles with engineered magnetic characteristics, can be used as magnetic fillers [2]. When these fillers are incorporated into a polymer matrix, MPCs are born with the ability to display variable magnetic behaviors. MPCs stand out from the crowd thanks to their adaptability in manufacture. Solution casting, melt blending, and electrospinning are only a few of the methods that can be used to create them [3]. By using these methods, the distribution of magnetic fillers inside the polymer matrix may be precisely manipulated, which in turn affects the composite's magnetic behavior and performance [4]. The production of MPCs continues to depend critically on the even distribution of fillers, which in turn affects the material's magnetic characteristics and potential uses. There is a complicated interaction between the type and concentration of magnetic fillers, particle size, and characteristics of the polymer matrix that determines the magnetic properties of MPCs. Saturation magnetization, coercivity, and remanence are all important magnetic factors that fall under the umbrella of these characteristics [5]. As a result, MPCs can be engineered to display a wide spectrum of magnetic behaviors, from soft behaviors characterized by high permeability and low coercivity to hard behaviors characterized by substantial remanence and coercivity. Because of their exceptional mix of magnetic and polymer properties, MPCs find use in a wide variety of fields [6]. Applications of MPCs in biomedicine include bioimaging, magnetic hyperthermia for cancer treatment, and targeted medication delivery. They are very suitable for regulated medication release and targeted therapeutic interventions due to their responsiveness to external magnetic fields. In addition to being useful in medicine, MPCs have been indispensable in the research and development of EMI shielding materials [7]. The need for reliable EMI shielding has skyrocketed as the number of electronic gadgets in use has increased exponentially. MPCs are effective at shielding electronic components because their magnetic characteristics allow for the absorption and dissipation of electromagnetic energy [8]. MPCs have also become a platform for developing high-tech sensors and actuators. Researchers have developed novel sensor technologies, such as magnetic field sensors and strain

DOI: 10.1201/9781003454236-1

sensors, with increased sensitivity and selectivity by taking advantage of the distinctive magnetic responses of these composites to external stimuli. The world of MPCs, however, is not without its difficulties. There is further work to be done in the areas of optimal dispersion of magnetic fillers, preservation of magnetic characteristics throughout production, and compatibility between the fillers and the polymer matrix [9]. However, these difficulties present exciting new opportunities for discovery and development.

In recent years, MPCs have emerged as a promising new class of materials thanks to their ability to combine magnetic and polymer characteristics [10]. The wide range of fields in which MPCs have been implemented demonstrates their critical role in solving difficult problems and inspiring new approaches. The following paragraphs explore the significance and varied uses of MPCs across disciplines.

Targeted medication delivery, imaging, and cancer therapy are just some of the medical applications that have benefited from MPCs. Because of their sensitivity to magnetic fields, they can be used to precisely target specific areas of the body and regulate the release of medicines [11]. Targeting specific locations allows drug concentration to be increased at the desired sites while minimizing unwanted effects, all thanks to magnetic nanoparticles incorporated in polymer matrices. Magnetic particle heaters (MPHs) are used in magnetic hyperthermia, a cancer treatment that makes use of the MPCs' ability to produce heat in response to alternating magnetic fields in order to preferentially kill tumor cells. Combining diagnostic imaging with therapeutic therapies improves the accuracy and efficiency of modern medicine.

Electromagnetic interference (EMI) problems have emerged as a result of the explosion of electronic gadgets and wireless technology [12]. Because of their exceptional magnetic properties, MPCs may absorb and discharge electromagnetic radiation, providing a viable option [13]. To ensure optimal device performance while minimizing interference from other electronics, these composites can be incorporated into device enclosures or packing materials to greatly increase the effectiveness of EMI shielding.

By taking advantage of MPCs' magnetic reaction to stimuli, the sensor and actuator industries have been completely transformed. MPCs are fundamental components of magnetic field sensors that can detect extremely small magnetic field fluctuations. The malleability of polymer matrices allows for the use of magnetic fillers, resulting in extremely responsive and versatile strain and pressure sensors. These can be used in fields where high levels of precision in monitoring and control are required, such as the automobile, aircraft, and robotics sectors.

Researchers are looking at novel energy harvesting techniques to meet the rising need for green energy. The magnetostrictive effect, which MPCs use to transform mechanical vibrations into electrical energy, is a key component of this effort. Energy can be captured and used for a variety of purposes, such as powering sensors or low-energy electronics, by incorporating these composites into structures that suffer mechanical oscillations, such as bridges or wearable devices.

Water purification and cleanup are two environmental issues that can be addressed with the help of MPCs. In order to improve the efficiency of water purification and recovery procedures, magnetic nanoparticles contained in polymer matrices can be engineered to selectively adsorb pollutants. This technique may help solve problems related to water contamination, leading to safer and cleaner water supplies.

MPCs' qualities are very useful in the study of soft robotics. In order to function as soft actuators, these materials can be designed to show unique magnetic responses when subjected to external fields [14]. Researchers can build more versatile and adaptive robots by adding these actuators into soft robotic systems, allowing devices to perform controlled movements, shape changes, and gripping activities.

MPCs are useful in aircraft engineering because they may be included into lightweight, multipurpose materials. Materials with increased mechanical qualities, including stiffness or damping characteristics, can be created by incorporating magnetic nanoparticles into polymer matrices. These materials may find use in aeroplane parts, where saving weight and maintaining strength are paramount [15].

The development of new items with improved functionality is facilitated by MPCs in the consumer goods sector. For instance, MPCs can be used to create "smart fabrics" that adjust to variations in temperature and humidity. The versatility and comfort of these materials make them useful in a wide variety of settings, from athletic clothing to medical scrubs [16].

The purpose of this chapter is to introduce readers to magnetic polymer composites and help them become familiar with their basic characteristics, fabrication processes, and magnetic behaviors. It highlights MPCs' significance in fields as varied as medicine, electronics, energy, and materials science [17]. This chapter's goal is to shed light on the many applications of MPCs, as well as the difficulties and opportunities that exist within this exciting field of materials science.

1.2 BASICS OF MAGNETIC POLYMERS

A wide range of engineering materials, as well as living organisms, exhibit a heterogeneous composition. The amalgamation of organic and inorganic substances at the colloidal level yields novel and occasionally unexpected characteristics [18]. Fillers are commonly utilized as solid additives within polymers in order to alter their physical properties. Fillers can be categorized into three distinct groups. The first category comprises fillers that serve to enhance the polymer's structural integrity and its mechanical properties [19]. Another category comprises those that serve the purpose of occupying space, thereby decreasing the overall expenditure on materials. The third category, which is less prevalent, involves the inclusion of filler particles within the material to enhance its responsive characteristics.

The magnetic particles utilized in contemporary magnetic polymer composites are typically categorised into three types: (i) powdered neodymium–iron–boron magnet (NdFeB) particles; (ii) iron oxide (Fe_3O_4) particles with varying diameters ranging from nanometers to micrometers, exhibiting a diverse range of magnetic profiles, including superparamagnetic particles and larger particles with multiple magnetic domains per particle [20]; and (iii) carbonyl iron [$Fe(CO)_5$] particles, also known as iron pentacarbonyl, which are particularly intriguing for biomedical applications [21]. There are two main approaches to the process of preparing composites. The first approach involves achieving a homogeneous dispersion of magnetic particles within a polymer matrix. Current research is focused on optimizing the mixing conditions and utilizing suitable solvents and surfactants to enhance the homogeneity of the composite and improve the reproducibility of its magnetic and mechanical properties. The second approach involves aligning the magnetic particles along a specific axis to reinforce both the mechanical and the magnetic properties of the composite in an anisotropic manner [22]. This alignment results in the definition of a magnetic easy axis. The objective is attained through the application of a uniform magnetic field in a predetermined orientation throughout the polymerization or gelation stage of the composite material. During the liquid phase or when the viscosity is low, the polymer matrix and magnetic particles exhibit mobility, allowing for their rearrangement within the composite material under the influence of an applied magnetic field. This rearrangement persists until the polymerization process concludes, resulting in a definitive configuration that is subsequently frozen in place. The film's resultant anisotropy can be effectively utilized in a wide range of applications, with a particular emphasis on actuators.

Magnetically active soft materials are elastomeric materials that are polymer-based and exhibit a responsive behavior when subjected to an external magnetic field, resulting in deformation or mechanical stress [23]. These materials are frequently referred to by different terms such as magnetostrictive polymers, magnetorheological polymers, or magnetoelasts. Magnetic polymer gels are a subset of magnetic elastomers, characterized by their flexible cross-linked polymers that incorporate magnetizable particles and a substantial volume of swelling liquid. The advancement of magnetically active polymer systems is closely linked to the progress made in the field of magnetic nanoparticles and magnetic fluids [24]. The utilization of magnetic nanoparticles has garnered significant attention due to their potential applications in high-density memory devices, spintronics, and diagnostic medicine. Ferrofluids, also known as magnetic liquids, are colloidal systems consisting of single domain magnetic nanoparticles that are dispersed in either aqueous or organic liquids. The particles typically exhibit sizes ranging from 5 to 15 nm and are retained within the sol through the utilization of specific stabilizers. These stabilizers serve the purpose of preserving the individual stability of the particles and preventing their coagulation [25]. Magnetorheological fluids consist of suspensions containing particles that are significantly larger, ranging from one to three orders of magnitude larger than the particles found in colloidal ferrofluids. The magnetorheological particles are composed of numerous magnetic domains. These fluids exhibit significant alterations in their rheological properties when subjected to an externally imposed magnetic field.

Extensive research has been conducted on the applications of magnetic micro- and nanospheres, as well as monolith gels composed of cross-linked polymers [4]. The enhanced physical and mechanical properties of polymer-encapsulated magnetic filler particles, which have diameters smaller than 1 µm, have garnered attention in various industries such as pharmaceuticals, cosmetics, and paint production [26]. The most significant examples include the employment of magnetic separation techniques for the isolation of labeled cells and other biological entities; the delivery of therapeutic drugs, genes, and radionuclides; the utilization of radio frequency methods to induce hyperthermia for tumor degradation; and the use of contrast enhancement agents for applications in magnetic resonance imaging.

1.3   MAGNETIC FILLERS AND NANOPARTICLES

In recent times, there has been a notable increase in the progress and utilization of sophisticated materials—namely, composites—across several industries. Composites are a type of engineered material that consists of two or more separate phases, which are joined in order to produce improved mechanical, thermal, electrical, or other qualities [27]. The utilization of magnetic fillers and nanoparticles in composites has attracted significant interest within several academic disciplines, mostly due to their distinct properties and prospective utilization in diverse domains such as electronics, medicine, and energy [28]. Magnetic fillers and nanoparticles are particles exhibiting magnetic characteristics, typically on a nanoscale, which can be uniformly distributed inside a non-magnetic matrix, hence giving rise to magnetic composites. The magnetic components encompass a range of materials, including metallic, ceramic, and organic substances, all of which demonstrate magnetic properties that can be effectively utilized for diverse applications [29]. The incorporation of magnetic components into composites presents the potential for innovative capabilities and creates opportunities for technological advancement. The inclusion of magnetic fillers and nanoparticles in composites offers a significant advantage due to their capacity to alter and react to external magnetic fields. This characteristic facilitates a diverse array of applications, including those in the field of electromagnetic interference (EMI) shielding [30]. In this context, these composites possess the ability to redirect or absorb electromagnetic waves, thereby safeguarding vulnerable electronic devices. Furthermore, magnetic composites are employed in the production of sensors, actuators, and transducers due to their ability to detect physical alterations and convert them into quantifiable signals by responding to magnetic fields.

The promise of magnetic composites has also been grasped by the medical community. The utilization of magnetic nanoparticles, when enveloped with biocompatible substances, enables their manipulation through external magnetic fields for precise localization within the human body [31]. This capability facilitates targeted administration of therapeutic agents, and thus magnetic nanoparticles may be employed as contrast agents in magnetic resonance imaging (MRI) [32]. Additionally, it should be noted that these nanoparticles have the capability to produce localized heat when exposed to alternating magnetic fields [33]. This characteristic holds significant potential for the advancement of hyperthermia treatments in the field of cancer therapy. The versatility of magnetic composites is further exemplified by their various energy-related applications. The composites have the potential to be utilized in the advancement of electromagnetic energy harvesting systems, wherein the conversion of mechanical vibrations into electrical energy is facilitated by the magnetic constituents [34]. Furthermore, the incorporation of magnetic nanoparticles into electrodes has been observed to augment the efficiency and durability of batteries and capacitors utilized in energy storage systems.

The incorporation and distribution of magnetic fillers and nanoparticles in composite materials pose various obstacles and opportunities [35]. Attaining a uniform dispersion of nanoparticles while simultaneously preventing their aggregation is of paramount importance in maintaining constant material characteristics. Several methodologies, including sol-gel procedures, co-precipitation, and chemical vapor deposition, have been effectively utilized to integrate these nanoparticles into composite matrices. Scholars additionally have investigated surface modification methodologies in order to augment the compatibility between magnetic particles and the matrix material, consequently enhancing the composite's mechanical and thermal properties holistically. Nevertheless, there are still obstacles that need to be surmounted. The issues around the possible toxicity of specific magnetic nanoparticles, particularly in the context of biomedical applications, are noteworthy. In order to have a thorough understanding of the enduring impacts of these substances on human health and the natural environment, it is imperative to conduct comprehensive investigations [36]. Furthermore, the capacity to accurately manipulate the dimensions, morphology, and magnetic characteristics of nanoparticles continues to pose a significant obstacle, hence impeding the reproducibility and scalability of these sophisticated composites.

Ferrites and iron oxides are frequently employed as magnetic fillers in composite materials, exhibiting unique magnetic characteristics and serving a wide range of purposes in multiple industrial sectors. Ferrites, which are frequently composed of compounds such as iron oxide and various metals, are widely recognized for their elevated magnetic permeability and their effectiveness for electrical insulation. The utilization of these materials is widespread in the manufacturing of electromagnetic devices, transformers, and inductors, owing to their highly effective electromagnetic interference shielding properties. Ferrite-based magnetic fillers are commonly used in composites to fulfill the need for regulated absorption and attenuation of electromagnetic waves in various applications. Iron oxides, specifically magnetite (Fe_3O_4) and hematite (Fe_2O_3), are magnetic nanoparticles that are extensively employed in composites due to their biocompatibility and adjustable magnetic characteristics. Within the field of biomedicine, nanoparticles are modified and utilized for the specific purpose of delivering drugs to particular areas, conducting hyperthermia treatments, and enhancing contrast in MRI. The variable magnetic properties and surface chemistry of these materials enable customized interactions with biological systems. The magnetic fillers have a wide range of applications, particularly in the field of energy. They have the ability to improve the efficiency of electromagnetic energy harvesting devices and play a significant role in the advancement of sophisticated energy storage systems. The incorporation of these materials into polymers, ceramics, and other matrices highlights their potential for generating multifunctional composites that exhibit magnetic responsiveness.

1.4   FABRICATION TECHNIQUES FOR MAGNETIC POLYMER COMPOSITES

Magnetic polymer composites, categorized as a class of sophisticated materials, amalgamate the distinctive characteristics of magnetic nanoparticles and polymers, yielding adaptable materials with an array of uses. The fabrication of these composites involves employing diverse approaches to achieve a uniform dispersion of magnetic nanoparticles inside a polymer matrix. The selection of the manufacturing procedure is a critical factor in defining the ultimate characteristics and effectiveness of these composite materials. Table 1.1 gives a brief overview of all the contemporary processes for fabrication.

1.4.1   In Situ Polymerization

In situ polymerization is considered a crucial technique in the fabrication of magnetic polymer composites. The synthesis of magnetic nanoparticles occurs within a polymer matrix through the process of polymerization. The utilization of this particular approach presents the benefit of attaining a uniform dispersion of nanoparticles inside the polymer matrix, a crucial factor for ensuring optimal material functionality. Furthermore, the utilization of in situ polymerization provides a heightened level of regulation over the concentration of nanoparticles, hence exerting an influence on the magnetic characteristics of the composite material.

Nevertheless, the technique of in situ polymerization does have certain drawbacks. The ability to endure the circumstances of the polymerization process sometimes limits the selection to specific types of polymers. In addition, the attainment of exact manipulation of nanoparticle dimensions and morphology presents a considerable challenge, which may have implications for the magnetic properties of the composite material. Notwithstanding these constraints, the utilization of in situ polymerization is observed across diverse domains, encompassing magnetic sensors and medication delivery systems, wherein the imperative requirement for regulated nanoparticle dispersion is evident.

1.4.2 SOLUTION MIXING

Solution mixing is a widely adaptable technology that is frequently employed in the fabrication process of magnetic polymer composites. In this methodology, magnetic nanoparticles are combined with a polymer solution, subsequently leading to the evaporation of the solvent, which results in the formation of the composite material. The utilization of solution mixing is a viable approach that combines ease of use and adaptability, rendering it suitable for a diverse array of polymer materials. The technique enables the integration of diverse magnetic nanoparticles and provides the ability to manipulate the magnetic characteristics of the composite by modifying the concentration of nanoparticles. Nevertheless, the process of solution mixing presents several difficulties. The mixing method poses a potential danger of nanoparticle agglomeration, resulting in non-uniform dispersion throughout the composite material. Furthermore, it is important to consider that the technique may possess certain limits in terms of attaining a substantial nanoparticle loading, which may be vital for specific applications. Notwithstanding these limitations, solution mixing is extensively employed in various domains, including electromagnetic interference shielding and flexible electronics, owing to its convenient implementation.

1.4.3 MELT MIXING

The process of melt mixing is considered a scalable method for the manufacturing of magnetic polymer composites on a big scale. The proposed technique involves the incorporation of magnetic nanoparticles into a liquid polymer matrix, which would be followed by the solidification of the resulting mixture to yield a composite material. Melt mixing possesses the advantageous characteristic of exhibiting compatibility with a diverse array of polymers, hence rendering it a useful technique for a multitude of applications. This feature facilitates a substantial loading of nanoparticles, which has the potential to result in improved magnetic characteristics. Nevertheless, it is important to acknowledge that melt mixing does possess certain restrictions. The potential deterioration of magnetic nanoparticles and its subsequent impact on their properties can be attributed to the elevated processing temperatures involved in their production.

Hence, the careful selection of appropriate processing conditions and materials is of utmost importance in order to achieve optimal performance of the composite. The utilization of melt mixing is prevalent in various industries, particularly in the automotive sector and the production of magnetic seals. The scalability and versatility of this technique offer significant benefits in these applications.

1.5 INCLUSION POLYMERIZATION

The process of inclusion polymerization entails the polymerization of monomers in the presence of pre-synthesized magnetic nanoparticles. This approach allows a significant level of manipulation of the organization of nanoparticles inside the polymer matrix, resulting in a homogeneous distribution of nanoparticles. The deliberate organization of components within the composite material can exert a substantial impact on its magnetic properties. The process of inclusion polymerization facilitates the utilization of diverse monomers, hence facilitating the development of composites with customized features. In contrast to alternative methodologies, inclusion polymerization is characterized by its intricate nature, necessitating meticulous consideration of monomer choice and reaction parameters. The selection of monomers that can efficiently interact with magnetic nanoparticles is typically restricted. Notwithstanding

TABLE 1.1

Different Types of Fabrication Techniques for Preparing Magnetic Polymer Composites

Fabrication Technique	Description	Advantages	Disadvantages	Applications
In Situ Polymerization	Magnetic nanoparticles are synthesized within a polymer matrix during polymerization.	Homogeneous distribution of nanoparticles Enhanced control over nanoparticle concentration	Limited to specific polymers Difficulty of controlling nanoparticle size and shape precisely	Magnetic sensors Drug delivery systems
Solution Mixing	Magnetic nanoparticles are mixed with a polymer solution, followed by solvent evaporation.	Relatively simple and versatile Can be used with various polymers	Potential nanoparticle agglomeration Limited nanoparticle loading	Electromagnetic interference shielding Flexible electronics
Melt Mixing	Magnetic nanoparticles are blended with a molten polymer and then solidified.	Scalable for mass production Can be used with a wide range of polymers	Possible nanoparticle degradation at high temperatures	Automotive components Magnetic seals
Inclusion Polymerization	Monomers are polymerized in the presence of pre-synthesized magnetic nanoparticles.	High control over nanoparticle arrangement Uniform dispersion of nanoparticles	Complex process Limited to specific monomers	Biomedical applications Magnetic data storage

these limitations, the utilization of inclusion polymerization is observed in various domains such as biological materials and magnetic data storage, wherein the careful regulation of nanoparticle distribution is of paramount importance.

1.6 MAGNETIC PROPERTIES OF MAGNETIC POLYMER COMPOSITES

Magnetic polymer composites are a category of materials that arise from the combination of magnetic particles or nanomaterials embedded within a polymer matrix. The combination of these elements creates a fusion of characteristics that are not present in either component on its own (Figure 1.1). Magnetic particles, frequently in the nanoparticle configuration, contribute to the magnetic characteristics of the composite, while the polymer matrix imparts flexibility, processability, and customizable functionality. The magnetic phase observed in MPCs can exhibit a range of characteristics, including ferromagnetism, where particles maintain their magnetic properties without the need for an external magnetic field, and superparamagnetism, where particles display magnetic behavior solely when subjected to an external magnetic field. The diverse array of magnetic properties observed can be attributed to variations in the size and configuration of magnetic particles. In general, smaller particles tend to display superparamagnetic behavior as a result of their diminished thermal stability. The magnetic characteristics of MPCs are affected by various parameters, such as the dimensions, configuration, composition, and arrangement of magnetic particles embedded in the polymer matrix. Nanoparticles composed of iron oxides, such as magnetite and maghemite, are frequently employed in various applications owing to their simple synthesis, compatibility with biological systems, and adjustable magnetic characteristics. Frequently, a stabilizing coating is applied to these particles in order to inhibit aggregation and improve compatibility with the polymer matrix.

The production of MPCs necessitates precise manipulation techniques to achieve consistent dispersion of magnetic particles throughout the polymer matrix. Various techniques, including solution mixing, melt blending, and in situ polymerization, are utilized in order to attain a state of homogeneity. The selection of the fabrication technique has a substantial influence on the ultimate characteristics of the composite, affecting factors such as the distribution, size, and alignment of particles. The attainment of a uniformly distributed and enduring blend is imperative in order to effectively utilize the intended magnetic properties. The comprehension of the magnetic characteristics of MPCs necessitates an exploration of the complex interrelationship between magnetic interactions and the dynamics of polymers. In the context of ferromagnetic MPCs, the interplay of magnetic particle interactions can give rise to notable phenomena, including magnetic hysteresis and remanence. These features have practical uses in the fields of memory devices and electromagnetic shielding. In contrast, superparamagnetic magnetite nanoparticles, known as SPIONs, exhibit significant utility in several fields, such as medication delivery and hyperthermia, due to their ability to respond to an external magnetic field in a controlled manner without retaining any residual magnetization. The utilization of MPCs encompasses a wide range of disciplines, with each field using unique magnetic properties and polymer characteristics of these materials. Within the field of biomedicine, mesenchymal stem/progenitor cells (MSCs) are significantly transforming the landscape of drug delivery systems. The feasibility of targeted medication delivery to specific areas within the body is enhanced by functionalizing magnetic nanoparticles and utilizing their responsiveness to external magnetic fields. Furthermore, the utilization of superparamagnetic particles in the production of thermal energy upon exposure to alternating magnetic fields is employed in the hyperthermia therapy of malignant cells. Within the field of electronics, microprocessors are employed for the purpose of mitigating electromagnetic interference. This is of utmost importance because of the ongoing trend of miniaturization and widespread adoption of electronic devices. These composite materials have the capability to efficiently divert and absorb electromagnetic radiation, thereby improving the performance and durability of delicate electronic components. Energy harvesting is a field that is seeing significant advantages as a result of microprocessor chips (MPUs). Through the exploitation of the alterations in magnetic flux generated by motion within the composites, the generation of electrical power can be achieved via the mechanism of electromagnetic induction. The aforementioned approach is now being investigated for its potential applications in the fields of self-powered sensors and energy-efficient devices. MPCs have the potential to significantly benefit environmental remediation efforts. These composite materials have the potential to be utilized for the purpose of extracting heavy metals and contaminants from various water sources. The composites containing magnetic nanoparticles exhibit the ability to selectively adhere to pollutants, hence facilitating their extraction from water by means of an externally applied

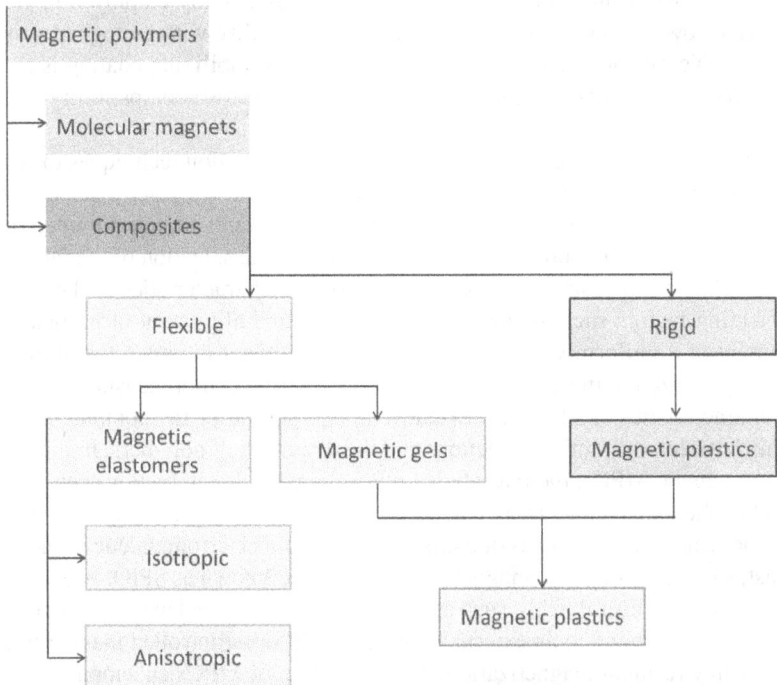

FIGURE 1.1 Magnetic polymer composites and their different types of fabrication techniques.

magnetic field. The concept of magnetic polymer composites signifies a notable amalgamation of magnetic materials and polymers, leading to the emergence of a novel epoch of functional materials that fit a wide range of applications. The magnetic characteristics of these materials are influenced by factors such as particle size, arrangement, and interactions. These qualities interact with the flexibility and processability of polymers, allowing for the customization of materials to suit particular applications. Multi-principal component systems are exerting a significant influence on diverse areas ranging from biomedicine and electronics to energy and the environment. As the research in this domain progresses, it is expected that there will be more developments in the field of MPCs, resulting in more advanced systems with improved characteristics. This will create opportunities for the emergence of innovative technologies that were previously considered unattainable.

1.7 APPLICATIONS OF MAGNETIC POLYMER COMPOSITES

1.7.1 DRUG DELIVERY

Within the realm of healthcare management, use of controlled drug administration rather than traditional dose forms plays a crucial role in enhancing effectiveness, minimizing toxicity, and promoting patient adherence and convenience. Polymeric materials have emerged as a prominent component in this novel technology, presenting prospects for extended drug release while simultaneously ensuring that the medication's blood concentration remains within therapeutic thresholds. Numerous polymeric nanomaterials and polymer-metal nanocomposites are presently being studied as vehicles for sustained release. The sustained release mechanism of these materials relies on a range of parameters, including pH, temperature, solubility, and the impact of magnetic fields [37]. The process of thermally induced controlled release from a polymer is characterized by the thermoresponsive nature of the polymeric system. This characteristic leverages thermal energy to induce alterations in the polymer structure, hence facilitating the liberation of a confined molecule. Poly(N-isopropylacrylamide) (PNiPAAm) and polycaprolactone (PCL) are commonly employed in drug absorption and release systems. PNiPAAm is an aqueous hydrogel, while PCL is a nonaqueous polymer. These materials exhibit the ability to undergo swelling and collapse of their polymer structure, melting of polymer chains, or an increase in porosity when subjected to thermal induction around 37°C [38]. This property enables them to effectively facilitate drug absorption and release processes.

The process of magnetically induced controlled release entails subjecting the polymeric system to an alternating current magnetic field. The application of a magnetic field induces an elevation in the temperature of the magnetic nanoparticles within the immediate vicinity. When the temperature exceeds the glass transition temperature (T_g) of the polymer, it is possible for the polymer strands surrounding the nanoparticles to undergo relaxation, resulting in the formation of a porous environment that facilitates the release of the medication. There is a scarcity of literature documenting the utilization of polymeric materials containing magnetic nanoparticles for the purpose of releasing or absorbing molecules under the influence of an alternating current magnetic field [39]. A new paper highlights the successful exhibition of magnetically regulated permeability in polyelectrolyte microcapsules containing ferromagnetic gold-coated cobalt (Co@Au) nanoparticles with a diameter of 3 nm, which are implanted within the walls of the capsules. In a similar vein, it was shown that collagen gels with a thickness of 1 cm, which were made using magnetite particles measuring 10 nm in diameter (Figure 1.2), exhibited the release of rhodamine-labeled dextran (Dex-R) when subjected to an oscillating magnetic field [40].

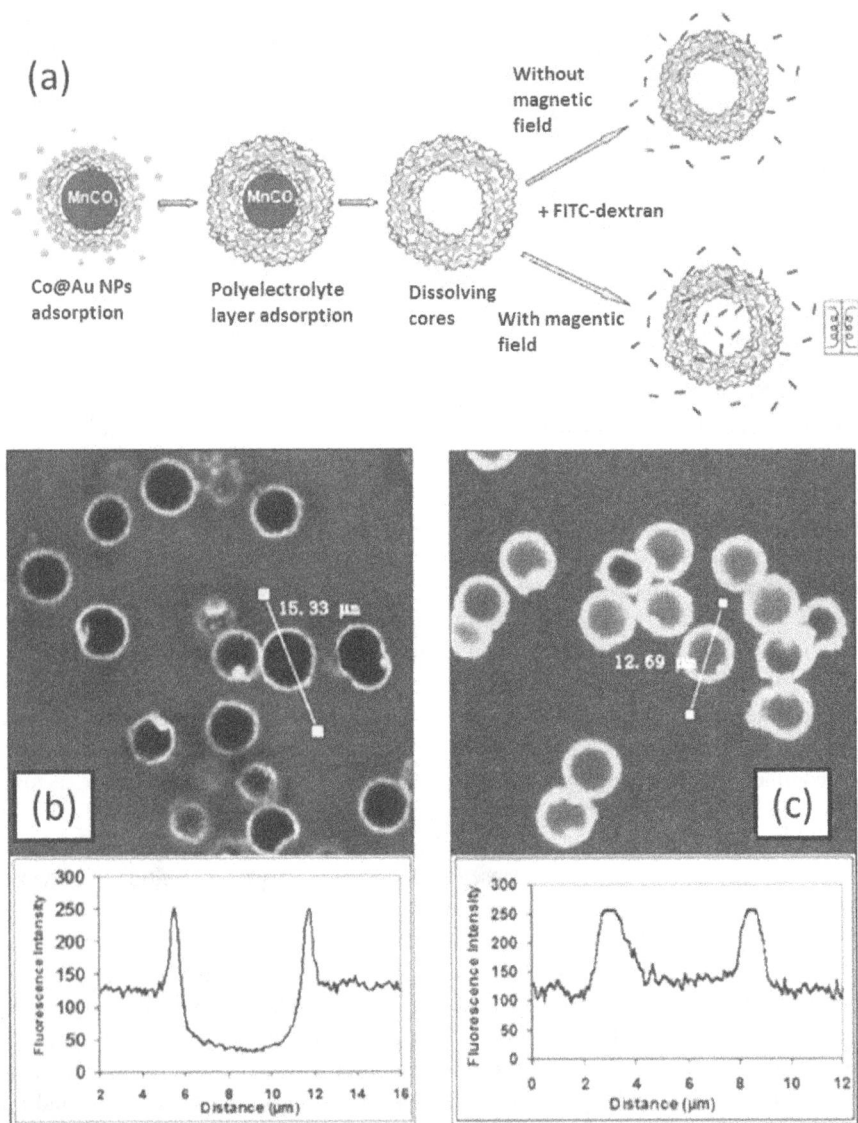

FIGURE 1.2 (a) Scheme of the assembly and permeability test for microcapsules embedded with Co@Au nanoparticles under an oscillating magnetic field. Confocal laser scanning microscopy (CLSM) image of magnetic capsules [(PSS/PAH)4(PSS/Co@Au)1(PSS/PAH)6] mixed with FITC–dextran (b) without applying an alternating magnetic field for 1 hour and (c) after applying an alternating magnetic field for 30 min. The corresponding optical density profiles are also shown. (Reproduced with the permission from ref. [40]. © 2005 American Chemical Society.)

In a separate study, the scientists proposed the use of a degradable superparamagnetic polymer composite. This composite enables the safe degradation of a device within the body and subsequent excretion of the degradation products. This approach offers a non-invasive alternative to the manual or surgical recovery of the device [41]. The composite material consisting of superparamagnetic hydrogel exhibits the ability to undergo cross-linking by both single- and two-photon polymerization (TPP). Additionally, it possesses the capacity to absorb and release chemicals that are biologically significant, while also degrading in aqueous environments (namely, those resembling physiological conditions). This method provides a single-step approach for the straightforward and effortless production of devices, effective activation, and device utilization, hence eliminating the need for device retrieval. The composite material consists of nanoparticles of magnetite (Fe_3O_4), poly(ethylene glycol) diacrylate (PEG-DA) with exceptional stealth properties, and pentaerythritol triacrylate (PE-TA). Hydrogel-based active biofunctional materials (ABFs) are produced through the process of thermal polymerization and are subsequently activated by the utilization of low-intensity oscillating magnetic fields. The study showcases the utilization of corkscrew propulsion for locomotion, tailored delivery of drugs to a specific model cell line (3T3 fibroblast cells), and the little cytotoxicity exhibited by the breakdown products. A study presents a highly effective technique for producing composite microfibers with magnetic responsiveness and biocompatibility, specifically designed for the purpose of controlled medication delivery. Polyvinyl alcohol solutions, containing Fe_3O_4 magnetic nanoparticles (MNPs) coated with citric acid, are treated with infusion gyration to produce composite fibers with diameters ranging from 100 to 300 nm [42]. The loading of MNPs may be controlled in these fibers. The fibers exhibit stability in polar solvents, such as ethanol, and do not demonstrate any leaching of MNPs for a duration exceeding 4 weeks. The efficacy of acetaminophen in drug immobilization and triggered release is exemplified, wherein the application of an external magnetic field induces these processes. The combination of remote actuation capability, biocompatibility, and light weight suggest significant potential for utilizing these fibers as an intelligent drug delivery agent. Polysaccharide-based magnetic nanocomposites exhibit notable characteristics that make them very suitable for use as carriers of anticancer drugs. A study employed rice straw–derived cellulose nanowire (CNW) as a solid substrate to incorporate Fe_3O_4 nanofillers, resulting in the synthesis of magnetic CNW. The cross-linked chitosan-coated magnetic cellulose nanowires used as a carrier for 5-fluorouracil can be shortened as CH/MCNW/5FU [43]. The nanocomposites exhibited desirable saturation magnetization and thermal stability, as well as great effectiveness in encapsulating drugs. Additionally, they demonstrated pH-dependent swelling and drug release capabilities. The nanocomposites composed of CH/MCNW/5FU exhibited significant cytotoxicity against colorectal cancer cells in both two-dimensional monolayer and three-dimensional spheroid models. The results of this study indicate that CH/MCNW may be a promising vehicle for delivering anticancer medicines due to its ability to effectively penetrate tumors. The influence of surface modification on the interactions between magnetic nanoparticles and cell membranes has been documented in previous studies. A newer study examines the interactions of polymer-functionalized Fe_3O_4 MNPs [namely, chitosan (CHI) and diethylaminoethyl dextran (DEAE-D)], medicines, and model cell membranes using Langmuir isotherms and adsorption studies [44]. All MNP-drug composites exhibit molecularly imprinted polymer (MIP) values that surpass the membrane pressure values usually acknowledged, suggesting the ability of MNPs:drugs to traverse a cellular membrane. The observation that the composite MNPs:drugs exhibit higher MIP values than those of individual compounds suggests that polymer-coated MNPs have the potential to serve as effective drug delivery platforms. In another study, the researchers employed a composite of paramagnetic Fe_3O_4 and carboxymethyl chitosan (CC) to augment the imprinting phenomenon and enhance the efficacy of drug delivery [45]. The preparation of MIP-doped Fe_3O_4-grafted CC (SMCMIP) involved the utilization of a binary porogen consisting of tetrahydrofuran and ethylene glycol. Salidroside is utilized as the template, methacrylic acid functions as the functional monomer, and ethylene glycol dimethacrylate (EGDMA) is employed as the cross-linker. The micromorphology of the microspheres was observed using scanning and transmission electron microscopy. The cytotoxicity trials yielded findings indicating that the SMCMIP composite did not have any detrimental impact on cellular proliferation. The study observed that the survival rates of IPEC-J2 intestinal epithelial cells exceeded 98%. The utilization of the SMCMIP composite enables the controlled release of medications, which has the potential to enhance therapeutic efficacy and mitigate adverse effects.

1.7.2   HYPERTHERMIA APPLICATIONS

The precise manipulation and non-cytotoxic properties of magnetic hydrogels and soft composites have significantly contributed to the advancement of biomimetic soft robots to the next generation. The regulation of bare magnetic nanoparticles by remote control poses challenges; however, when these nanoparticles are confined within a polymeric matrix; the resulting system exhibits characteristics similar to those of an integrated artificial soft mussel system [3]. The synthesis of magnetite (Fe_3O_4) nanoparticles was conducted by the co-precipitation method, involving the combination of Fe^{3+} and Fe^{2+} ions with an aqueous NaOH solution. The magnetic composite nanoparticles consisting of Fe_3O_4 and polyaniline (PANI) were synthesized using an in situ polymerization method. The nanoparticles exhibited a core-shell structure and had a diameter ranging from 30 to 50 nm. The synthesis process involved polymerizing aniline in an aqueous solution containing the Fe_3O_4 magnetic fluid. The investigation focused on examining the inductive heat behavior of Fe_3O_4/PANI composite nanoparticles when subjected to an alternating current (ac) magnetic field [46]. After being subjected to a 29-minute exposure in an alternating current magnetic field, the physiological saline solutions containing either Fe_3O_4 nanoparticles or Fe_3O_4/PANI composite nanoparticles exhibited temperatures of 63.6°C and 52.4°C, respectively. The utilization of Fe_3O_4/PANI composite nanoparticles has promise as an effective means of localized hyperthermia treatment for cancer. The co-precipitation method was considered for the synthesis of ferrite filler, while the in situ polymerization technique was employed to obtain polypyrrole. The incorporation of $Nd_{0.04}Fe_2O_4$ ferrites into the polypyrrole polymeric matrix facilitated the formation of nanocomposites. The examination of the created samples encompassed an analysis of their structural, morphological, and magnetic properties. The observation of ferromagnetic activity in the manufactured composites was achieved by analyzing the hysteresis loop [47]. An increase in the amount of ferrite filler within the matrix led to changes in the magnetic characteristics, specifically magnetization and coercivity. The elevated dielectric constants observed in these samples indicate potential suitability for utilization in microwave devices. Composite ferrogels were made by encapsulating magnetic nanoparticles at 2.0 and 5.0% w/v in mixed agarose/chitosan hydrogels containing 1.0, 1.5, and 2.0% agarose and 0.5% chitosan. Scanning electron microscopy found micrometer-sized pores in dried composite ferrogels [48]. Ferrogels degrade faster than blank chitosan/agarose hydrogels without magnetic nanoparticles, according to thermogravimetric studies. The elastic moduli of the composite ferrogels further showed that magnetic nanoparticles in the beginning aqueous solutions prevent agarose gelation by chilling chitosan/agarose solutions. Finally, composite chitosan/agarose ferrogels heat in response to an alternating magnetic field, making them ideal biomaterials for magnetic hyperthermia therapies. A separate team of researchers conducted an analysis of the methods employed in order to establish a routine that is both reproducible and accurate for evaluating the heating efficiency of magnetic scaffolds used as bone implants in the context of deep-seated hyperthermia tumor treatment. This study demonstrated the integration of a magnetic polymer filament production technique with the heating efficiency analysis of the resultant 3D-printed magnetic polymer scaffolds, building on prior research [49]. The study provides a comprehensive examination of the structural and magnetic properties of magnetic polymer filaments. Additionally, the specific absorption rate results of 3D-printed magnetic polymer scaffolds are analyzed to evaluate their effectiveness in magnetic hyperthermia treatment. The findings highlight the potential of magnetite MNPs as a guiding tool for the fabrication of magnetic polymer scaffolds that can successfully facilitate magnetic hyperthermia therapy. A composite material with extrinsic magnetic properties is synthesized using magnetite and poly(butylene succinate) (PBS) that has been modified. The magnetic composite is synthesized using the emulsion–solvent evaporation technique. The synthesis of the composites has been verified using multiple analytical testing [50]. The composite material exhibits magnetic force, enabling it to provide magnetic navigation toward a specific target region. Additionally, the magnetic induction heating test facilitates heating within the therapeutic range of 40 to 45 °C, which, as documented in the literature, has the potential to induce cell death in several forms of cancer. Hence, the material exhibits significant potential for extensive research as a prospective tool for cancer treatment. Poly-N-isopropylacrylamide (PNIPAM) is a polymer that is both biocompatible and thermosensitive. It demonstrates a reversible transformation in volume phase from a hydrophilic coil to a hydrophobic globule when the temperature reaches the lower critical solution temperature (LCST) of 305 K in an aqueous solution. In order to induce this conformational alteration only through magnetic heating, magnetite nanoparticles with an approximate diameter of 12 nm were integrated into the polymer. The PNIPAM/magnetite nanoparticle composites exhibited significant thermal response upon exposure to an alternating magnetic field. Another study documented the controlled release of the anti-cancer medicine mitoxantrone [51].

The release was achieved with a non-invasive in vitro method, utilizing a composite material. Notably, the release mechanism was driven exclusively by the heating effects generated by an external magnetic field. Stable suspensions at physiological pH were obtained by synthesizing superparamagnetic nanoparticles (NPs) through chemical precipitation and subsequently stabilizing them with either oleic acid (OA) or dimercaptosuccinic acid (DMSA). The confirmation of the integration of NPs into a cellulose acetate (CA) matrix was achieved through the utilization of scanning and transmission electron microscopy techniques [52]. The findings of the study indicate that the adsorption of magnetic nanoparticles on the surface of fibers leads to the formation of composite membranes with heating capabilities superior to those of membranes formed by including magnetic NPs in the fibers. Nevertheless, the presence of magnetic nanoparticles on the surface of fibers might lead to cytotoxic effects, which are dependent on the concentration of these NPs.

1.7.3 Soft Robotics

In contrast to organisms possessing solid skeletal structures, soft-bodied animals such as caterpillars and earthworms have a reduced constraint on their locomotion and the ability to modify their form and morphology. These organisms possess the ability to manipulate their pliable and flexible anatomical structures with a considerable range of motion, hence exhibiting the utilization of robust muscle force. One instance of such an organism with a soft body is the inchworm. The organism possesses individual muscles that are innervated and lacks segmented body regions with constriction. The organism moves by use of sequential undulations that propagate in a wave-like manner from the posterior to the anterior region. The design of our soft robot was influenced by the body structure of the inchworm [53]. Our robot features a soft monolithic body without segmented parts and utilizes a push-pull locomotion method. Similar to the inchworm, our robot maintains surface contact solely through its legs rather than its entire body. This phenomenon is also observed in the locomotion of inchworms, which exhibit a relatively slow pace compared to that of mollusks and worms [54]. Th soft robot primarily draws inspiration from the movement patterns and soft-bodied structure exhibited by an inchworm [55]. The subsequent sections delineate the crawling mechanism and intrinsic friction control of the soft robot, drawing inspiration from the locomotion of inchworms

The domain of soft robotics is experiencing a swift expansion due to the development of soft robots that possess novel capabilities, designs, and materials. Soft robots possess a wide array of potential applications, encompassing domains such as biomedical and surgical instruments, assistive healthcare apparatuses, tissue engineering, and aerospace technology [56]. The incorporation of stimuli-responsive particles into polymer matrices facilitates the activation of composites, enabling their application in the field of soft robotics and conferring intricate functionalities. The majority of soft robots exhibit responsive behavior because of their composition of elastomers, which possess a mechanical reaction characterized by the absence of hysteresis. Responsive materials exhibit the ability to react to external stimuli and afterwards revert to their original form in the absence of this stimulus. The desirability of soft robots that demonstrate sequential and programmable reactions arises from their potential to carry out intricate tasks in remote settings, such as in vivo applications or space exploration [14].

Artificial bacterial flagella (ABFs) are microrobots that exhibit magnetic actuation and swimming capabilities, drawing inspiration from the propulsion mechanism observed in Escherichia coli bacteria, whereby a helical tail is employed. The described ABFs are constructed using an MPC including iron-oxide nanoparticles within an SU-8 polymer matrix [57]. The MPC is then molded into a helical structure by the process of direct laser writing. The locomotion model incorporates the fluidic drag experienced by the microrobot, which is determined through the application of the resistive force theory. The magnetization of the robot is estimated using an analytical model that represents it as a soft-magnetic ellipsoid. The fluidic and magnetic characteristics of the robot are affected by the helicity angle. It has been observed that robots with weak magnetization tend to exhibit corkscrew-like motion when the helicity angle is minimal. Joyee and Pan designed a soft robot featuring a novel grooved body design, which enhances its capacity to execute various tasks in dynamic environments [58]. This design also improves the robot's multi-modal locomotion capability, providing it with three degrees of freedom (DOF). The fabrication of this soft robot involves the use of elastic polymer and magnetic nanoparticles, resulting in a multi-material, monolithic structure. The robot exhibits exceptional bidirectional locomotion capability. The distribution of magnetic nanoparticles within the polymer material, which is controlled at a local level, contributes to the magnetic actuation capabilities of the robot. This enables the robot to have several modes of locomotion, such as linear crawling and rotating locomotion. The body construction of the robot has compliant properties that enable it to bend in both the xy plane and the z axis, facilitating its movement within confined areas. This flexibility grants the robot three degrees of freedom for maneuverability. Furthermore, the robot's deformation profile has been enhanced to closely resemble that of an actual inchworm, resulting in improved crawling speed and increased tilt angles. The implemented design enhancements facilitate the robot's navigation in complex situations, such as the intricate rugae found in the human gut. Furthermore, in order to highlight the diverse range of capabilities of the soft robot, a controlled simulation of a drug delivery application was shown. The utilization of magnetic field–guided self-assembly of magnetic particles into chains holds significant potential in the development of directionally sensitive materials, particularly for the field of soft robotics. The utilization of materials with higher levels of complexity enables the implementation of advanced functionalities while simultaneously employing straightforward device topologies. Elastomer films incorporating interconnected magnetic microparticles were fabricated using the solvent casting technique and afterwards shaped into lifters, accordions, valves, and pumps that may be activated by magnetic forces [59]. Cantilevers employed as lifting mechanisms demonstrated the capability to elevate masses of the polymer film that were up to 50 times greater. The concept of specific torque is proposed as a metric to evaluate and compare the efficiency of lifters and similar apparatuses. It represents the torque exerted per unit of magnetic field strength and mass of magnetic particles.

1.8   CHALLENGES AND FUTURE DIRECTIONS

In the culmination of this investigation into MPCs, it becomes evident that although researchers have achieved notable advancements, they also have encountered several obstacles in their pursuit of enhanced scientific comprehension and pragmatic implementation. Moreover, with the identification of these obstacles, it becomes possible to foresee potential future research lines within this particular sector.

The development of scalable fabrication techniques poses a substantial hurdle for the general deployment of MPCs. Although there are several techniques available for producing these composites in a laboratory setting, the challenge lies in successfully adapting these processes for large-scale industrial manufacturing while maintaining consistent levels of quality. It is imperative for researchers to effectively tackle the challenges pertaining to the even distribution of magnetic particles, precise control over particle dimensions, and the mitigation of particle agglomeration throughout the process of synthesizing materials on a large scale.

The practical applications of MPCs heavily rely on their long-term stability and durability. It is necessary to conduct comprehensive investigations and optimizations regarding factors such as thermal stability, resistance to degradation, and the capacity to sustain ideal magnetic characteristics over an extended period. This holds particular significance in scenarios where MPCs are exposed to challenging conditions or undergo repetitive instances of external stimulation.

Although empirical research has provided useful insights into the behavior of magnetic particle composites, there is still no comprehensive theoretical framework that can reliably predict their magnetic properties. The development of models that incorporate the intricate interactions among magnetic particles, the polymer matrix, and external stimuli will facilitate a more comprehensive comprehension of the behavior of magnetically responsive polymer composites. Moreover, such models will serve as valuable tools for the creation of materials with properties tailored to meet specific requirements.

Numerous applications require the utilization of MPCs that possess multi-functional capabilities extending beyond their inherent magnetic characteristics. The incorporation of properties such as electrical conductivity, optical activity, or catalytic activity into these composites poses a range of obstacles and opportunities. The difficulty lies in attaining these features while maintaining the magnetic properties and overall performance of the MPCs.

The utilization of improved characterization techniques will be crucial in the ongoing research efforts to understand the complex structure-property interactions present in MPCs. Methods such as in situ imaging, sophisticated spectroscopy, and computational simulation will yield more profound understandings of the properties and dynamics of these materials across several spatial and temporal dimensions.

The future prospects of MPCs are contingent upon their customized applications. By comprehending the distinctive demands of various businesses and sectors, researchers have the ability to create MPCs with tailored characteristics that effectively tackle distinct issues. In the field of medicine, the advancement of mesenchymal progenitor cells for the purpose of targeted drug administration, incorporating precise release mechanisms, has the potential to significantly transform treatment approaches. The growing significance of sustainability has led to a heightened focus on the advancement of fabrication techniques that are environmentally friendly, as well as the development of biocompatible MPCs. These areas of research show great potential to address the pressing need for sustainable solutions. The future trajectory of these materials will be influenced by research endeavors that prioritize the utilization of renewable resources, waste reduction, and the establishment of safe practices for the application of MPCs in biological surroundings.

The issues connected with MPCs are multi-faceted, necessitating the involvement of experts from various disciplines including materials science, chemistry, physics, and engineering. The acceleration of progress and the exploration of novel paths for innovation can be facilitated by collaborative endeavors that involve the integration of expertise from many fields.

In summary, magnetic polymer composites exhibit an intriguing amalgamation of magnetic and polymeric characteristics, showing significant promise in a wide range of applications. Despite the numerous hurdles that exist, it is evident that the future of MPCs holds great promise. Ongoing research endeavors are strategically positioned to surmount existing limits and fully unleash the vast potential of these systems. Through the exploration of manufacturing issues, the improvement of stability, the advancement of theoretical models, and the pursuit of specialized applications, researchers have the potential to propel the field of MPCs toward significant progress. This progress, in turn, can result in the development of transformational technologies that will have a profound impact on our society.

REFERENCES

1. Filipcsei, G., Csetneki, I., Szilágyi, A., and Zrínyi, M. (2007). Magnetic Field-Responsive Smart Polymer Composites, in Oligomers - Polymer Composites - Molecular Imprinting. Advances in Polymer Science, Vol. 206. 2006, Springer. p. 104. https://doi.org/10.1007/12
2. Thevenot, J., et al., *Magnetic responsive polymer composite materials*. Chemical Society Reviews, 2013. **42**(17): pp. 7099–7116.
3. Ganguly, S., and S. Margel, *3D printed magnetic polymer composite hydrogels for hyperthermia and magnetic field driven structural manipulation*. Progress in Polymer Science, 2022. **131**: p. 101574.
4. Wizel, S., et al., *The preparation of metal-polymer composite materials using ultrasound radiation*. Journal of Materials Research, 1998. **13**: pp. 211–216.
5. Liu, H., et al., *Fabrication and functionalization of dendritic poly (amidoamine)-immobilized magnetic polymer composite microspheres*. The Journal of Physical Chemistry B, 2008. **112**(11): pp. 3315–3321.
6. Alicandro, R., M. Cicalese, and M. Ruf, *Domain formation in magnetic polymer composites: An approach via stochastic homogenization*. Archive for Rational Mechanics and Analysis, 2015. **218**: pp. 945–984.
7. Nguyen, V.Q., A.S. Ahmed, and R.V. Ramanujan, *Morphing soft magnetic composites*. Advanced Materials, 2012. **24**(30): pp. 4041–4054.
8. Thomas, S., et al., *Polymer Composites, Macro-and Microcomposites*. Vol. 1. 2012: John Wiley & Sons.
9. Kimura, T., et al., *Polymer composites of carbon nanotubes aligned by a magnetic field*. Advanced Materials, 2002. **14**(19): pp. 1380–1383.

10. Dollig, L.M., et al., *3D printed magnetic polymer composite transformers.* Journal of Magnetism and Magnetic Materials, 2017. **442**: pp. 97–101.

11. Stabik, J., et al., *Magnetic induction of polymer composites filled with ferrite powders.* Archives of Materials Science and Engineering, 2010. **41**(1): pp. 13–20.

12. Ganguly, S., et al., *Polymer nanocomposites for electromagnetic interference shielding: A review.* Journal of Nanoscience and Nanotechnology, 2018. **18**(11): pp. 7641–7669.

13. Simplício, S., et al., *Thermal resistance of magnetic polymeric composites based on styrene, divinylbenzene, and Ni and Co particles.* Journal of Thermal Analysis and Calorimetry, 2014. **117**: pp. 369–375.

14. Liu, J.A.-C., et al., *Photothermally and magnetically controlled reconfiguration of polymer composites for soft robotics.* Science Advances, 2019. **5**(8): p. eaaw2897.

15. Balakrishnan, P., et al., *Natural fibre and polymer matrix composites and their applications in aerospace engineering*, in *Advanced Composite Materials for Aerospace Engineering.* 2016, Elsevier. pp. 365–383.

16. Singh, M.K., *Textiles functionalization-a review of materials, processes, and assessment.* Textiles for Functional Applications, 2021. pp. 43–374. DOI:10.5772/intechopen.96936

17. Alivisatos, P., et al., *From molecules to materials: Current trends and future directions.* Advanced Materials, 1998. **10**(16): pp. 1297–1336.

18. Watanabe, Y., L. Warmington, and N. Gopishankar, *Three-dimensional radiation dosimetry using polymer gel and solid radiochromic polymer: From basics to clinical applications.* World Journal of Radiology, 2017. **9**(3): p. 112.

19. Ganguly, S., et al., *Photopolymerized thin coating of polypyrrole/graphene nanofiber/iron oxide onto nonpolar plastic for flexible electromagnetic radiation shielding, strain sensing, and non-contact heating applications.* Advanced Materials Interfaces, 2021. **8**(23): p. 2101255.

20. Ganguly, S., and S. Margel, *Design of magnetic hydrogels for hyperthermia and drug delivery.* Polymers, 2021. **13**(23): p. 4259.

21. Das, T.K., et al., *A facile green synthesis of silver nanoparticle-decorated hydroxyapatite for efficient catalytic activity towards 4-nitrophenol reduction.* Research on Chemical Intermediates, 2018. **44**: p. 1189–1208.

22. Ganguly, S., *13 challenges prospects of and magnetic future quantum dots.* Magnetic Quantum Dots for Bioimaging, 2023: p. 269. DOI: 10.1201/9781003319870-13

23. Ziv-Polat, O., et al., *Novel magnetic fibrin hydrogel scaffolds containing thrombin and growth factors conjugated iron oxide nanoparticles for tissue engineering.* International Journal of Nanomedicine, 2012: pp. 1259–1274.

24. Boguslavsky, Y., and S. Margel, *Synthesis and characterization of poly (divinylbenzene)-coated magnetic iron oxide nanoparticles as precursor for the formation of air-stable carbon-coated iron crystalline nanoparticles.* Journal of Colloid and Interface Science, 2008. **317**(1): pp. 101–114.

25. Galperin, A., et al., *Radiopaque iodinated polymeric nanoparticles for x-ray imaging applications.* Biomaterials, 2007. **28**(30): p. 4461–4468.

26. Marcus, M., et al., *Iron oxide nanoparticles for neuronal cell applications: Uptake study and magnetic manipulations.* Journal of Nanobiotechnology, 2016. **14**(1): p. 1–12.

27. Ganguly, S., *Preparation/processing of polymer-graphene composites by different techniques*, in *Polymer Nanocomposites Containing Graphene.* 2022, Elsevier. p. 45–74.

28. Trache, D., et al., *Microcrystalline cellulose: Isolation, characterization and bio-composites application—A review.* International Journal of Biological Macromolecules, 2016. **93**: p. 789–804.

29. Merkoçi, A., et al., *New materials for electrochemical sensing VI: Carbon nanotubes.* TrAC Trends in Analytical Chemistry, 2005. **24**(9): pp. 826–838.

30. Mondal, S., et al., *High-performance carbon nanofiber coated cellulose filter paper for electromagnetic interference shielding.* Cellulose, 2017. **24**: pp. 5117–5131.

31. Ganguly, S., et al., *Mussel-inspired polynorepinephrine/MXene-based magnetic nano-hybrid for electromagnetic interference shielding in x-band and strain-sensing performance.* Langmuir, 2022. **38**(12): pp. 3936–3950.

32. Ganguly, S., et al., *Layer by layer controlled synthesis at room temperature of tri-modal (MRI, fluorescence and CT) core/shell superparamagnetic IO/human serum albumin nanoparticles for diagnostic applications.* Polymers for Advanced Technologies, 2021. **32**(10): pp. 3909–3921.

33. Chen, Y., et al., *AI-based reconstruction for fast MRI—A systematic review and meta-analysis.* Proceedings of the IEEE, 2022. **110**(2): pp. 224–245.

34. Fernandes, M.C., M.J. Gollub, and G. Brown, *The importance of MRI for rectal cancer evaluation.* Surgical Oncology, 2022. **43**: p. 101739.

35. Deoni, S.C., et al., *Development of a mobile low-field MRI scanner.* Scientific Reports, 2022. **12**(1): p. 5690.

36. Montiel Schneider, M.G., et al., *Biomedical applications of iron oxide nanoparticles: Current insights progress and perspectives.* Pharmaceutics, 2022. **14**(1): p. 204.

37. Kumar, C.S., *Nanomaterials for Medical Diagnosis and Therapy.* 2007: John Wiley & Sons.

38. Shamim, N., et al., *Thermosensitive polymer (n-isopropylacrylamide) coated nanomagnetic particles: Preparation and characterization.* Colloids and Surfaces B: Biointerfaces, 2007. **55**(1): pp. 51–58.

39. De Paoli, V.M., et al., *Effect of an oscillating magnetic field on the release properties of magnetic collagen gels.* Langmuir, 2006. **22**(13): pp. 5894–5899.

40. Lu, Z., et al., *Magnetic switch of permeability for polyelectrolyte microcapsules embedded with Co@ au nanoparticles.* Langmuir, 2005. **21**(5): pp. 2042–2050.

41. Peters, C., et al., *Degradable magnetic composites for minimally invasive interventions: Device fabrication, targeted drug delivery, and cytotoxicity tests.* Advanced Materials, 2016. **28**(3): pp. 533–538.

42. Perera, A.S., et al., *Polymer–magnetic composite fibers for remote-controlled drug release.* ACS Applied Materials & Interfaces, 2018. **10**(18): pp. 15524–15531.

43. Yusefi, M., et al., *Chitosan coated magnetic cellulose nanowhisker as a drug delivery system for potential colorectal cancer treatment.* International Journal of Biological Macromolecules, 2023. **233**: p. 123388.

44. Moya Betancourt, S.N., C.I. Cámara, and J.S. Riva, *Interaction between pharmaceutical drugs and polymer-coated Fe3O4 magnetic nanoparticles with Langmuir monolayers as cellular membrane models.* Pharmaceutics, 2023. **15**(2): p. 311.

45. Ma, X., et al., *Development of an Fe3O4 surface-grafted carboxymethyl chitosan molecularly imprinted polymer for specific recognition and sustained release of salidroside.* Polymers, 2023. **15**(5): p. 1187.

46. Zhao, D.-L., et al., *Inductive heat property of Fe3O4/polymer composite nanoparticles in an ac magnetic field for localized hyperthermia.* Biomedical Materials, 2006. **1**(4): p. 198.

47. Qindeel, R., N.H. Alonızan, and L.G. Bousiakou, *Magnetic behavior of ferrite-polymer composites for hyperthermia applications.* Journal of Materials Science: Materials in Electronics, 2020. **31**: pp. 19672–19679.

48. Zamora-Mora, V., et al., *Composite chitosan/agarose ferrogels for potential applications in magnetic hyperthermia.* Gels, 2015. **1**(1): pp. 69–80.

49. Makridis, A., et al., *Composite magnetic 3D-printing filament fabrication protocol opens new perspectives in magnetic hyperthermia.* Journal of Physics D: Applied Physics, 2023. **56**(28): p. 285002.

50. Moraes, R.S., et al. *Hyperthermia system based on extrinsically magnetic poly (butylene succinate).* in *Macromolecular Symposia.* 2018. Wiley Online Library.

51. Regmi, R., et al., *Hyperthermia controlled rapid drug release from thermosensitive magnetic microgels.* Journal of Materials Chemistry, 2010. **20**(29): pp. 6158–6163.

52. Matos, R.J., et al., *Electrospun composite cellulose acetate/iron oxide nanoparticles non-woven membranes for magnetic hyperthermia applications.* Carbohydrate Polymers, 2018. **198**: pp. 9–16.

53. Joyee, E.B., and Y. Pan, *A fully three-dimensional printed inchworm-inspired soft robot with magnetic actuation.* Soft Robotics, 2019. **6**(3): pp. 333–345.

54. Kim, Y., and X. Zhao, *Magnetic soft materials and robots.* Chemical Reviews, 2022. **122**(5): pp. 5317–5364.

55. Dong, G., Q. He, and S. Cai, *Magnetic vitrimer-based soft robotics.* Soft Matter, 2022. **18**(39): pp. 7604–7611.

56. Laschi, C., B. Mazzolai, and M. Cianchetti, *Soft robotics: Technologies and systems pushing the boundaries of robot abilities.* Science Robotics, 2016. **1**(1): p. eaah3690.

57. Peyer, K.E., et al., *Magnetic polymer composite artificial bacterial flagella.* Bioinspiration & Biomimetics, 2014. **9**(4): p. 046014.

58. Joyee, E.B., and Y. Pan, *Additive manufacturing of multi-material soft robot for on-demand drug delivery applications.* Journal of Manufacturing Processes, 2020. **56**: pp. 1178–1184.

59. Schmauch, M.M., et al., *Chained iron microparticles for directionally controlled actuation of soft robots.* ACS Applied Materials & Interfaces, 2017. **9**(13): pp. 11895–11901.

2 Various Approaches to the Synthesis of Magnetic Nanoparticles

Christian Chapa González

2.1 INTRODUCTION TO MAGNETIC NANOPARTICLES

Nanoparticles exhibit interesting properties depending on their nature, size, and composition [1, 2]. For this reason, nanoparticles have been developed for application in several areas such as environmental remediation [3–5], artificial lighting [6, 7], catalysis [8, 9], semiconductors [10, 11], and biomedicine [12–16]. Magnetic particles consist of elements such as iron, nickel, and cobalt or some chemical compounds. The magnetic behavior largely depends on the particle size.

2.1.1 MAGNETISM IN NANOSCALE PARTICLES

Magnetism arises because of the spin property of electrons, and it is precisely the combination of this property and the magnetic moments of each electron that causes the atom to have an overall magnetic moment. A global magnetic moment will result if the valence layer contains unpaired electrons. Depending on their response to an applied magnetic field, materials can be classified as diamagnetic (opposed to the magnetic field), paramagnetic (magnetized in the direction of the applied field), or in some ordered magnetic state such as ferromagnetic, ferrimagnetic, or antiferromagnetic.

Ferromagnetic and ferrimagnetic substances are strongly attracted to a magnetic field. They contain unpaired electrons whose moments result from interactions between neighboring spins, at least partially aligned, even in the absence of a magnetic field. The coupling of energetic spins is positive. In a ferromagnetic material, magnetic moments tend to align in the same direction and sense. These materials consist of domains that, upon the application of a magnetic field, tend to align. As a result, domains in which dipoles are oriented in the same direction and along the field's orientation will expand. In a ferromagnetic substance, the alignment of electronic spins is parallel. These substances have a net magnetic moment, long magnetic permeability, and susceptibility (0.01 to 10^6). With increasing temperature, spin alignment decreases due to thermal fluctuations of individual magnetic moments, and susceptibility rapidly decreases.

In contrast, antiferromagnetic substances exhibit electronic spins with equal magnetic moments aligned in an antiparallel manner. These substances possess a global magnetic moment of zero, positive permeability, and a very small positive susceptibility (ranging from 0 to 0.1). Elevating the temperature typically leads to an increase

DOI: 10.1201/9781003454236-2

in susceptibility as the antiparallel alignment is disrupted. Ferrimagnetic substances, like antiferromagnetic materials, entail an alignment of antiparallel spins. However, in a ferrimagnetic substance, the different spins have unequal moments, resulting in a net magnetic moment for the material.

Ferromagnetic, antiferromagnetic, and ferrimagnetic materials exhibit a domain structure, with particles within the size range of a few nanometers typically forming a single domain. Within a domain, spins align either in a parallel or an antiparallel configuration, while different domains display varying spin orientations. To eliminate domains in a ferro- or antiferromagnetic material, a sufficiently high magnetic field is necessary; with the increasing magnetic field, spins in the domains gradually align. At a critical magnetic field strength, saturation magnetization is achieved, signifying that all spins within the domains have become parallel.

Achieving a state of magnetic saturation, the material's magnetization reaches its maximum value, M_s. When the magnetic field decreases to zero, the material still has residual magnetic moment, and the value of magnetization when the magnetic field is zero is called remanent magnetization, M_r. To bring the material to a zero-magnetization value after it has been magnetized to saturation, a magnetic field in the negative direction must be applied, and the magnitude of the field is called the coercive field, H_c (Figure 2.1). This parameter allows measurement of the resistance of a ferromagnetic material to demagnetization.

2.1.2 SUPERPARAMAGNETISM

In this context, when referring to "magnetic nanoparticles," the discussion typically centers around materials characterized by the property of superparamagnetism. This form of magnetism manifests in small, single-domain ferromagnetic and ferrimagnetic nanoparticles. In a simplified conceptualization, the collective magnetic moments of a nanoparticle can be envisioned as a consolidated, giant magnetic moment, encompassing all individual moments of the constituent atoms. In the absence of an applied magnetic field, the net moment is zero. Upon the application

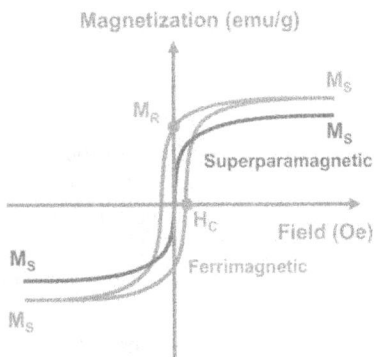

FIGURE 2.1   Hysteresis curve of applied field (H)–magnetization (M) for ferrimagnetic and superparamagnetic materials.

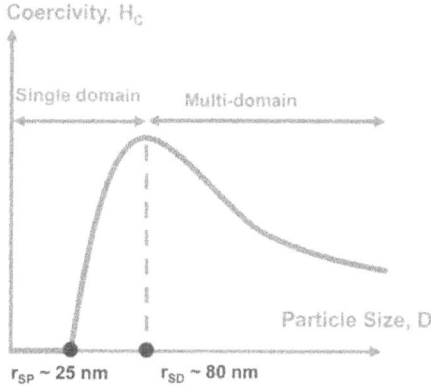

FIGURE 2.2   Coercivity behavior as a function of particle sizes and domains.

of a magnetic field, nanoparticles exhibit behavior akin to that of a paramagnet, with the noteworthy distinction that their magnetic susceptibility is significantly extended (hence the term "super"). In the realm of nanoparticles, theoretical findings indicate that, due to thermal fluctuations potentially hindering the establishment of stable magnetization, H_c tends to approach zero, as depicted in Figure 2.2. In consequence, superparamagnetism is characterized by two experimental criteria: the lack of hysteresis in the magnetization curve and the overlap of magnetization curves at different temperatures.

Very often, nanoparticles show a certain preference for a direction during magnetization alignment. It is said that these nanoparticles have anisotropy in these directions. If nanoparticles mainly align in a single direction, it is called uniaxial anisotropy. Superparamagnetism arises because of magnetic anisotropy. Nanoparticles with uniaxial anisotropy randomly flip in the direction of their magnetization, behaving as if they did not have a net magnetic moment. This effect is due to thermal energy. The average time to carry out the direction change is given by the relaxation time:

$$\tau = \tau_0 \exp\left(\frac{\Delta E}{k_B \cdot T}\right)$$

ΔE represents the energy barrier during inversion, and $k_B \cdot T$ stands for thermal energy. In the case of non-interactive particles, the pre-exponential factor τ_0 falls within the range of 10^{-10} to 10^{-12} s and exhibits weak dependence solely on temperature. The energy barrier is defined as $\Delta E = K \cdot V$, where K signifies the anisotropy energy density and V corresponds to the particle's volume. With small particles, ΔE becomes comparable to $k_B \cdot T$ at room temperature, underscoring the significance of superparamagnetism for particles of diminutive dimensions, also called nanoparticles. It is important to note that the size of magnetic nanoparticles is closely related to synthetic approaches.

2.2 OVERVIEW OF BIOMEDICAL APPLICATIONS OF MAGNETIC NANOPARTICLES

Magnetic nanoparticles have demonstrated significant utility across diverse scientific disciplines, particularly in the biomedical domain, offering the capability of remotely manipulating entities labeled with magnets through the application of magnetic fields. Noteworthy advancements have been achieved through the integration of magnetic particles in diagnostic, therapeutic, and medical imaging applications. Furthermore, magnetic particles have exhibited excellence in tasks such as bioseparation, purification, and the identification of molecules, cells, and microorganisms adhered to these particles. Regarding biosensors, magnetic particles are very useful in remote manipulation of components in samples and as transduction labels in magnetic sensors.

Magnetic nanoparticles find diverse applications in biomedical engineering, particularly in the development of drug delivery systems and hyperthermia for cancer treatment. Consequently, a comprehensive understanding of the optimal magnetic nanoparticle properties is essential for advancing studies in cancer treatment applications. Moreover, magnetic nanoparticles serve as contrast agents in magnetic resonance imaging (MRI), leveraging characteristics like superparamagnetism and low toxicity. The synthesis of iron oxide nanoparticles has demonstrated significant sensitivity in imaging and detecting ovarian cancer. Furthermore, the modulation of the coating composition of iron oxide nanoparticles allows for the synthesis of particles with distinct contrasts and low cytotoxicity. Magnetic nanoparticles, specifically superparamagnetic iron oxide nanoparticles (SPIONs), have emerged as a groundbreaking discovery in the field of tissue engineering, showcasing diverse applications. These nanoparticles, when placed in proximity to damaged tissue, have been shown to bind proteins by inducing a temperature increase of up to 50°C. Moreover, SPIONs play a crucial role in tracking and locating stem cells within damaged tissue. Beyond tissue engineering, SPIONs find regular application in magnetic fluid hyperthermia (MFH) therapy, particularly as a treatment for cancer. In this context, nanoparticles are intravenously injected at the site of a cancerous tumor, followed by the application of an alternating magnetic field. The resultant temperature elevation of the nanoparticles selectively affects cancer cells, exploiting their heightened sensitivity to higher temperatures compared to the surrounding healthy cells.

2.2.1 PARAMETERS DETERMINING THE APPLICATIONS OF MAGNETIC NANOPARTICLES

Magnetic nanoparticles are synthesized in such a way that their properties are suitable for the desired application. In the context of biomedical applications, special care must be given to properties such as size, shape, and surface charge. For example, if the biomedical application is a matter of delivering drugs in the body, it must be borne in mind that size is linked to drug circulation and biodistribution. A size ranging from 1 to 100 nm allows for a longer half-life. Simultaneously, smaller sizes enable better cellular acceptance compared to macroparticles. Concerning nanoparticle elimination from the body, those smaller than 10 nm can pass through blood vessels and be excreted by the kidneys, while larger ones are captured and eliminated by the mononuclear phago-cytic system [17]. Another factor influenced by size is drug release; smaller NPs, with a larger surface area, release the drug more rapidly. Hence, the ideal size for targeted drug administration ranges from 2 to 200 nm [18], and spherical particles within this size range enter cells through clathrin-mediated endocytosis [19]. In the same way, shape impacts nanoparticles' performance, depending on the cell and cancer type being treated. Shape affects cellular acceptance, biodistribution, and blood-brain barrier pen-etration efficiency. For example, rod-shaped nanoparticles show better acceptance in brain tissue compared to spherical ones [20]. It also determines the adhesion pattern of nanoparticles, with non-spherical NPs exhibiting higher carrier efficiency. In the same sense, surface charge affects biodistribution, cellular acceptance, and interaction with other biological environments. Positively charged nanoparticles are generally more easily internalized, especially through adsorption-mediated transcytosis. Positively charged nanoparticles are more readily accepted by proliferative cells, whereas nega-tively charged nanoparticles diffuse more rapidly in tumor tissue [18, 20]. To enhance cellular acceptance, polymeric coatings such as chitosan are employed. This method works by disrupting tight junction integrity to facilitate the passage of more nanopar-ticles to the specific site [17]. Surface charge also influences cytotoxicity, as cationic nanoparticles tend to interact with negatively charged cell membranes, and a positively charged and small-sized particle leads to greater cytotoxicity in tumor cells [18, 21].

Indubitably, superparamagnetism is a highly sought-after property for specific biomedical applications. This characteristic defines the behavior of nanoparticles, showcasing their elevated magnetic susceptibility and saturation when subjected to a magnetic field. But magnetic nanoparticles must also have colloidal stability. The colloidal stability of nanoparticles is intricately linked to their capacity to prevent agglomeration, enabling them to persist as a colloidal, monodispersed entity within the solution or suspension. Agglomeration adversely impacts the superparamagnetic properties of nanoparticles, necessitating surface modification. Surface modification agents, comprising organic molecules (such as citric acid, ascorbic acid, and antibod-ies, among others) or biopolymers (like dextran, chitosan, and PEG, among others), must adhere to the criteria of being non-toxic or biocompatible.

Furthermore, the property of magnetic nanoparticles known as magnetic suscep-tibility is defined as the variation in magnetization due to the applied field strength. The susceptibility of ferromagnetic materials tends to change with temperature. In applications of magnetic nanoparticles for use in magnetic hyperthermia, the mag-netic field to be applied should be of low radio frequency, in the range of 100–300 kHz, so that the temperature achieves the therapeutic effect and the cancer cells are destroyed. All these properties are strongly influenced by the method of obtaining magnetic nanoparticles.

2.3 SYNTHETIC APPROACHES TO OBTAIN MAGNETIC NANOPARTICLES

Identifying optimal synthesis processes for magnetic nanoparticles in spe-cific applications is crucial for researchers conducting in vitro or in vivo experi-ments. Moreover, the preparation and synthesis of magnetic nanoparticles, such as magnetite, maghemite, or others, must consider the diverse morphologies and surfaces of each nanoparticle, making these processes crucial. Despite positive findings reported in various articles and reviews regarding the use of magnetic nanoparticles in biomedical applications, there is a pressing need to refine preparation and synthesis methods. Improving these methods holds the potential to enhance outcomes in many applications. In recent years, considerable attention has been dedicated to studying the relationship among the size, shape, composition, and surface modification of nanoparticles to identify optimal properties for specific applications. The ongoing progress in the development of magnetic nanoparticles, particularly in biomedical applications, is driven by their ability to induce a thermal effect in an alternating magnetic field (AMF) [22, 23]. In both clinical and experimental applications, the heat generated by magnetic nanoparticles is influenced by the frequency (f) and field strength (H) of the AMF, with experiments constrained by a range ensuring biological safety and physiological tolerance ($H \cdot f < 4.85 \times 10^8$ Am^{-1}s^{-1}). Additionally, beyond field parameters, magnetic nanoparticles dissipate heat through relaxation losses [24]. Among other properties, such as anisotropy energy and rate of magnetic relaxation, the heating efficiency of magnetic nanoparticles depends on their saturation magnetization (M_s). Thus, several research groups have established versatile methods to develop superparamagnetic nanoparticles [25–27]. Researchers have reported the variance in the magnetic properties of nanoparticles when they change the particle size, composition, and coatings. Hence, there is a need to exercise control over the properties of nanoparticles through experimental variables in the synthesis method.

Controlling experimental variables in the synthesis of magnetic nanoparticles is paramount for achieving desired outcomes, especially when targeting specific compounds like magnetite. The importance of carefully considering the stoichiometry of the reaction cannot be overstated. In the case of iron oxides, approximately 16 different compounds can form because of the intricate interplay between oxygen and iron. Failure to meticulously control the synthesis process may lead to undesired products, with common outcomes being the formation of phases like maghemite (γ-Fe$_2$O$_3$) and hematite (α-Fe$_2$O$_3$) [28–30]. Therefore, a precise understanding and careful management of experimental parameters are crucial to guide the synthesis toward the desired magnetic nanoparticle and avoid unintended outcomes.

There are various methods for the preparation and synthesis of magnetic nanoparticles, which are classified into physical and chemical methods. The methods within each group yield different results in nanoparticle production. Understanding the available options for nanoparticle preparation is crucial for determining the best alternative for a study or research.

2.3.1 PHYSICAL METHODS

2.3.1.1 The Gas-Phase Synthesis Method

The gas-phase synthesis technique can be performed through physical vapor deposition (PVD) or chemical vapor deposition (CVD). While it allows the synthesis of particles in large quantities with higher purity than liquid-based methods do, maintaining the nanoscale size throughout the experiment is challenging. Magnetron sputtering is a versatile method for synthesizing magnetic nanoparticles through inert-gas cooling and gas-phase condensation. This technique, marked by advancements like improved geometry and multiple-target usage, provides control over magnetic nanoparticles' size and crystallinity. In a high-vacuum gas-aggregation magnetron-sputter deposition system, a target material undergoes bombardment by Ar ions, generating free atoms for cluster formation. The growth process involves rapid cooling, seed formation, and subsequent coalescence. Control parameters, including pressure differentials and mass selection, enable precise manipulation of magnetic nanoparticles' properties, making magnetron sputtering a flexible and promising approach for tailored magnetic nanoparticle synthesis, with applications in various fields.

2.3.1.2   Laser-Induced Pyrolysis

Laser-induced pyrolysis involves heating a gaseous mixture of iron precursors using a laser. This method stimulates a substance that absorbs laser light and transfers it to a reaction medium, where it causes a sharp increase in temperature. Reactants that are heated to a high temperature break down. These substances, known as precursors, dissociate to generate nanoparticles, which exit the flame and experience a quench, or abrupt cooling, effect. The particles' ability to develop is halted by this abrupt reduction in temperature. This yields particles that are nanoscale. In the end, the differentiable factors (precursor kinds, pressures and flow rates; laser power) yield a variety of products with a broad range of sizes, crystallinities, and chemical compositions, as well as nanopowders with good physical and chemical homogeneity.

The synthesis of magnetic nanoparticles from a hydrolysable precursor using laser pyrolysis exhibits several advantages. The flame stability is remarkable, characterized by a sharp and consistent flame without pulsation or flashing, ensuring a reliable reaction. Raman spectroscopy confirms the superior stability of laser pyrolysis compared to pyrosol, emphasizing its robust performance. Reproducibility is a key strength, as the nanostructures obtained exhibit perfect consistency from one pyrolysis to another. Under similar pyrolysis conditions, the specific surface area and particle diameters demonstrate high reproducibility, emphasizing the precision and reliability of the synthesis process. The technique also ensures a narrow grain size dispersion, typically below 40%. Even under different precursor flow rate conditions, the powders maintain a uniform morphology, as depicted in histograms of particle size measurements. This narrow dispersion, around 35%, indicates a high degree of control over the synthesis process, contributing to the overall reliability and precision of the method. Laser pyrolysis technique for synthesizing magnetic nanoparticles offers exceptional advantages, including outstanding flame stability, high production efficiency, excellent reproducibility, and a narrow grain size dispersion. These attributes collectively make the method a robust and controlled approach for nanoparticle synthesis [31–33].

2.3.1.3   High-Energy Ball Milling

Utilizing balls within a mill, high-energy ball milling (HEBM) forms uniform, small-sized magnetic nanoparticles. In this process, the sample preparation involves subjecting the starting material to mechanical forces generated by high-energy ball collisions in a milling apparatus. The milling process is conducted in controlled air conditions using a ball mill machine, with a ball-to-powder weight ratio and a determined milling speed. Various samples are milled for durations ranging from minutes to hours. Subsequently, the as-milled samples could undergo annealing under flowing controlled atmosphere to manipulate grain size and minimize microstrain, dislocation density, and oxygen contamination [34–36].

The HEBM synthesis process unfolds through distinct stages. In the initial stage, powder particles experience flattening due to compressive forces resulting from ball collisions, inducing changes in particle shapes. As the process progresses to the intermediate stage, cold welding becomes pronounced, causing significant alterations in the powder constituents' mixture. Fracturing and cold welding dominate this stage, accompanied by possible dissolution. Moving to the final stage, a noticeable refinement and reduction in particle size occurs, rendering the microstructure more homogeneous at a microscopic scale. It is likely that true alloys have formed by this point. In the completion stage, powder particles exhibit an extremely deformed metastable structure. Lamellae are no longer discernible by optical microscopy, and further mechanical alloying fails to enhance dispersoid distribution. A genuine alloy, akin to the starting constituents' composition, is ultimately formed. Particle size depends on the time and rotation speed of the balls in the mill. However, nanoparticle agglomeration is a drawback, which can be addressed using coatings or surfactants.

2.3.2 Chemical Methods

2.3.2.1 Solution Combustion Method

Also known simply as the combustion method, the solution combustion method is commonly used for ferrites and other oxides. In this method, precursor reactants are prepared in a homogeneous mixture and exposed to temperatures of up to 1500°C using fuels. The resulting powder is crystalline and nanosized, yielding different iron oxides depending on the precursor used. Combustion synthesis stands out as an eco-friendly and energy-efficient technique, offering scalability and continuous production capabilities. This method has been widely explored for creating various nanoscale materials, including nitrides, sulfides, metals, and alloys. Among them, of course, are magnetic nanoparticles for diverse applications. The combustion approach has increased prominence for its simplicity, speed, and cost-effectiveness. The process involves self-propagating and exothermic reactions, leading to the formation of high-purity, homogeneous powders with fine particle sizes. Remarkably, the choice of organic compounds as fuels, such as glycine, urea, citric acid, and alanine, influences the properties of the synthesized magnetic nanoparticles. This method facilitates the uniform distribution of metal ions within the resulting nanoparticles, ensuring single-phase synthesis without the need for additional calcination steps. The advantages of combustion synthesis become apparent when it is compared to other methods like wet chemical processes. Unlike multi-step wet chemical techniques, combustion synthesis operates as a single-step process, simplifying the production procedure significantly. Furthermore, this method offers excellent scalability, ensuring high yields and continuous large-scale production. The ease of scalability and the ability to achieve substantial quantities of nanoparticles make combustion synthesis particularly attractive for industrial applications. The versatility of this approach is underscored by the variety of organic fuels employed, with urea being a standout choice due to its wide availability, low cost, safety, and high purity. Recent reviews and comparisons with wet-based methods highlight the efficiency and practicality of combustion synthesis, making it a promising avenue for nanoparticle production, particularly in laboratory and semi–pilot scale settings.

The synthesis of nanoparticles by combustion represents a fundamental technique in nanotechnology, noted for its efficiency and versatility in the production of nanometric materials with diverse applications. The process starts with the careful preparation of a precursor solution. This solution is composed of specific reagents, such as metallic nitrates like ferric nitrate and organic compounds like citrate, or urea [37]. The selection of these reagents and their molar ratio (represented by ϕ) are crucial, dictating the final properties of the nanoparticles. The next step involves the application of heat using a suitable device, such as a heating mantle or a microwave oven. The choice of device depends on the type of synthesis required. During the combustion process, the reaction should be closely monitored. The release of gases and changes in the color of the solution indicate the onset and completion of the exothermic reaction. After completion of the combustion reaction, the resulting powders are collected and can be transformed into colloidal suspensions by the addition of distilled water [38–42].

2.3.2.2   Chemical Co-precipitation

One of the most often used methods for synthesizing iron oxide nanoparticles such as magnetite, chemical co-precipitation has proven to be a minimally damaging and easily implementable method. The chemical co-precipitation method stands out as a robust and efficient approach for synthesizing magnetic nanoparticles from aqueous salt solutions. This method involves the controlled addition of a base as a precipitating agent, either at ambient or elevated temperature. Despite its apparent simplicity, this method provides the advantage of producing a substantial quantity of nanoparticles and achieving precise control over the particle size distribution. The procedure employs a carefully arranged chemical co-precipitation reaction, utilizing precursor solutions containing Fe^{3+} and Fe^{2+} ions in a stoichiometric 2:1 ratio. For example, the experimental procedure commences with the precise weighing of 1.623 g of ferric chloride ($FeCl_3$) and 0.760 g of ferrous sulfate [$Fe_2(SO_4)$]. Subsequently, each weighed substance is placed in a separate 100 mL volumetric flask. Deionized water is added to each flask to achieve a precisely controlled total volume of 100 mL. The solutions from individual flasks are then combined into a 500 mL beaker, and the mixture is subjected to continuous magnetic stirring and controlled heating until it reaches the prescribed temperature of 50°C. At this point, 100 mL of ammonium hydroxide (NH_4OH) is rapidly injected into the mixture, leading to a noticeable color change (Figure 2.3). The process continues with magnetic stirring and controlled heating until reaching 80°C, followed by a cooling phase to room temperature.

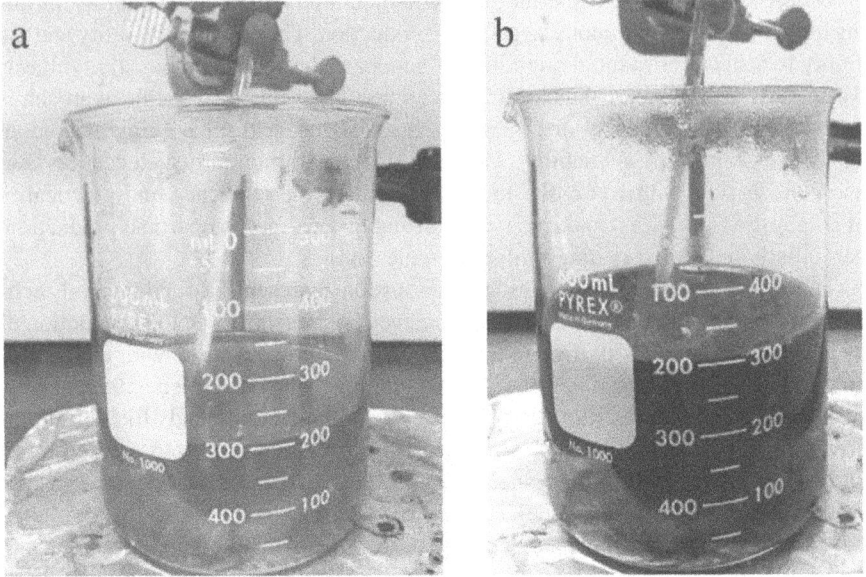

FIGURE 2.3 Transformation process: (a) the precursor salt mixture undergoes magnetic agitation and heat, displaying an orange hue, and (b) a notable change in color occurs in the mixture upon the addition of 100 mL of ammonium hydroxide at 50°C, resulting in the formation of a black precipitate.

The subsequent step involves centrifuging the mixture in 15 mL tubes at 3000 rpm for 5 minutes. This step is repeated several times, until the mixture reaches pH 7. The magnetite nanoparticles obtained are then transferred to a crucible and placed in an oven set at 50°C for precisely 24 hours for controlled drying (Figure 2.4).

The co-precipitation method is also employed to obtain different kind of ferrite nanoparticle—for example, nickel or cobalt ferrites [43–45]. For instance, maintaining a molar ratio of 2:1, as per the stoichiometric formula of magnetite, precursor solutions of iron(III) chloride (0.1 M), iron(II) sulfate (0.05 M), and nickel sulfate (0.05 M) are dissolved in distilled water, each with a volume of 100 mL. The synthesis process involves the mixing of iron(III) and iron(II) solutions to produce magnetite. For nickel ferrite, the iron(II) solution is replaced with the nickel solution. Nickel-doped magnetite nanoparticles are synthesized, where the precursors include ferric chloride, ferrous sulfate, and nickel sulfate. The ferric chloride solution remains consistent across all substitutions (0.1 M, 100 mL). The stoichiometric ratio of 2:1 [Fe^{3+}:(Ni/Fe)$^{2+}$] is achieved by preparing a mixture of ferrous and nickel sulfate solutions.

2.3.2.3 Sol-Gel Method

Because of its features such as good uniformity, low cost, and high purity, the sol-gel approach stands out among alternative chemical synthesis methods for metal oxides.

FIGURE 2.4   Result of drying the black precipitate formed during the co-precipitation reaction at a temperature of 50°C for 24 hours.

This process has recently been developed for the synthesis of magnetite nanoparticles from metallorganic precursors. Despite the capacity to produce highly crystalline and evenly sized magnetite nanoparticles, conventional synthetic processes have limits for large-scale and cost-effective manufacturing, frequently requiring expensive and hazardous reagents and intricate synthetic steps.

Ferrite-based nanocomposites can be produced effectively and economically using the sol-gel method, which gives you control over the material's structure and characteristics. One of the most important parts of the process is the development of stable sols, which is followed by casting or shape formation, gelation of the sols, aging, drying, calcination, and, if required, sintering. The sol-gel process is noteworthy for its ability to handle a wide range of precursors, not just alkoxides.

Sol-gel is a unique chemical synthesis method for metal oxides that has several advantages over other processes, including high purity, cheap cost, and good uniformity. Recently, metallorganic precursors have been used to adapt this process for the synthesis of magnetite nanoparticles. While some synthetic methods can provide highly crystalline and evenly sized magnetite nanoparticles, their large-scale and cost-effective production is limited. These methods frequently include complex synthetic stages and costly, hazardous reagents. The sol-gel process is a bottom-up synthesis technique that forms its final products through irreversible chemical processes. Research investigated the wet gel's drying process and the mechanism and selection of appropriate catalysts. It is difficult to keep the gel structure intact when the solvent is being removed. For solvent separation, xerogel and supercritical techniques are used. While the supercritical approach reduces changes in the solid network, producing aerogel with low strength and porous networks, the xerogel method involves spontaneous drying in the atmosphere after synthesis.

In that regard, the technique for making magnetite nanoparticles entails dissolving 0.2 mol ferric nitrate in 100 mL ethylene glycol, which results in the production of a brown gel. After aging and drying, the xerogel is annealed under a vacuum at diverse temperatures, giving various-sized magnetite nanoparticles. This complete technique provides for exact control over the final structure and attributes of sol-gel products based on the development of a colloidal structure known as a sol via hydrolysis processes and metal precursor polycondensation. The sol is subsequently dehydrated chemically, resulting in the creation of a liquid phase known as a gel. The sol-gel approach has proven to be an effective way of manufacturing nanoparticles with precise control over size and structure [46–51].

2.4 CONCLUDING REMARKS

The synthesis of magnetic nanoparticles plays a central role in various industries, with a significant impact on biomedical applications, wastewater treatment, and electronics. The unique properties of magnetite nanoparticles, including their non-toxic nature and superparamagnetic characteristics, have fueled increasing demand, particularly in the medical sector. The ability to manipulate these nanoparticles using a magnetic field opens avenues for applications in drug delivery, microfluidics, and magnetic resonance imaging. Moreover, their application in diagnosing and treating life-threatening and chronic diseases, such as cancer and brain tumors, adds another dimension to their significance in the medical industry.

The versatility of magnetic nanoparticles extends to material science, magnetic fluid recording, catalysis, and biomedicines, offering a wide range of applications. Various synthesis approaches, mentioned in this chapter, provide avenues for tailoring the shape, size, magnetic properties, and surface chemistry of these nanoparticles. In conclusion, the synthesis of magnetic nanoparticles not only opens new frontiers in scientific research but also addresses the growing demands in diverse industries, making them a key player in advancing technologies and applications with significant societal implications.

REFERENCES

1. A. L. Tiano et al., "Correlating size and composition-dependent effects with magnetic, Mössbauer, and pair distribution function measurements in a family of catalytically active ferrite nanoparticles," Chemistry of Materials, vol. 27, no. 10, pp. 3572–3592, May 2015. doi: 10.1021/acs.chemmater.5b00767.
2. F. J. Carrillo-Pesqueira, R. C. Carrillo-Torres, M. E. Álvarez-Ramos, and J. Hernández-Paredes, "Synthesis and characterization of silica nanoparticles obtained by additions of amino acids into the Stöber reaction," Microscopy and Microanalysis, vol. 24, no. S1, pp. 1098–1099, 2018. doi: 10.1017/S1431927618005974.
3. D. M. de Santiago Colín et al., "Sonochemical coupled synthesis of Cr-TiO2 supported on Fe3O4 structures and chemical simulation of the degradation mechanism of malachite green dye," Journal of Photochemistry and Photobiology A: Chemistry, vol. 364, pp. 250–261, Sep. 2018. doi: 10.1016/J.JPHOTOCHEM.2018.06.004.

4. B. I. Kharisov, O. V. Kharissova, and H. V. R. Dias, *Nanomaterials for Environmental Protection*, Wiley Blackwell, 2015. doi: 10.1002/9781118845530.
5. B. E. Monárrez-Cordero, A. Sáenz-Trevizo, L. M. Bautista-Carrillo, L. G. Silva-Vidaurri, M. Miki-Yoshida, and P. Amézaga-Madrid, "Simultaneous and fast removal of As3+, As5+, Cd2+, Cu2+, Pb2+ and F− from water with composite Fe-Ti oxides nanoparticles," *Journal of Alloys and Compounds*, vol. 757, pp. 150–160, Aug. 2018. doi: 10.1016/J.JALLCOM.2018.05.013.
6. D.-L. Flores, E. Gutierrez, D. Cervantes, M. Chacon, and G. Hirata, "White-light emission from Y2SiO5:Ce3+, Tb3+ and Sr2Si5N8:Eu2+ phosphor blends: A predictive model," *Micro & Nano Letters*, vol. 12, no. 7, pp. 500–504, 2017. doi: 10.1049/mnl.2017.0154.
7. I. V. García-Amaya *et al.*, "Influence of Eu2O3 on phase crystallization and nanocrystals formation in tellurite glasses," *Journal of Non-Crystalline Solids*, vol. 499, pp. 49–57, Nov. 2018. doi: 10.1016/J.JNONCRYSOL.2018.07.018.
8. J. Juárez, A. Cambón, S. Goy-López, A. Topete, P. Taboada, and V. Mosquera, "Obtention of metallic nanowires by protein biotemplating and their catalytic application," *Journal of Physical Chemistry Letters*, vol. 1, no. 18, pp. 2680–2687, 2010. doi: 10.1021/jz101029f.
9. V.-F. Ruiz-Ruiz *et al.*, "Mechanochemically obtained Pd-Ag nanoalloys. Structural considerations and catalytic activity," *Materialia (Oxf)*, vol. 4, pp. 166–174, Sep. 2018. doi: 10.1016/J.MTLA.2018.09.031.
10. H. Rojas-Chávez, J. L. González-Domínguez, R. Román-Doval, J. M. Juárez-García, N. Daneu, and R. Farías, "ZnTe semiconductor nanoparticles: A chemical approach of the mechanochemical synthesis," *Materials Science in Semiconductor Processing*, vol. 86, pp. 128–138, Nov. 2018. doi: 10.1016/J.MSSP.2018.06.029.
11. P. V. Kamat, K. Murakoshi, Y. Wada, and S. Yanagida, "Semiconductor nanoparticles," *Handbook of Nanostructured Materials and Nanotechnology*, pp. 291–344, Jan. 2000. doi: 10.1016/B978-012513760-7/50037-X.
12. R. Esquivel, J. Juárez, M. Almada, J. Ibarra, and M. A. Valdez, "Synthesis and characterization of new thiolated chitosan nanoparticles obtained by ionic gelation method," *International Journal of Polymer Science*, vol. 2015, pp. 1–18, Dec. 2015. doi: 10.1155/2015/502058.
13. S. Correa-Espinoza, C. A. Rodríguez-González, S. A. Martel-Estrada, and J. F. Hernández-Paz, and I. Olivas-Armendáriz, "Synthesis of Ag2S quantum dots and their biomedical applications." http://cathi.uacj.mx/20.500.11961/3927
14. K. Maldonado-Lara *et al.*, "Preparation and characterization of copper chitosan nanocomposites with antibacterial activity for applications in tissue engineering," *Revista Mexicana De Ingeniería Biomédica*, vol. 38, no. 1, pp. 306–313, 2017.
15. J. Roacho-Perez, H. Gallardo-Blanco, M. Sanchez-Dominguez, P. Garcia-Casillas, C. Chapa-Gonzalez, and C. Sanchez-Dominguez, "Nanoparticles for death-induced gene therapy in cancer (review)," *Molecular Medicine Reports*, vol. 17, no. 1, pp. 1413–1420, 2017. doi: 10.3892/mmr.2017.8091.
16. T. Ruthradevi *et al.*, "Investigations on nickel ferrite embedded calcium phosphate nanoparticles for biomedical applications," *Journal of Alloys and Compounds*, vol. 695, pp. 3211–3219, Feb. 2017. doi: 10.1016/J.JALLCOM.2016.11.300.
17. G. Adamo, S. Campora, and G. Ghersi, "Chapter 3 - Functionalization of Nanoparticles in Specific Targeting and Mechanism Release," in *Micro and Nano Technologies,*, Elsevier, 2017, pp. 57–80. doi: https://doi.org/10.1016/B978-0-323-46142-9.00003-7.
18. D. H. Jo, J. H. Kim, T. G. Lee, and J. H. Kim, "Size, surface charge, and shape determine therapeutic effects of nanoparticles on brain and retinal diseases," *Nanomedicine*, vol. 11, no. 7, pp. 1603–1611, 2015. doi: https://doi.org/10.1016/j.nano.2015.04.015.

19. J. Zhang, H. Tang, Z. Liu, and B. Chen, "Effects of major parameters of nanoparticles on their physical and chemical properties and recent application of nanodrug delivery system in targeted chemotherapy," *International Journal of Nanomedicine*, vol. 12, pp. 8483–8493, Nov. 2017. doi: 10.2147/IJN.S148359.

20. J. Li *et al.*, "Nanoparticle drug delivery system for glioma and its efficacy improvement strategies: A comprehensive review," *International Journal of Nanomedicine*, vol. 15, pp. 2563–2582, Apr. 2020. doi: 10.2147/IJN.S243223.

21. J. Basso *et al.*, "Sorting hidden patterns in nanoparticle performance for glioblastoma using machine learning algorithms," *International Journal of Pharmaceutics*, vol. 592, p. 120095, 2021. doi: https://doi.org/10.1016/j.ijpharm.2020.120095.

22. S. R. Gunakala, V. M. Job, P. V. S. N. Murthy, and S. Sakhamuri, "Influence of alternating magnetic field on non-newtonian blood perfusion and transport of nanoparticles in tissues with embedded blood vessel during hyperthermia," *Ain Shams Engineering Journal*, vol. 14, no. 1, Feb. 2023. doi: 10.1016/J.ASEJ.2022.101831.

23. M. Dan, Y. Bae, T. A. Pittman, and R. A. Yokel, "Alternating magnetic field-induced hyperthermia increases iron oxide nanoparticle cell association/uptake and flux in blood-brain barrier models," *Pharmaceutical Research*, vol. 32, no. 5, pp. 1615–1625, 2015. doi: 10.1007/s11095-014-1561-6.

24. P. Ilg, and M. Kröger, "Dynamics of interacting magnetic nanoparticles: Effective behavior from competition between Brownian and Néel relaxation," *Physical Chemistry Chemical Physics*, vol. 22, no. 39, pp. 22244–22259, 2020. doi: 10.1039/D0CP04377J.

25. D. Ling, and T. Hyeon, "Chemical design of biocompatible iron oxide nanoparticles for medical applications," *Small*, vol. 9, no. 9–10, pp. 1450–1466, 2013. doi: 10.1002/smll.201202111.

26. R. V. Mehta, "Synthesis of magnetic nanoparticles and their dispersions with special reference to applications in biomedicine and biotechnology," *Materials Science and Engineering: C*, vol. 79, pp. 901–916, Oct. 2017. doi: 10.1016/J.MSEC.2017.05.135.

27. A. Sohail, Z. Ahmad, O. A. Bég, S. Arshad, and L. Sherin, "A review on hyperthermia via nanoparticle-mediated therapy," *Bulletin du cancer*, vol. 104, no. 5, pp. 452–461, 2017. doi: 10.1016/J.BULCAN.2017.02.003.

28. M. Sanna Angotzi, V. Mameli, S. Khanal, M. Veverka, J. Vejpravova, and C. Cannas, "Effect of different molecular coatings on the heating properties of maghemite nanoparticles," *Nanoscale Advances*, vol. 4, no. 2, pp. 408–420, 2022. doi: 10.1039/d1na00478f.

29. M. A. Zayed, M. A. Ahmed, N. G. Imam, and D. H. El Sherbiny, "Preparation and structure characterization of hematite/magnetite ferro-fluid nanocomposites for hyperthermia purposes," *Journal of Molecular Liquids*, vol. 222, pp. 895–905, Oct. 2016. doi: 10.1016/J.MOLLIQ.2016.07.082.

30. C. L. Snow, C. R. Lee, Q. Shi, J. Boerio-Goates, and B. F. Woodfield, "Size-dependence of the heat capacity and thermodynamic properties of hematite (α-Fe2O3)," *Journal of Chemical Thermodynamics*, vol. 42, no. 9, pp. 1142–1151, 2010. doi: 10.1016/J.JCT.2010.04.009.

31. C. Spreafico, D. Russo, and R. Degl'Innocenti, "Laser pyrolysis in papers and patents," *Journal of Intelligent Manufacturing 2021 33:2*, vol. 33, no. 2, pp. 353–385, 2021. doi: 10.1007/S10845-021-01809-9.

32. I. I. Lungu *et al.*, "laser pyrolysis of iron oxide nanoparticles and the influence of laser power," *Molecules*, vol. 28, no. 21, p. 7284, Oct. 2023. doi: 10.3390/MOLECULES28217284.

33. O. Bomati, M. P. Morales, C. J. Serna, and S. Veintemillas, "Magnetic nanoparticles prepared by laser-induced pyrolysis," *INTERMAG Europe 2002 - IEEE International Magnetics Conference*, 2002. doi: 10.1109/INTMAG.2002.1001479.

34. J. C. Aphesteguy, G. V. Kurlyandskaya, J. P. De Celis, A. P. Safronov, and N. N. Schegoleva, "Magnetite nanoparticles prepared by co-precipitation method in different conditions," *Materials Chemistry and Physics*, vol. 161, pp. 243–249, Jul. 2015. doi: 10.1016/j.matchemphys.2015.05.044.

35. P. A. Calderón Bedoya, P. M. Botta, P. G. Bercoff, and M. A. Fanovich, "Influence of the milling materials on the mechanochemical synthesis of magnetic iron oxide nanoparticles," *Journal of Alloys and Compounds*, vol. 939, Apr. 2023. doi: 10.1016/j.jallcom.2023.168720.

36. J. F. De Carvalho, S. N. De Medeiros, M. A. Morales, A. L. Dantas, and A. S. Carriço, "Synthesis of magnetite nanoparticles by high energy ball milling," *Applied Surface Science*, vol. 275, pp. 84–87, Jun. 2013. doi: 10.1016/J.APSUSC.2013.01.118.

37. I. A. Flores Urquizo, H. Sanchez Correa, F. T. Montes De Oca Ayala, J. Rivera De La Rosa, and T. C. Hernandez Garcia, "Synthesis of La-Sr-Mn-O and La-Sr-Ca-Mn-O perovskites through solution combustion using urea at fuel deficient conditions," *IEEE Transactions on NanoBioscience*, vol. 19, no. 2, pp. 183–191, 2020. doi: 10.1109/TNB.2019.2963703.

38. H. Gyulasaryan, A. Kuzanyan, A. Manukyan, and A. S. Mukasyan, "Combustion synthesis of magnetic nanomaterials for biomedical applications," *Nanomaterials*, vol. 13, no. 13, p. 1902, 2023. doi: 10.3390/NANO13131902.

39. O. Seibert, J. Grégr, and P. Kejzlar, "The preparation of iron oxide nanoparticles by a self-combustion method," *Manufacturing Technology*, vol. 19, no. 4, pp. 680–684, 2019. doi: 10.21062/UJEP/355.2019/A/1213-2489/MT/19/4/680.

40. F. Gu, K. Jung, A. A. Martinez-Morales, Z. K. Karakaş, R. Boncukçuoğlu, and İH. Karakaş, "The effects of fuel type in synthesis of NiFe2O4 nanoparticles by microwave assisted combustion method," *Journal of Physics: Conference Series*, vol. 707, no. 1, p. 012046, 2016. doi: 10.1088/1742-6596/707/1/012046.

41. N. H. Sulaiman, M. J. Ghazali, J. Yunas, A. Rajabi, B. Y. Majlis, and M. Razali, "Synthesis and characterization of CaFe2O4 nanoparticles via co-precipitation and auto-combustion methods," *Ceramics International*, vol. 44, no. 1, pp. 46–50, 2018. doi: 10.1016/J.CERAMINT.2017.08.203.

42. A. Căpraru, E. A. Moacă, C. Păcurariu, R. Ianoş, R. Lazău, and L. Barbu-Tudoran, "Development and characterization of magnetic iron oxide nanoparticles using microwave for the combustion reaction ignition, as possible candidates for biomedical applications," *Powder Technology*, vol. 394, pp. 1026–1038, Dec. 2021. doi: 10.1016/J.POWTEC.2021.08.093.

43. I. A. F. Urquizo *et al.*, "Effect of aminosilane nanoparticle coating on structural and magnetic properties and cell viability in human cancer cell lines," *Particle & Particle Systems Characterization*, vol. 39, p. 2200106, 2022. doi: 10.1002/PPSC.202200106.

44. C. Chapa González, I. A. Flores Urquizo, D. I. Máynez Tozcano, L. E. Valencia Gómez, and J. A. Roacho Pérez, "Enhancing the cytocompatibility of cobalt-iron ferrite nanoparticles through chemical substitution and surface modification," *Advanced Materials Interfaces*, vol. 10, no. 18, p. 2300206, 2023. doi: 10.1002/ADMI.202300206.

45. I. A. Flores-Urquizo, P. García-Casillas, and C. Chapa-González, "Desarrollo de nanopartículas magnéticas Fe+32X+2IO4(X= Fe, Co y Ni) recubiertas Con amino silano," *Revista Mexicana de Ingenieria Biomedica*, vol. 38, no. 1, pp. 402–411, 2017. doi: 10.17488/RMIB.38.1.36.

46. D. Bokov *et al.*, "Nanomaterial by sol-gel method: Synthesis and application," *Advances in Materials Science and Engineering*, vol. 2021, 2021. doi: 10.1155/2021/5102014.

47. M. Benamara *et al.*, "Study of the magnetic properties of Mg, Gd, and Co doped maghemite (γ-Fe2O3) nanoparticles prepared by sol–gel," *Journal of Magnetism and Magnetic Materials*, vol. 569, p. 170479, Mar. 2023. doi: 10.1016/J.JMMM.2023.170479.

48. L. Kopanja, I. Milosevic, M. Panjan, V. Damnjanovic, and M. Tadic, "Sol-gel combustion synthesis, particle shape analysis and magnetic properties of hematite (α-fe 2 o 3) nanoparticles embedded in an amorphous silica matrix," *Applied Surface Science*, vol. 362, pp. 380–386, Jan. 2016. doi: 10.1016/j.apsusc.2015.11.238.

49. S. M. Hashemi, Z. Ataollahi, S. Hasani, and A. Seifoddini, "Synthesis of the cobalt ferrite magnetic nanoparticles by sol–gel auto-combustion method in the presence of egg white (albumin)," *Journal of Sol-Gel Science and Technology*, vol. 106, no. 1, pp. 23–36, 2023. doi: 10.1007/S10971-023-06073-2.

50. K. K. Patankar, D. M. Ghone, V. L. Mathe, and S. D. Kaushik, "Structural and physical property study of sol–gel synthesized CoFe2-xHoxO4 nano ferrites," *Journal of Magnetism and Magnetic Materials*, vol. 454, pp. 71–77, May 2018. doi: 10.1016/j.jmmm.2018.01.039.

51. J. Xu *et al.*, "Preparation and magnetic properties of magnetite nanoparticles by sol–gel method," *Journal of Magnetism and Magnetic Materials*, vol. 309, no. 2, pp. 307–311, Feb. 2007. doi: 10.1016/J.JMMM.2006.07.037.

3 Magnetism, Metamaterials, and Other Physical Properties of Magnetic Nanoparticles

Javeed Iqbal, Sonia Maqbool, Nadia Bhatti, and Aziz-ur-rehman

3.1 HISTORY OF MAGNETISM

The first magnetic substance known to man was a mineral called magnetite, which is where the history of magnetism begins. Its early history is hazy, but 2500 years ago, people were aware of its ability to draw iron. Magnetite is found widely. The district of Magnesia was home to the most abundant deposits in ancient times. Our word magnet is derived from a related Greek word, which is said to have originated from the name of this district. The Greeks also understood that if a piece of iron was contacted or rubbed with magnetite, it would turn magnetic on its own. Later, but at an unknown time, it was discovered that a properly formed piece of magnetite would point roughly north and south if supported such that it might float on water. If previously scraped with magnetite, a pivoting iron needle would as well. Thus, the mariner's compass came into being. The old English name for magnetite, lodestones, which means "way stone" because it shows the way, derives from its north-pointing characteristic. William Gilbert, an Englishman, conducted the first true scientific investigation into magnetism and published his seminal work, "On the Magnet," in 1600. After Gilbert, no fundamental discoveries were made for more than a century and a half, but there was much useful advancement achieved in the production of magnets. Compound steel magnets, which were created in the seventeenth century from several magnetized steel strips joined together, were able to lift 28 times as much iron as they weighed. The fact that rubbing iron or steel with a lodestone was the only method for creating magnets makes this feat all the more amazing. There was no alternative method until the first electromagnet was created in 1825 as a result of Hans Christian Oersted's groundbreaking discovery in 1820 that an electric current generates a magnetic field. It can be stated that research on magnetic materials began with the development of the electromagnet, which made available far stronger fields than those created by lodestones.

DOI: 10.1201/9781003454236-3

3.2 FUNDAMENTAL TERMINOLOGY OF MAGNETISM

3.2.1 MAGNETIC POLES

Almost everyone experienced the enigmatic forces of attraction and repulsion between magnets when playing with them as a youngster. The poles, or areas close to the ends of the magnet, appear to be the source of these forces. The north-seeking pole, or simply the North Pole, is the end of a pivoted bar magnet that roughly faces the north geographic pole of the Earth. According to this convention, there is a zone of south polarity close to the north geographic pole, since unlike poles attract and similar poles repel. Michael Faraday came up with the brilliant notion of using "lines of force" to symbolize a magnetic field. A single north pole would travel along these lines, or a tiny compass needle would be tangent to them. It appears that there is just one North Pole from which lines of force emanate. The lines of force exit a bar magnet at the North Pole and re-enter at the South Pole. By scattering iron filings or powder on a card positioned immediately above the magnet, one may see the resultant field in two dimensions (Figure 3.1). Each iron atom becomes magnetized, and with its long axis parallel to the lines of force, it behaves like a tiny compass needle [1].

3.2.2 MAGNETIC DIPOLE AND MAGNETIC MOMENT

Magnetic poles only appear in pairs; they have never been seen alone. A magnetic dipole is a pair like this. A dipole's magnetic moment is described by equation 3.1.

$$\mu = ml \tag{3.1}$$

where l is a vector heading in the direction of the N pole from the S pole and μ is represented in A m^{-2}.

3.2.3 MAGNETIC FIELD *H*

The magnetizing force operating on a unit test pole ($m_2 = 1$) positioned at that location is used to quantify the strength of the magnetic field H produced by either pole

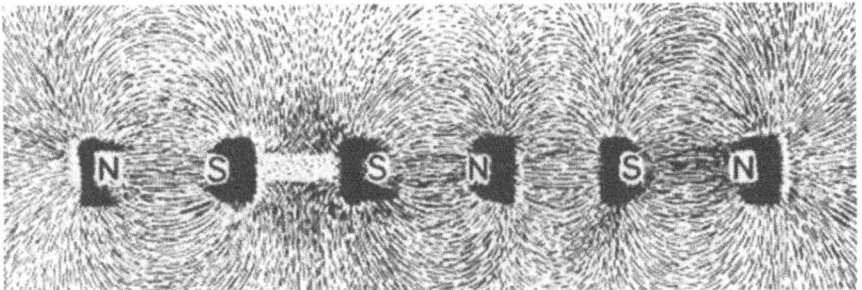

FIGURE 3.1 Fields of bar magnets revealed by iron filings.

of a bar magnet with pole strength m_1, at a distance **r**. Magnetic field H is represented by equation 3.2:

$$H = \frac{F}{m_2} = K \frac{m_1}{r_2} r_0 \qquad (3.2)$$

The unit of H in the CGS system is the widely used oersted. In that system, one oersted corresponds to a repulsive force of one dyne exerted on a unit test pole by another unit pole of the same sort at a distance of 1 centimeter since K has the justified value of unity. The fact that an electric current frequently results in a magnetic field is the foundation for the SI unit of H. The magnetic field generated within a long solenoid with N turns per meter and carrying I amperes of current is presented by equation 3.3:

$$H = NI \qquad (3.3)$$

As a result, the ampere-turn per meter, or simply the ampere per meter, is the SI unit of H. One oersted is equal to around 80 A m^{-1}, and one ampere per meter is equal to 0.0126 oersted.

3.2.4 MAGNETIZATION M AND MAGNETIC INDUCTION B

Magnetization or, more specifically, the degree of magnetization M is the total magnetic moment of all dipoles in a given volume, measured in amperes (A) per m. Consider a bar magnet with a cross-sectional area of A, a length of l, and a pole strength of m. The vector quantity **M** has a direction that points from the S pole to the N pole, and its magnitude is provided by equation 3.4:

$$\mathbf{M} = \mu|V = m|A \qquad (3.4)$$

Since 10^{-3} gauss in the CGS system is comparable to A m^{-1} in terms of pole strength per unit area, magnetization is also expressed in those units. The flux per unit area, also known as magnetic induction or magnetic flux density B, is measured in watts per square meter or tesla. Flux is the number of right-angle induction lines that pass a specific region. The Weber is the magnetic flux that connects a one-turn circuit and generates a one-volt electromotive force as the flux is uniformly lowered to zero over the course of one second. M. Faraday was the first to propose the use of infinite flow lines between the magnetic poles of a magnet to symbolize magnetic induction. The magnetic induction pattern may be seen when a magnet is placed beneath a piece of cardboard on which iron filings have been scattered. The magnet will cause the filings to create infinite flow lines that start at the N pole and converge at the S pole [2].

3.2.5 MAGNETIC SUSCEPTIBILITY

The process of applying a magnetic field to a material and measuring the magnetization that is generated by the field is arguably the most widespread

magnetic experiment [3]. The susceptibility of the material is the indicator of
how well a magnetic dipole is produced by an applied field and is described by
equation 3.5:

$$\kappa = M \mid H \tag{3.5}$$

3.3 DIFFERENT CATEGORIES OF MAGNETISM

3.3.1 PARAMAGNETISM

When exposed to an external magnetic field, some materials show paramagnetism,
a particular kind of magnetism. Because the individual atoms or molecules in paramagnetic materials have unpaired electrons, each one of them has a net magnetic
moment. There is a weak pull to the magnetic field as a result of these magnetic
moments aligning with the external field. Titanium dioxide, chromium, and magnesium all exhibit paramagnetism.

3.3.2 FERROMAGNETISM

A significant permanent magnetization is shown by some materials as a result of
the phenomenon known as ferromagnetism. Ferromagnetic materials are what
these substances are. Ferromagnetism is one of the three forms of magnetism,
along with paramagnetism and diamagnetism. A macroscopic magnetic field is
produced in ferromagnetic materials when individual or groups of atoms' magnetic moments spontaneously align in a parallel pattern. Neighboring atom interactions that result in a collective alignment of their magnetic moments cause this
pattern to happen. Ferromagnetic materials display spontaneous magnetization
even in the absence of an external magnetic field below a certain temperature
known as the Curie temperature. The substance loses its ferromagnetic characteristics above the Curie temperature and starts to behave more like a paramagnetic
substance. Fe, Ni, and Co exhibit ferromagnetism. The susceptibility of ferromagnetic materials depends on the field strength; the relationship between magnetization and the field strength is not one-valued; and Weiss domains, which
are elementary regions of magnetization, exist in the material below the Curie
temperature [4].

3.3.3 DIAMAGNETISM

Diamagnetic behavior is shown by atoms or compounds with no effective magnetic
moment or by substances with fully filled electronic shells, such as noble gases,
ionic solids, or semiconductors with strong covalent bonds. The electrical charges in
all such materials have a propensity to partially hide the body's interior from magnetic fields. Because of this, the susceptibility is minimal, has a negative sign, and
is caused by the progression of the electronic orbits (Figure 3.2). Gold, silver, and
carbon all exhibit diamagnetism.

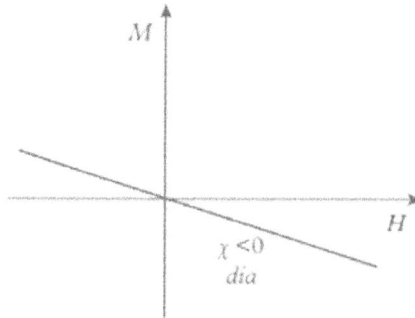

FIGURE 3.2   Magnetism M as a function of the applied field H of a diamagnet.

3.3.4   ANTI-FERROMAGNETISM

Anti-ferromagnetic materials might be thought of as consisting of atoms with permanent magnetic moments but with a strong negative contact between them; as a result, nearby magnetic moments tend to be arranged in an antiparallel fashion. Furthermore, there is no net or spontaneous magnetization when the two interpenetrating but antiparallel lattices of the crystal are precisely aligned, since only one magnetic species is involved and it is distributed equally throughout both. At ambient temperature, several of these materials are shown to display paramagnetic behavior. However, as the temperature drops below a crucial level called the Neel temperature, they suddenly shift susceptibility (Figure 3.3). Furthermore, below T_N susceptibility depends on the orientation of the field and is a function of whether the field is applied perpendicular to H or parallel to H, the preferred direction of the spin-lattice in this material. The temperature dependence of an antiferromagnet's susceptibility when $T > T_N$ is presented by equation 3.6:

$$X = \frac{C}{T - \theta} = \frac{C}{T + T_N} \tag{3.6}$$

where T_N is the Neel temperature and $\theta = -T_N$, when projected to the temperature axis, is the intercept of $1/x$ for $T > T_N$.

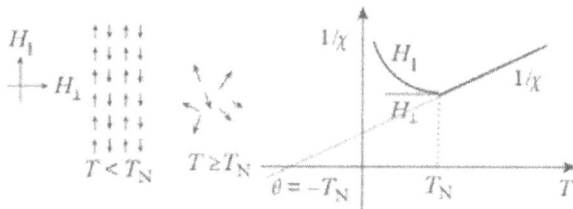

FIGURE 3.3   Magnetization behavior of an antiferromagnet as a function of temperature.

3.3.5 FERRIMAGNETISM

A tendency toward ferrimagnetism is often seen in materials when two or more magnetic species are present at places on various sublattices and have magnetic moments that differ from one another. The two sublattices interact negatively and strongly, resulting in an antiparallel configuration [5]. When the two sublattices are organized, ferrimagnets, unlike antiferromagnets, show a spontaneous magnetism below Curie temperature T_C. Ferrimagnetism is shown by ferrites, yttrium iron garnet, cobalt ferrite, and magnetite (Figure 3.4).

3.3.6 NANOMAGNETISM

A subfield of nanotechnology called nanomagnetism investigates the magnetic characteristics of nanoparticles. Different magnetic states, including single-domain superparamagnetism, super ferromagnetism, and super spin glasses, have been seen in a system of nanoparticles. The spectrum of applications for magnetic nanoparticle assemblies is determined by the distinctive characteristics of each of these magnetic states. While the core spins of nanoparticles are virtually normal, the shell is made up of canted spins. Numerous characteristics, such as various anisotropies, interactions between the particles, relaxation durations, coercivity, remanent magnetization, and saturation magnetization, must be explored and characterized in the study of nanoparticle systems. Magnetic nanoparticles have undergone extensive research into their properties. By expanding the use of nanoparticles in business, health, and daily life, research in the subject of nanomagnetism advances [6]. Nanomagnetism encompasses a wide spectrum of studies on the magnetism and magnetic characteristics of low-dimensional systems, including both theoretical modeling and simulations as well as experimental approaches for sample production and characterization [7]. A fascinating new class of organic nanomagnets, graphene-based nanostructures, including graphene nanoribbons, nanodots, and nanoholes with zigzag edges, have recently been demonstrated to

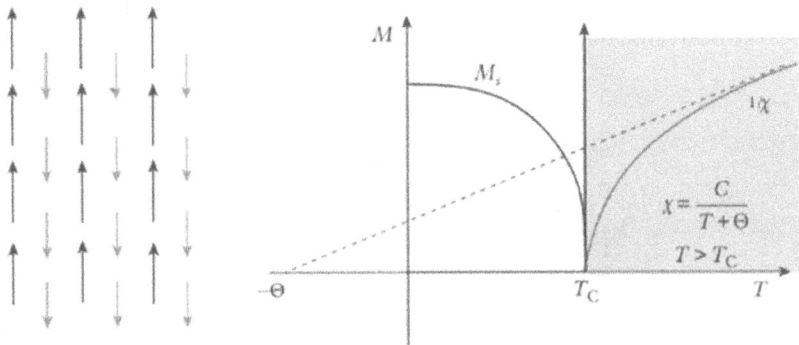

FIGURE 3.4 Magnetization behavior of a ferrimagnet as a function of temperature.

display magnetism [6]. Magnetic nanoparticle application in vivo is generating a lot of interest as a way to deliver personalized therapy [9].

3.4 APPLICATIONS OF MAGNETISM

3.4.1 ROLE OF MAGNETISM IN TECHNOLOGY

The fundamentals of contemporary life, such as electric power, communications, and information storage, are permeated by uses of magnetism. Magnetism has a financial effect of more than 1% of the gross national product, despite the fact that the general public is largely ignorant of its function. The market for direct magnetic materials is now expected to be worth 650 million dollars annually. Electrical steels and recording tape share the top spot in terms of revenue, followed by soft ferrites, magnetic disc packs and drums, permanent magnets, nickel-iron alloys, and soft ferrites. The market for magnetism is also discussed in terms of research's contribution to quality improvement and the expansion of new applications. About 5% of the market for direct magnetics is thought to be spent on fundamental and applied research in the United States. After reaching a peak in the 1950s, the global production of papers on magnetics has surged six-fold during the past ten years. The breadth of magnetism's study, discoveries, and applications has expanded over this time, and its popularity among scientists as a whole has increased [10]. In technical applications including novel information storage technologies, four-state logic devices, magnetoelectric sensors, and low-power magnetoelectric devices, manipulation of magnetization by an electric field at room temperature is crucial. Additionally, it can miniaturize devices and reduce power consumption, both of which are highly helpful for practical applications [11]. The construction of numerous kinds of sensors for mechanical and electromagnetic hydrodynamic measurements is made possible by magnetic fluids. Inertial and gravity sensors are an important use of magnetic fluids technology. The potential for creating virtually all sorts of sensors required for motion measurement is provided by magnetic fluids [12]. Permanent magnets have been employed in electrical equipment for more than a century, but due to recent substantial advancements in their availability and qualities, their employment in electromechanical and electronic devices is presently expanding quickly [13].

3.4.2 MAGNETISM IN MEDICINE

It has been reported since the 17th century that bar magnets may be used to remove iron from the body. The electromagnet has been used in medicine since the end of the 19th century. Alnico magnets have been utilized in artificial eyes. The invention of alnico made it possible to dramatically remove hazardous materials from previously inaccessible bodily channels. The most common use of magnetism in medicine is for the treatment of hardware, also known as traumatic gastritis, in cattle. The potential of magnetic resonance as a tool for biochemical and medical investigation is only now starting to become clear. The investigation of the recombination rate of free radicals generated by radiation in liver cells is an illustration. The method in which starch-coated iron particles with a micron-sized diameter are captured by white blood cells demonstrates a distinct type of use of magnetism in the analysis. The white cells can then be divided by utilizing a gradient magnetic field. Researchers in Edison's Laboratory conducted the first conclusive investigation of the direct biological effects of the magnetic field in 1892 [14]. There are well-established uses of magnetism in human medicine, such as magnetic resonance imaging, a cutting-edge diagnostic tool [15]. To carry out various functions, magnetic elements are purposefully introduced into the body. Catheters that employ magnetic guidance have been used to navigate the blood vessels' complicated interiors. Barium sulphate may be replaced with ferrite particles in some applications for gastrointestinal X-ray diagnostics. Additionally, magnetic fields have been employed for a variety of measures, utilizing both those created by the body's own electrical or electromagnetic activity and those produced by little quantities of material that have been intentionally injected into the body [16].

3.4.2.1 Magnetic Micromachines for Medical Applications

As driving mechanisms for micromachines, magnetic systems have outstanding properties. A magnetic field from outside the machine can be used to apply energy for work. As a result, in the field, the machine may operate without power supply wires. Because a magnetic field can easily be supplied to the small region where the micromachine operates, utilizing a magnetic field to deliver energy makes sense. This technique makes it possible to create wireless micromachines. The magnetic micromachine is operated using two fundamental concepts. One is the force generated by the gradient magnetic field acting on the magnetic pole. The other is the torque that the uniform magnetic field's magnetic moment is subjected to. Magnetic micromachines have been researched using these ideas. The benefits of magnetic micromachines point to potential uses in the medical industry. The spiral-type magnetic micromachine, for instance, may swim in both a liquid and a gel. Additionally, this device features a heating feature. This device is appropriate for applying localized heat to the body. At the guide wire's tip, a small magnet is connected to provide active bending. This is straightforward and practical for placing a catheter into a blood vessel or lung [17].

3.4.2.2 Applications of Magnetic Particle Imaging (MPI)

A recent advancement in medical imaging is magnetic particle imaging (MPI). High spatial and temporal resolution quantitative pictures of the magnetic material's local distribution may be acquired using MPI. Its sensitivity far exceeds that of other techniques, including magnetic resonance imaging, that are used to identify and quantify magnetic materials. On the basis of an intravenous injection of magnetic particles, MPI has the potential to be a significant advancement in sectors of exploration like cell labeling and tracking as well as in fields of medicine like cardiology and cancer. MPI has a number of significant benefits over other methods for measuring tissue perfusion now in use, such as nuclear methods, MRI, CT, and ultrasound. The approach enables non-invasive quantitative assessments of tissue perfusion, expressed as milliliters of blood per gram of tissue, due to the linear relationship between the tracer concentration and the MPI signal [18].

3.4.2.3 Use of Magnetic Nanoparticles in Biotechnology and Medicine

Due to their unusual properties, such as magnetic quantum tunneling and quantum size effects, magnetic nanoparticles have garnered a lot of attention [19]. Iron oxide, cobalt ferrite, iron platinum, and manganese ferrite are the magnetic nanoparticles that have received the most research [20]. MNPs are utilized in biotechnology and medicine for the separation, purification, and immobilization of numerous biomolecules from bacteria, viruses, and other biological fluids, including ribonucleic acid and deoxyribonucleic acid. To obtain highly purified genetic material for further analysis using methods like polymerase-chain reaction amplification and DNA hybridization, the majority of commercially available DNA extraction kits necessitate time-consuming pipetting, centrifugation, or filtration processes. However, target analytes might be lost or undesirable interferents could separate during a high-speed spin. MNPs have certain advantages over traditional methods, including automation, a reduction in the number of hazardous chemicals used, and a shorter time required for analyte separation [21]. A recently created groundbreaking bioseparation technique called nano-sized magnetic support is particularly useful for cell isolation and purification, ligand fishing, and protein, enzyme, DNA, and RNA purification. Different terms, such as magnetic beads or micro- and nano-sized magnetic beads, are used to describe magnetic nanoparticles. They can also be colloidal suspensions of magnetic particles in a liquid carrier and are then referred to as ferrofluids or magnetic fluids. These particles are often a component of nanotechnology, which is the creation of practical systems at the molecular level. It has the potential to produce a large number of novel materials and gadgets with a diverse range of uses in the domains of biomedicine, chemistry, electronics, and mechanics. Magnetic nanoparticles may be used to isolate any target and are connected to a variety of human and automated applications because of their superparamagnetic characteristics and micro- and nano-dimensions [22].

3.4.3 APPLICATIONS OF MAGNETIC NANOPARTICLES FOR WASTEWATER PURIFICATION

Precipitation, evaporation, solvent extraction, ion exchange, reverse osmosis, membrane separation, and other treatment methods have all been employed to remove metallic ions from wastewater. The majority of these approaches have certain disadvantages, such as high treatment and disposal costs for the remaining metal sludge and high capital and operating expenses. As a result, work has been done to create inexpensive materials that can purge pollutants from aqueous solutions. Fortunately, recent advances in nanotechnology have illuminated this area. Because of their distinct characteristics and prospective uses, nanoparticles, which frequently contain a large number of surfaces, have attracted a lot of attention. In this regard, in recent decades magnetic Fe_3O_4 nanoparticles have garnered interest not only in the field of magnetic recording but also in the fields of magnetic sensing and medical care. According to theory, magnetic nanoparticles show the finite-size effect or have a high surface-to-volume ratio, which increases their capacity for metal adsorption. An external magnetic field can also be used to separate metal-loaded magnetic adsorbent from solution easily [23].

3.4.4   APPLICATIONS OF MAGNETIC TECHNOLOGY IN AGRICULTURE

As a result of the magnetic field being a component of the environment and a source of energy, it now influences both regular metabolisms and meristem cell proliferation. Additionally, MF has an impact on water ionization, preservation, and absorption. Magnetophoresis in macromolecules may be brought on by MF forces. Plant photosynthetic pigments and other metabolic components may be impacted by MF. According to research, plants' chemical processes accelerate under MF, which benefits the plants' photochemical activity, respiration ratio, and enzyme activity. Utilizing magnetic water in agricultural production will allow for more intensive, high-quality and quantity output [24].

3.5   METAMATERIALS

3.5.1   INTRODUCTION TO METAMATERIALS

In the term "metamaterial," "meta" denotes anything above and beyond the ordinary, altered, transformed, or advanced. Thus, a metamaterial is an artificial substance created to acquire physical qualities not found in natural substances. Rodger M. Walser is the author of the phrase "metamaterial," which refers to composite structures with rationally planned building units built of one or more basic bulk materials. They are claimed to possess useful qualities above and beyond those of ordinary material. For instance, Veselago presented them theoretically as the material required for the negative refraction of light in optics. Later in 2000, Pendry advocated utilizing a thin metal layer to create such negative refractive index slabs and create the ideal lens [25]. Because of the possibilities presented by their odd characteristics, such as a negative index of refraction and magnetism at optical frequencies, metamaterials have received a lot of interest recently. The majority of research so far has been on the propagation of plane waves or Gaussian beams in metamaterials [26]. Metamaterials were first developed in the discipline of electromagnetics and afterward used in the fields of acoustics and solid mechanics [27]. Since the start of the field, there has been a lot of research on metamaterials that can actively modify electromagnetic radiation. The geometrical layout and material composition of conventional metamaterials determine their resonant characteristics [28]. Metamaterials and intricate transmission lines that facilitate the propagation of reverse waves at microwave frequencies have attracted increasing attention in recent years. These metamaterials go by a variety of names, including backward-wave media, left-handed media, Veselago media, media with a negative refractive index, media with a negative permittivity and permeability, and lines with a negative dispersion [29].

3.5.2   HISTORY OF METAMATERIALS

The word "meta" was selected to indicate that such composites transcend the qualities of natural materials; it is said that R. Walser first used the phrase during a workshop on composite materials. Metamaterials can be broadly categorized as structures made of subwavelength arrangements of metallic or dielectric inclusions designed to produce unusual or otherwise inaccessible macroscopic properties not found in the natural materials they are made of, despite the fact that definitions abound and continually change. The first instance of negative refraction naturally sparked a frenzy of activity. The field was intrinsically interdisciplinary, incorporating ideas from plasmonic and nanotechnology to microwave engineering and optics to solid-state physics and materials research. As a result, it garnered instant and ongoing interest from a number of communities, each encouraged by the virtually endless potential for learning about one or more novel properties. The ongoing contributions of Pendry, in particular the discovery that a properly constructed negative-index flat lens could theoretically offer an infinite resolution, breaking the classical diffraction limit, and his later prediction that metamaterials could be used to precisely alter optical space in order to make objects invisible, both of which were ultimately proven in experiments, provided periodic energy to this frenetic pace. Beyond the SRR medium, this period of discovery also saw the development of several metamaterial implementations inspired by metamaterial properties, including ultracompact phase shifters, power dividers, and other components as well as leaky-wave antennas capable of back-fire to end-fire radiation through broadside [30]. Metamaterials today span a wide range of academic fields, including physics, electrical engineering, material science, mathematics, mechanics, acoustics, fluid dynamics, and chemistry.

3.5.3 Different Metamaterials

3.5.3.1 Terahertz Metamaterials

Metamaterials are made up of subwavelength metallic components that are organized in regular patterns and have special electromagnetic features including cloaking, subdiffraction restricted focusing, and negative refraction. Additionally, the gap structures of the metamaterials are characterized by intensely localized and amplified fields, allowing for sensitive detection of incredibly minute quantities of chemical and biological constituents. Because the size of fungus and bacteria is consistent with the gap size, metamaterials operating in the THz frequency band in particular have a micro-sized gap and may thus be used as an appropriate platform for the sensitive detection of these microorganisms. THz metamaterials are also very sensitive to materials close to the surface, which is advantageous for sensing in an aquatic environment, since it enables us to employ a thin water layer without experiencing a sizable reduction in the THz wave transmission. A potential method that permits the label-free, non-contact, and non-destructive assessment of chemical and biological compounds is terahertz spectroscopy. Particularly, on-site detection and identification of these compounds with a high signal-to-noise ratio is now possible thanks to recently developed, portable THz spectroscopic equipment. Due to its applicability to food and security inspection, THz frequency range detection of microorganisms including fungi, bacteria, and viruses has attracted a lot of attention. However, as they are mostly transparent to THz waves, the microorganisms are frequently nonresponsive in the THz frequency range. Additionally, a typical microbe has a size of about 1/100, which results in a low scattering cross-section [31].

3.5.3.2 Acoustic Metamaterials

Humans have traditionally found noise to be annoying. According to the mass density law, thick and heavy blocking materials must be utilized for low-frequency sound insulation. Therefore, a thin, lightweight insulating material that resists the mass rule is very desirable for noise reduction. Low-frequency sound insulation has recently been moving in a positive direction thanks to the invention of acoustic metamaterials. Acoustic metamaterials achieve sound insulation in the form of complete reflection based on the local resonance process. A thin membrane with one or more stiff masses connected to it, for instance, makes up membrane-type acoustic metamaterials. They can display greater acoustic attenuation at a deeper subwavelength distance compared to standard insulating materials. Sound insulation and energy recovered from these noises are compatible from a practical standpoint and should be preferred. Due to the fact that structural resonance often results in the localization and augmentation of the deformation energy, which is easily collected by energy conversion devices, the realization of this dual functionality is achievable [32].

3.5.3.3 Tunable Metamaterials

The unique potential that metamaterials have for modifying macroscopic characteristics by carefully selecting and arranging their structural constituents make them stand out. By doing so, it is feasible to integrate additional modification capabilities at the level of assembly, in addition to designing a metamaterial for a desired capability. Metamaterials are distinct from ordinary materials due to this feature, which also creates new prospects for multi-functionality through tunability. Metamaterials having an innate mechanism for tunability are said to be tunable if an external force or signal can continually modify their characteristics. Naturally, changing the system in such a way as to alter the resonance's characteristics constitutes the primary method of tuning resonant metamaterials. As a result, a metamaterial's properties may be changed, allowing for features like adjustable transmission [33].

3.5.3.4 Nonlinear Tunable Magnetic Metamaterials

Figure 3.5 shows a nonlinear metamaterial made of two split-ring resonators that each have a varactor diode added so that the magnetic resonance may be dynamically controlled by changing the input power [34]. Higher powers cause the metamaterial's transmission to become power-dependent, which allows this type of metamaterial to exhibit a variety of nonlinear features.

3.5.3.5 Photonic Metamaterials

Victor Veselago explored the way that matter may interact with the magnetic field of light 40 years ago. The materials were then tabulated based on the sign of their magnetic permeability and electric permittivity. He demonstrated that the Maxwell equations' solution led to an index of refraction with a negative sign when both the electric and the magnetic properties were negative. Veselago had to wait more than 30 years for the engineering of a new class of structures known as metamaterials before his theoretical hypotheses were empirically verified. Split ring resonators, which display $\mu < 0$ at particular resonance frequencies, are metallic-based resonant structures that were made possible by Sir John Pendry's pioneering work.

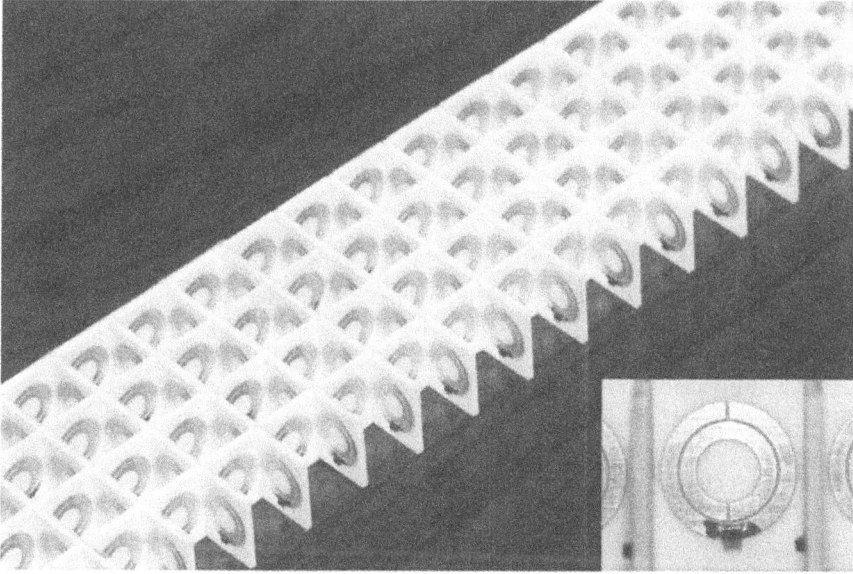

FIGURE 3.5   Nonlinear tunable magnetic metamaterial.

Smith et al. developed double-negative composite metamaterials by combining an array of SRRs and an array of metallic wires. Then, they were put to use in a wedge-shaped construction to show negative refraction experimentally at microwave frequencies. The well-known "right-handed rule" between the electric and the magnetic fields becomes left-handed in certain circumstances. As a result, these materials are sometimes referred to as negative index materials or left-handed materials. Scientists have shown that metamaterials may theoretically be utilized for magical purposes as well, including subwavelength imaging, super lenses, perfect lenses, and cloaking. It only took a few years for scientists to downscale these structures to optical frequencies, despite the fact that the initial tests were conducted at microwave frequencies. "Photonic metamaterials" is the new name for these nanoscale metamaterials. Typical 2D manufacturing methods for photonic metamaterials include focused ion beam etching, evaporation of metal sheets, and electron beam lithography. One of the main drawbacks of photonic metamaterials, which has long been cited, is the optical losses connected to the metallic inclusions. It may be possible to create loss-less photonic metamaterials using active media, where the gains from the active media will make up for the losses in metallic structures [35].

3.5.3.6   Mechanical Metamaterials

Recently, mechanical metamaterials, human-made materials whose mechanical characteristics are mostly determined by their structures rather than the characteristics of each component, have garnered a lot of interest. Mathematicians, scientists, and engineers have transformed origami to take advantage of the deformability and compactness of folded objects in applications ranging from space exploration to automotive safety, biomedical devices, and extremely foldable and stretchable electronics. By folding along creases, origami creates 3D structures from 2D sheets [36]. Mechanical metamaterials' intriguing mechanical characteristics are a result of their intentional architecture, not of the bulk substance that makes them up. Because of the intricate designs required, three-dimensional printing is now the primary method used to create mechanical metamaterials [37].

3.5.3.7 Auxetic Metamaterials

In auxetic metamaterials, the Poisson's ratios are negative. In 1987, Lakes published the first account of a practical technique for creating re-entrant foam with a Poisson's ratio of −0.7. Since then, auxetic metamaterials have been thoroughly researched by academics, and other varieties have been suggested and looked at. Auxetic metamaterials exhibit a wide range of advantageous mechanical characteristics, including the ability to absorb energy and withstand shear, fracture, and indentation. They, therefore, have enormous promise for applications in the real world, such as cushion covers, esophageal stents, nails, intelligent filters, and intelligent sensors [38]. Auxetic metamaterials come in many different forms in nature, including iron pyrite, several cubic elemental metals, cancellous bone, and cat skin. Auxetic metamaterials that have structural features including re-entrant, chiral, crumpled, and perforated geometries have been extensively researched [39]. There has been a lot of interest in recent years in the creation of unique techniques that can be used to quickly and cheaply create auxetic mechanical metamaterials. One such technique, which is gaining popularity, involves adding certain perforation patterns to a block or sheet of material that is conventional or has a positive Poisson's ratio. To put it another way, this entails cutting material into pieces from a block or sheet, resulting in a system with a geometric arrangement like that seen in a variety of systems having the ability to display auxeticity. The auxetic perforated systems that have been made so far using this technology resemble spinning constructions made of squares, triangles, and cubes, among other shapes [40]. The most extensively researched class of mechanical metamaterials is auxetic materials, which are recognized by a negative Poisson's ratio. Similar to other kinds of mechanical metamaterials, the topology of their nanoarchitecture is typically a direct cause of the negative Poisson's ratio of auxetics [41].

3.5.4 Properties of Metamaterials

The primary factors influencing a material's sensitivity to electromagnetic waves are its electrical permittivity and magnetic permeability. Both of these material properties are negative in metamaterials. The metamaterials' refractive index is thus negative. As a result, when light reaches the metamaterials from a vacuum, it bends "the wrong way." When light enters from a rarer medium to a denser medium, it often bends toward the normal. However, in the case of metamaterials, light deviates from the usual (Figure 3.6). When we modify the sign of ε and μ in the four Maxwell's equations, the immediate result of negative ε and μ in metamaterials is a left-handed triplet of the electric field E, magnetic field H, and phase vector $\mathbf{k}$. It is a triplet of

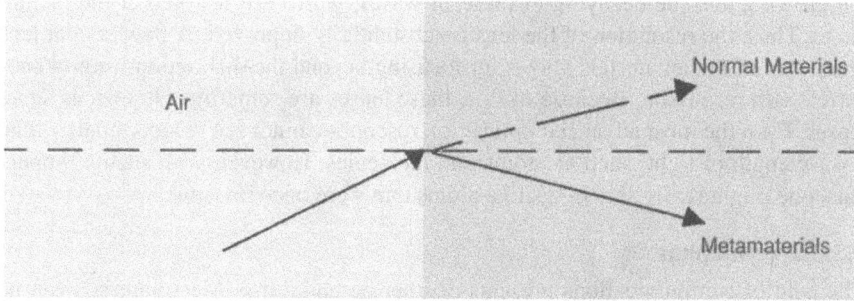

FIGURE 3.6   Light bends the "wrong way" when it enters from the vacuum to the metamaterial.

right-handed materials. This suggests that for metamaterials, the phase vector **k** and the group velocity or Poynting vector **S** are in opposing directions. As a result, the waves appear to be moving in a backward manner. The phase vector **k** indicates the direction of the phase velocity, whereas the Poynting vector **S** shows the direction of the energy or group velocity's propagation.

The reverse Doppler effect is yet another peculiar characteristic of metamaterials. It causes emission of a train-whistle sound that becomes louder as it gets closer and softer as it gets farther away. How such a phenomenon may be used in mobile wireless communication is still being researched. The reverse Cerenkov effect is another possibility. In contrast to "normal" materials, a charged particle moving through a medium emits light from a cone behind the particle. When a charged particle travels through a substance faster than the speed of light, Cerenkov radiation is produced. In particle physics, building a Cerenkov detector to recognize charged particles of varying velocity is one potential use.

3.5.5   APPLICATIONS OF METAMATERIALS

3.5.5.1   Super Lenses

Due to their negative refractive index, the first potential uses of such intentionally constructed metamaterials are in optics. A lens composed of metamaterials, according to John B. Pendry, a scientist at Imperial College London, might concentrate light for objects smaller than wavelength to a geometric point. Light cannot be focused onto an area smaller than the square wavelength of the light used to study it by any natural lens now in existence. For instance, the wavelength of visible light is smaller than atoms. Therefore, they are invisible to optical microscopes. Plane waves make up the EM field that a line source emits. While some of them are attenuating, others are growing. We refer to them as evanescent waves. A typical lens can only focus on waves that are already in motion. As a result, the picture of the item is imperfect. The evanescent waves' significant attenuation as they pass from the item to the picture causes the finer spatial information of the object that they carry to be lost. The diffraction limit's inception may be traced back to this loss of the evanescent spectrum. However, the metamaterials-based lens provides flawless imaging by supporting both the growing and the decaying evanescent waves, which are restored at the picture plane. Thus, the resolution of the lens is substantially improved. A flawless flat lens composed of metamaterials allows for focusing beyond the diffraction limit or sub-wavelength resolution. Because of this, these lenses are sometimes known as super lenses. Even the most advanced optical microscopes cannot see objects smaller than a wavelength of light, such as atoms and molecules. However, with such a "super" lens, one might really see things like atoms that were once invisible.

3.5.5.2   Antennas

The field of communications presents another potential use. Metamaterials can be used to create leaky-wave antennas with backward-to-forward scanning. The disadvantage of conventional leaky-wave antennas is that they can only do forward scanning or half-space scanning. We also have the capacity to scan backward using metamaterials. It is feasible to create backfire-to-end-fire leaky-wave antennas that can scan in all directions by combining conventional and metamaterials-based leaky-wave antennas [42].

3.5.5.3   Improve Antenna Parameters

In order to accommodate modern living, scientists in recent decades have put a lot of time and effort into investigating novel materials and physical phenomena. Since the discovery of several new artificial materials to replace the previous natural ones, many different facets of life have benefited greatly. The arrangement of metal structures on the surface of dielectric substrates accounts for one of the novel materials that include metamaterials. As a result, metamaterials' physical characteristics are more dependent on their structures than on the individual parts that make them up. Recently, there has been a lot of interest in the need to miniaturize and integrate a variety of telecommunications equipment functions, particularly for items that are used often in daily life, such as mobile communication systems, cell phones, portable tablets, GPS receivers, and wireless Internet devices. The mobile device's components must be portable and able to operate across many frequency bands in order to meet this criterion. One of them is an antenna, which is required to be smaller, adhere to the device's body, and be able to operate at various mobile communication system frequencies while being integrated into a single so-called smart gadget. To meet such criteria, a variety of technical solutions are being used in antenna building. Through the use of high-permittivity dielectric substrates, shorting walls, shorting pins, and other techniques, microstrip antenna technology has been miniaturized. Nevertheless, these techniques have drawbacks such as limited bandwidth and poor gain. The use of electromagnetic metamaterials for antenna design is a novel approach that is of significant interest to designers. In addition to substantially reducing the size of the antenna, the use of metamaterials in antenna design can also improve other antenna characteristics, such as boosting bandwidth, raising gain, or creating multi-band frequencies for antenna operation [43].

3.5.5.4   Laser Sources

The concept of merging gain media with metamaterials has become more and more popular among researchers. The creation of new laser sources will heavily relay on this new generation of metamaterials. In fact, it now has been scientifically shown that combining a gain medium with a plasmonic metamaterial may improve their photoluminescence's intensity by a factor of many and restrict their spectral ranges. The quantum Purcell effect is displayed clearly in the luminescence enhancement, which is controllable by the design of a metamaterial. This is a crucial step in the creation of gain media boosted by metamaterials as well as the "lasing spacer," a flat laser whose emission is driven by plasmonic excitations in a cluster of coherently emitting metamolecules. The emission wavelength of the laser may be altered by meta-molecule design, in contrast to conventional lasers, which function at wavelengths of appropriate natural atomic or molecular transitions.

3.5.5.5 Sensor Applications

Another fast-expanding area of metamaterials study is sensor applications. For example, metamaterial arrays of nanoscale antennas or asymmetrically split ring resonators supporting high-quality Fano resonances are ideally adapted to detecting low-concentration analytes like hydrogen and sugar through fluctuations in their transmission and reflection properties. For instance, a single molecule sheet of graphene has the ability to alter a metamaterial's transmission by many orders of magnitude. Plasmonic metamaterial nanostructures can also be utilized to enhance light-harvesting methods, allowing solar photovoltaic absorber layers to be considerably thinner and more effective while maintaining their optical properties.

3.5.5.6 Superconducting and Quantum Metamaterials

Recently developed superconducting metamaterials promise to provide a fundamentally different paradigm for information technology and data processing. The extraordinary sensitivity of the superconducting state to outside stimuli and the unique nonlinearity of superconductors will be accessible, resulting in a huge decrease of losses. Additionally, superconductors are a fascinating plasmonic medium because of their strong kinetic resistance and negative dielectric constants. Superconductivity also causes a fundamental shift in the makeup of information carriers. It will be feasible in certain implementations to flip between the classical excitations of traditional plasmonic and metamaterial devices and quantum excitations supported by flux quantization and quantum interference effects. Indeed, the typical split-ring meta molecule, the traditional focus of metamaterials research, has a lot of similarities with the Josephson junction ring, the basic building block of superconductivity [44].

3.6 PHYSICAL PROPERTIES OF MAGNETIC NANOPARTICLES

3.6.1 Types of Magnetic Nanoparticles

3.6.1.1 Oxides: Ferrites

The most researched magnetic nanoparticles to date are ferrite or iron oxide nanoparticles. Since they exhibit their magnetic behavior only when a magnetic field from the outside is applied, superparamagnetic ferrite particles, which are smaller than 128 nm, avoid self-agglomeration. By carefully arranging a number of separate superparamagnetic nanoparticles into superparamagnetic nanoparticle groups, such as magnetic nanobeads, the magnetic field of ferrite nanoparticles can be significantly boosted. Remanence returns to zero when the outside magnetic field is turned off. The exterior of ferrite nanoparticles is frequently altered by surfactants, silica, silicones, or phosphoric acid derivatives to boost their stability in solution, analogous to non-magnetic oxide nanoparticles [45].

3.6.1.2 Ferrites with a Shell

Maghemite or magnetite magnetic nanoparticles typically do not allow for strong covalent interactions with functionalization molecules due to their relatively inert surface. However, by adding a layer of silica to the surface of the magnetic nanoparticles, their reactivity can be increased. Organo-silane molecules and the silica shell can form covalent bonds that make it simple to add different surface functional groups to the silica shell. The customized silica shell can also be covalently linked to various fluorescent dye molecules. In comparison to metallic nanoparticles, ferrite nanoparticle groups with small size distributions, made of superparamagnetic titanium oxide nanoparticles 80 maghemite is superparamagnetic nanoparticle per bead encapsulated with a silica shell, have the following advantages:

- Increased chemical stability, which is essential for biological applications
- Higher stability of colloids because they do not mechanically agglomerate
- Narrow size distribution (essential for biological applications)
- Magnetic moment that can be adjusted based on the size of the nanoparticle cluster
- Silica surface that permits simple covalent functionalization (retained superparamagnetic characteristics regardless of nanoparticle cluster size) [46].

3.6.1.3 Metallic Nanoparticles

Due to their stronger magnetic moments, metallic nanoparticles may be useful for some technological applications, although oxides (maghemite, magnetite) might be more useful for biological applications. This means that nanoparticles made of metal can be produced in a shorter time than their oxide equivalents, resulting in them being physically shorter.. On the other hand, nanoparticles of metal have the significant drawback of being highly pyrophoric and to varying degrees reactive to oxidizing chemicals. This makes them harder to handle and encourages unintended side effects, which makes them less suitable for biomedical applications. Metallic particle colloids are also significantly more difficult to create.

3.6.1.4 Metallic Nanoparticles with a Shell

Magnetic nanoparticles' metallic cores can be passivated using surfactants, polymers, precious metals, and light oxidation. Co nanoparticles create an anti-ferromagnetic CoO surface on their surface when exposed to oxygen. Recently, research examined the processes of synthesis as well as the exchange bias impact in these gold-coated Co core, CoO shell nanoparticles. Npw it has been possible to create nanoparticles with a core that is magnetic, formed of either elemental iron or cobalt, and a nonreactive graphene shell. When contrasted to ferrite or elements nanoparticles, the benefits of this composition are

- Increased magnetization
- Increased stability in organic solvents and acidic and basic solutions
- Chemical reactions on the graphite surface using techniques for carbon nanotubes [47].

3.6.2 Physical Properties of Magnetic Nanoparticles

Particles with both mass and electrical charges move, which results in magnetic phenomena. Protons, electrons, holes, and both negative and positive ions make up nanoparticles. A magnetic dipole, or "magneton," is produced by a rotating electric-charged particle. Magnetons are grouped together in ferromagnetic materials. The area of ferromagnetic substance that occurs when all magnetons align themselves in the same orientation by the exchange of forces is referred to as a magnetism domain (also known as a Weiss domain). The domain theory separates ferromagnetism from paramagnetism. The size dependence of a ferromagnetic material's magnetic behavior is governed by its domain structure. A ferromagnetic substance becomes an individual domain when its size is decreased below a certain value. Size impacts, which depend on the magnetic zone structure of ferromagnetic materials, are what cause fine particle magnetism. It is predicated that the situation of minimum free energy of particles that are ferromagnetic has nonuniform magnetism for bigger particles and homogeneous magnetization for particles less than a particular threshold size. Single-domain particles are the first kind, and multi-domain particles are the second kind. The magnitude of the magnetic saturation point, the intensity of the crystal heterogeneity and exchange forces, the surface or sector-wall energy, and the form of the particles are some of the variables that have an impact on the critical size of a single domain, according to the magnetism domain theory. A hysteresis loop, which is defined by the two key parameters of remanence and coercivity, accurately captures the response of ferromagnetic materials to an applied field. The latter has to do with the curve's "thickness." The most important attribute when coping with fine particles is coercivity, which is substantially size-dependent. It has been discovered that the coercivity climbs to a maximum as the number of particles lowers and then declines toward zero (Figure 3.7).

Single-domain particles have 0% coercivity when their size falls below a certain diameter, which causes them to become superparamagnetic. Thermal processes are the cause of superparamagnetism. Superparamagnetic nanoparticles have zero coercivity and no hysteresis because thermal oscillations are sufficiently strong to independently demagnetize a recently saturated assembly. When an external magnet is present, nanoparticles become magnetic; however, when the external magnet is withdrawn, the nanoparticles return to their nonmagnetic state. This prevents the particles from acting actively when they have no supplied field. Particles introduced into live systems are only 'magnetic" in the presence of a magnetic field from the outside, giving them a distinct advantage while operating in biological settings.

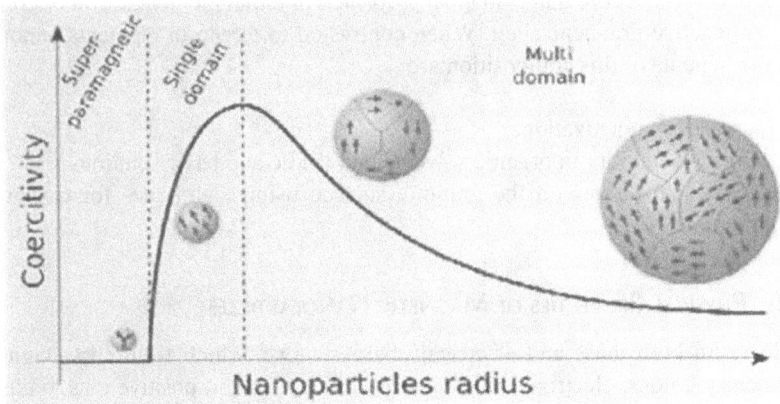

FIGURE 3.7   Schematic illustration of the coercivity-size relations of small particles.

Ferromagnetism can be found in a variety of crystalline minerals, including Fe, Co, and Ni. Being the most magnetic of all minerals that occur naturally on Earth, ferrite oxide–magnetite (Fe_3O_4) is frequently employed in the shape of superparamagnetic nanoparticles for a variety of biological purposes [48–52]. Magnetic nanoparticles have many properties that are described below:

3.6.2.1   Size and Shape and Properties of Nanoparticles

More and more work has been put into creating innovative nanomaterials with a wide range of uses as nanotechnology has advanced. Among the transition metal magnetic particles, Ni nanoparticles have received the most attention due to their wide range of applications involving magnetic sensor memory devices and separation of bio-molecules. The final efficiency of substances and devices formed of Ni nanoparticles is significantly influenced by their purity, structure, size, and shape. Therefore, it is important to create high-quality Ni nanoparticles utilizing a practical and affordable technology. Ni nanoparticles have recently been synthesized under controlled conditions using a variety of techniques, including anodic spark plasma, chemical reduction pyrolysis, reversed micelle, and polyol process. Pure Ni nanoparticles, however, are challenging to create since they are rapidly oxidized. Numerous procedures are used in organic media to create pure Ni nanoparticles in order to prevent the development of oxide or hydrolysis. For instance, Ni particles are created using $Ni(COD)_2$ by spontaneous breakdown in CH_2Cl_2 when polyvinylpyrrolidone is present. More intriguingly, trigonal in Ni nanoparticles were created by combining the reductant and stabilizer tetra-n-octyl ammonium carboxylates with Ni $(COD)_2$ in tetrahydrofuran. To create Ni nanoparticles with precise control over size and form, the magnetic characteristics, chemical stability, and microstructure of the so-produced Ni particles are carefully examined. The effect of particle size and shape on saturate magnetization, coercivity, and Curie temperature is the main topic of study (Figure 3.8). Particularly, a theoretical model may be used to explain the size dependency of the Curie temperature [53].

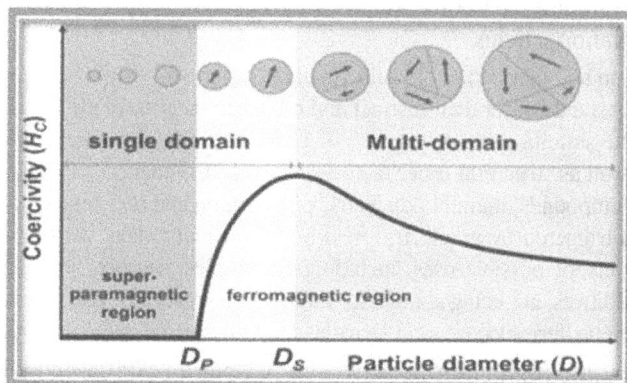

FIGURE 3.8 A schematic showing the coercivity (H_c) behavior of a magnetic particle as a function of its diameter (D).

3.6.2.2 Highly Crystalline Properties of Magnetic Nanoparticles

Due to their appealing qualities, such as their high magnetism moment, biological compatibility, chemical resistance, and magnetoelectric properties, magnetite (Fe_3O_4) nanoparticles (NPs) have garnered a lot of interest. Numerous articles have discussed developments in their nanostructures and the creation of a variety of uses include printer toners, wastewater treatment, catalysis, Li batteries, and medicinal applications. Controlling the crystalline shape, dimension, and domain's structure is crucial for defining the chemical and physical characteristics of these NPs, which have a wide range of applications. This was shown in research on photoluminescent nanoparticles8, where it was shown that the domain size, rather than the particle size, was responsible for the performance. The particle size, which results from the magnetism domain structure, is now acknowledged as a crucial element in the use of magnetic NPs. The coercivity (H_c) of Fe_3O_4 NPs, one of its magnetic characteristics, and particle size have a well-documented connection. The modification in H_c regarding particle size was used to assess the essential dimensions of magnetic NPs, which infer the transformation from a single- to a multi-domain structure. Because the critical size for Fe_3O_4 NPs relies on the arrangement of the crystals, which might contain spherical, cubic, or many phases, this number has not yet been rigorously proven. Therefore, it is crucial to investigate the critical size based on the crystal structure. The necessary size for generating a multi-domain configuration has been theoretically calculated to be 76 nm for cubic along with 128 nm for sphere Fe_3O_4 NPs. The particle size sufficient to generate magnetic properties in Fe_3O_4 NPs is usually thought to be below 20 nm. The essential dimension for cubic Fe_3O_4 NPs, however, has been determined by testing to be more than 160 nm. For cubic Fe_3O_4 NPs, critical sizes of 30-46 nm have been found. Fe_3O_4 multi-granule NPs with diameters ranging from 16 to 512 nm had a transition size of around 120 nm [54].

3.6.2.3   Crystalline Structure Regarding Magnetic Properties of Nanoparticles

For the creation and fabrication of multi-function materials, multifarious materials hold promise. They stand out for their distinct and robust connection of electric and magnetic factors and the simultaneous emergence of ferroelectricity, ferromagnetism, and ferro elasticity, as well as structural order factors. Ferro electromagnets, or multiferroic magnetoelectric compounds, maintain dielectric polarization and magnetization that may be modified and triggered by an electric field and a magnetic field, respectively. For this reason, a variety of possible uses, including spintronics, sensors, magnetic recording media, and additives, are being studied for multiferroic materials. Because it serves as an antiferromagnetic, ferroelectric, and ferro-elastic multiferroic compound with electrical power, magnetism, and morphological ordering temperatures considerably beyond room temperature, bismuth ferrite ($BiFeO_3$) has been the focus of study. A spinning of the canting method away from perfect antiferromagnetic ordering results from the interplay of exchange and spin orbital forces. A spiral spin organization with a wavelength of 62 nm is superimposed on the motion of the resulting tiny moment, creating a helimagnetic order and diminishing magnetization in the bulk. Bismuth ferrite has thus far proved difficult to incorporate into practical devices because of leakage that results in poor resistivity, likely as a result of defect and nonstoichiometric related problems. There has thus been a compelling necessity to provide top-notch samples (Figure 3.9). Recent strategies have concentrated on creating innovative structural compositions such as $BiFeO_3$ materials with zero- and one- along with two-dimensional nanostructures [55].

3.6.2.4   Monodisperse Magnetite Properties of Magnetic Nanoparticles

Applications for magnetic iron oxide nanoparticles include ferro fluids, magnetic separations, and pharmaceuticals. Iron oxides are employed as magnetic materials

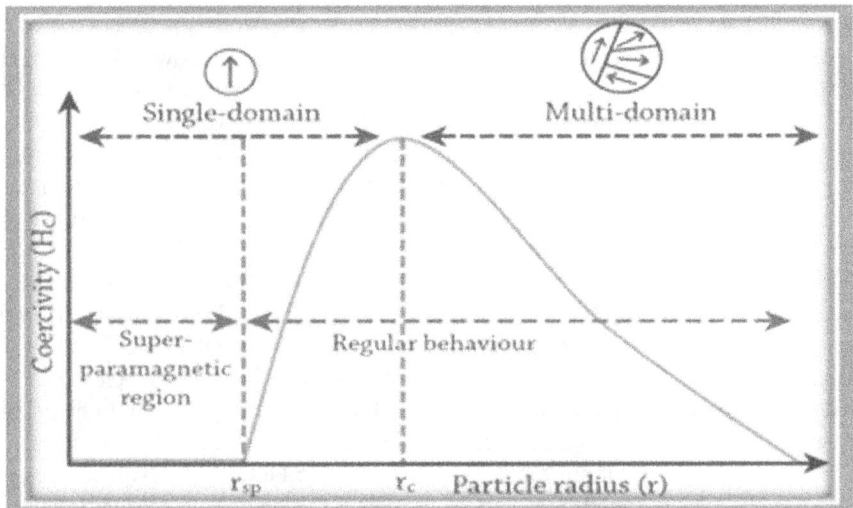

FIGURE 3.9   The change in magnetic coercivity of NPs as a function of particle radius.

in the clinic. Emerging uses of magnetic particles imaging (MPI) include medication delivery, hyperthermia treatment, contrast agents for magnetic resonance imaging (MRI), and treatment of iron deficiency. The nanoparticles of iron oxide are a crucial element in the investigation of magnetism. Iron oxide nanoparticle production methods have been established, and they may typically be categorized as water-based co-precipitation processes or non-polar solvent-based thermolysis reactions. The production of monodisperse nanoparticles by thermolysis results in programmable physical characteristics and variable size, making them very desirable for use in applications research. Naturally, a range of other iron precursors have been used in thermolysis processes, including iron pentacarbonyl, $Fe(acac)_3$, iron oxyhydroxide, and metal oleate. Typically, a substance called iron oleate (or similar iron-organic) is used in thermolysis. It is combined with octadecene (or another high-boiling solvent) and oleic acid (or another surfactant) and heated to a high temperature (250 to 360°C), where iron oxide nanoparticle production takes place. With a dispersity of between 5 and 10%, a variety of sizes may be produced depending on the solvent, surfactant, and temperature [56].

3.6.2.5   Optical Properties of Magnetic Nanoparticles

The creation of nanostructures necessitates a thorough knowledge of the physical processes occurring at this size. It has been demonstrated that low-dimensional quantum structures exhibit distinctive optical and electrical features. Specifically, the form and important factors that govern their physical qualities include the size of low-dimensional structures. The characterization of these factors, which range from growth and classification to device processing, is a crucial problem in both basic research and technical applications. Because several of their primary physical characteristics may be entirely different from the equivalent ones with respect to molecules or bulk materials, nanoparticles made of metallic material are significant among nanostructures. For instance, they could suppose crystal structures that do not match those of the solid's mass. Additionally, several of them have a considerable size and shape dependence on their catalytic activity. Typically, methods including atomic force microscopes (AFM), scanning tunneling microscopy (STM), electron microscopy with transmission (TEM), and reflection high-energy electron diffraction (RHEED) have been used to estimate the form and size of nanoparticles. By characterizing a limited number of nanoparticles at a time, these approaches reveal information on local attributes by producing a picture of a tiny portion of the sample. These approaches have led to the discovery of nanoparticles in a variety of forms, including spheres, spheroids, lenses, cones, pyramids with various facets, truncation pyramids, and various polyhedrals. These "structural-characterization techniques" have been used to gather a lot of data, but they still have significant restrictions. One drawback is that the growth and characterization are often done in distinct environments, which is a severe issue because nanoparticle characteristics depend on their environment. Some of these methods contact the sample when characterizing it, which can occasionally significantly change a nanoparticle's characteristics. Additionally, nanoparticles are often grown and characterized at separate periods, and this might introduce another unpredictable variable [57].

3.6.2.6   Hydrothermal Properties of Magnetic Nanoparticles

Since materials in the range of sizes of nanoparticles will enable exploring the fundamental elements of magnetic-ordering events in magnetic substances with reduced dimensions, the fabrication and characterization of nanoscale magnetic elements have received a lot of attention. This could also open up new opportunities for applications, such as high-density recording using arrays of magnetic nanoparticles. Multiple techniques have been noted in the research for the setup of nanotechnology Fe_3O_4 particles, such as a decrease of iron oxide Fe_2O_3 by CO as well as H_2, simultaneous precipitation of a solution of ferrous/ferric mixed alongside $NH_3 \cdot H_2O$ fluid, the oxidation of ferrous hydroxide gels employing KNO_3, g-ray sunlight, microwave plasma therapy creation, water with oil an emulsion manner using an inadequate the quantity of cyclohexane as the oil phase. However, compared to those of buck materials, the stated magnetic characteristics, such as saturation magnetization, are substantially lower, and the results also vary from method to method. According to certain theories, nanoparticles have a high surface-to-volume ratio, meaning that their surface area is larger than their bulk state. The material's magnetization behavior would be reduced by a disordered phase and non-magnetic layer. It has been discovered that the formation circumstances have a significant impact on the crystallization of nanosized particles. One of the effective ways to create crystals of many various minerals, including quartz and malachite, is through the hydrothermal process. This method has also been applied to create dislocation-free single crystal particles, and grains produced by this method may be more crystalline than grains produced by other methods. The goal of this research is to hydrothermally create nanotechnology Fe_3O_4 grains with higher crystallinity and assess the relationship between their magnetic characteristics and crystallinity. According to reports, iron powder and ferric chloride may be combined hydrothermally in an alkaline medium to create ultrafine magnetite powder. To prevent traces of metallic iron in the finished product, the precise quantity of iron pellets must be carefully regulated. In one study, a one-step hydrothermal method was established that produced well-crystallized nanoscale Fe_3O_4, a powder at low temperatures. A vibrating sample magnetometer was used to examine its magnetic characteristics [58].

3.6.2.7   Superparamagnetic Properties of Metal Nanoparticles

Spinel-shaped magnetic transition oxide nanostructures (MFe_2O_4; M = Mn, Fe, Co, Ni, etc.) have special magnetic characteristics, including moderate magnetization, and remarkable industrial and biological applications are produced by their coercivity, such as specific zone effects, superparamagnetism, and spin filtering. Due to its captivating magnetic and electromagnetic characteristics, $MnFe_2O_4$ (MFO) is a significant member of the ferrite family and has drawn significant study interest. Because $MnFe_2O_4$ is partly inverse spinal, 80% of the Mn^{2+} ions are concentrated at the tetrahedral (A) site, while only 20% are concentrated at the octahedral (B) site. The material $MnFe_2O_4$ exhibits a variety of features in the form of nanoparticles, nanostructures, and thin films, including a high anisotropic steady, size-dependent saturation magnetization, extreme spin glass state, and superparamagnetism, as well as a high Curie temperature. Numerous appealing uses, including magnetic recording, a microwave, MRI contrast, a ferrofluid, and a site-specific drug, are made possible by MFO's desirable features. $MnFe_2O_4$ magnetic nanoparticles have recently been discovered to be an effective adsorbent for the purification of the dye azo Acids Red B out of water. The composition, morphology, and size of manganese ferrite are largely dependent on the preparation circumstances, as are its size and shape [48, 59–61].

3.6.2.8 Magnetorheological (MR) Properties of Magnetic Nanoparticles

Magnetic fluids, also known as ferrofluids, are magnetic colloids in which nanoparticles are typically protected from agglomeration by a surfactant-like layer (satirical stabilization) the surfactant's grade and type will determine. As a result, you can balance the attracting and repellent interactions between particles by determining the effectiveness of the particle surface covering. The appealing interactions may result in a variety of aggregate forms, most frequently the form of linear chains that are nearly parallel to the magnetism or aggregates that resemble drops. Since most applications do not want the agglomeration process, a microstructural characterization of the samples based on their physical characteristics is required. These particles are appealing for controlling movement and separation among attached material because of the inherent coupling of nanoparticles of magnetic material with applied magnetic field gradients [62–64].

3.6.2.9 Controlling Transport Properties of Magnetic Nanoparticles

Spanning from nanoparticles to cells in size and employing optical techniques using nanoscale precious metals and semiconductor particles, nanotechnology is a potent means for interacting with materials at the nanoscale. For this reason, magnetic nanoparticles have attracted a lot of attention in recent years as researchers have sought to create and comprehend artificial techniques to regulate their size, magnetic behavior, and chemical reactivity. The creation of useful magnetic nanoparticles that may be used as tools for manipulating, tracking, and delivering attached cargo is made possible by simultaneously adjusting the surface chemical and physical characteristics. Nanoscale magnets are expected to have a wide range of biomedical uses, including magnetic field–aided separation and analysis. Multi-functional magnetic probes will be created as a result of expanding the chemistry used to add surface functionality, providing new chances to carry out a variety of tasks with a single particle. The size and composition–dependent magnetic moments, on the other hand, have the potential to be used in the creation of magnetic separation technology. The special characteristics seen in nanoscale magnets have potential uses in a wide range of fields, including hyperthermia therapy for cancer cells and MRI contrast [20].

3.6.2.10 Sensor and Biosensor Properties of Magnetic Nanoparticles

One of the most significant areas of study in the material sciences has been nanotechnology. Compared to non-NP materials, nanomaterials (NPs with sizes between 1 and 100 nm) exhibit striking changes in their physical and chemical characteristics. Due to their tiny size, they have unique optical, electrical, catalytic, thermal, and magnetic properties. Due to their unique benefits, including their size, physicochemical features, and cheap cost of manufacture, magnetic NPs (MNPs) have received a lot of attention in recent years. Due to supermagnetism, MNPs function

BIOSENSOR STRUCTURE

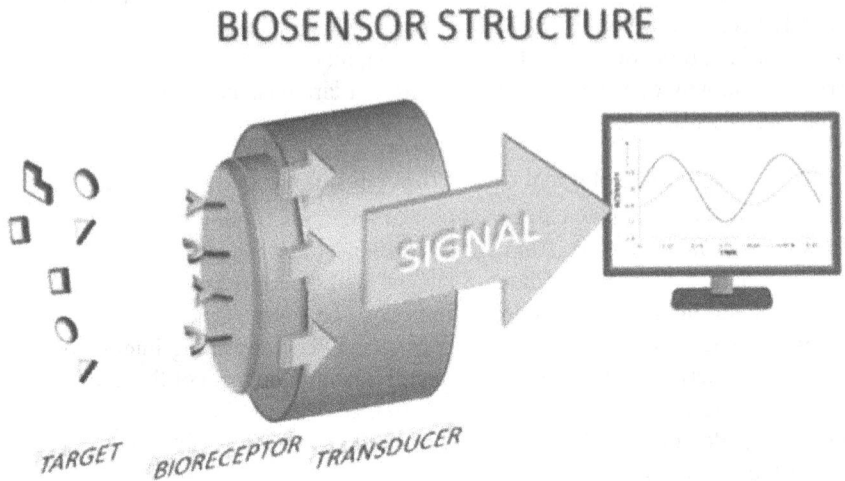

FIGURE 3.10   Schematic representation of a biosensor structure and transducing mechanism.

best at diameters between 10 and 20 nm, making them particularly ideal for applications where a quick reaction to applied magnetic fields is required. Additionally, MNPs have excellent mass transference and a huge surface area. MNP synthesis and preparation must be planned to produce particles with the right size-dependent physicochemical characteristics, since the physical features of MNPs largely depend on their dimensions. MNPs have a wide range of applications, including sample preparation, treatment of wastewater, purification of water, treatment of illnesses, medical diagnosis through magnetic resonance imaging, and cell labeling, as well as imaging, tissue engineering, and biosensors and other detection systems. Through physiochemistry, nanoparticles with customized surface characterization have been synthesized under precise conditions. The sensitivity and durability of sensors as well as biosensors have also been improved by MNPs for the detection of a variety of analytes in clinical, food, and environmental applications (Figure 3.10) [65–68].

REFERENCES

1. Cullity, B.D. and Graham, C.D., 2011. *Introduction to Magnetic Materials*. John Wiley & Sons.
2. Chen, C.W., 2013. *Magnetism and Metallurgy of Soft Magnetic Materials*. Courier Corporation.
3. Klabunde, K.J. and Richards, R.M. eds., 2009. *Nanoscale Materials in Chemistry*. John Wiley & Sons.
4. Heck, C., 2013. *Magnetic Materials and Their Applications*. Elsevier.
5. Krishnan, K.M., 2016. *Fundamentals and Applications of Magnetic Materials*. Oxford University Press.
6. Aslibeiki, B., Kameli, P. and Salamati, H., 2019. Nanomagnetism. *Iranian Journal of Physics Research*, 16(4), pp.251–272.
7. Schmool, D.S., 2022. Recent advances in nanomagnetism. *Magnetochemistry*, 8(9), p.110.

8. Yu, D., Lupton, E.M., Gao, H.J., Zhang, C. and Liu, F., 2008. A unified geometric rule for designing nanomagnetism in graphene. *Nano Research, 1*, pp.497–501.
9. Gould, P., 2006. Nanomagnetism shows in vivo potential. *Nano Today, 1*(4), pp.34–39.
10. Jacobs, I.S., 1969. Role of magnetism in technology. *Journal of Applied Physics, 40*(3), pp.917–928.
11. Wang, L., Wang, D., Cao, Q., Zheng, Y., Xuan, H., Gao, J. and Du, Y., 2012. Electric control of magnetism at room temperature. *Scientific Reports, 2*(1), p.223.
12. Piso, M.I., 1999. Applications of magnetic fluids for inertial sensors. *Journal of Magnetism and Magnetic Materials, 201*(1–3), pp.380–384.
13. Strnat, K.J., 1990. Modern permanent magnets for applications in electro-technology. *Proceedings of the IEEE, 78*(6), pp.923–946.
14. Freeman, M.W., Arrott, A. and Watson, J.H.L., 1960. Magnetism in medicine. *Journal of Applied Physics, 31*(5), pp.S404–S405.
15. Hickey, R. and Schibeci, R.A., 1999. The attraction of magnetism. *Physics Education, 34*(6), p.383.
16. Frei, E.H., 1969. Magnetism and medicine. *Journal of Applied Physics, 40*(3), pp.955–957.
17. Ishiyama, K., Sendoh, M. and Arai, K.I., 2002. Magnetic micromachines for medical applications. *Journal of Magnetism and Magnetic Materials, 242*, pp.41–46.
18. Borgert, J., Schmidt, J.D., Schmale, I., Rahmer, J., Bontus, C., Gleich, B., David, B., Eckart, R., Woywode, O., Weizenecker, J. and Schnorr, J., 2012. Fundamentals and applications of magnetic particle imaging. *Journal of Cardiovascular Computed Tomography, 6*(3), pp.149–153.
19. Ichiyanagi, Y., Moritake, S., Taira, S. and Setou, M., 2007. Functional magnetic nanoparticles for medical application. *Journal of Magnetism and Magnetic Materials, 310*(2), pp.2877–2879.
20. Latham, A.H. and Williams, M.E., 2008. Controlling transport and chemical functionality of magnetic nanoparticles. *Accounts of Chemical Research, 41*(3), pp.411–420.
21. Wierucka, M. and Biziuk, M., 2014. Application of magnetic nanoparticles for magnetic solid-phase extraction in preparing biological, environmental and food samples. *TrAC Trends in Analytical Chemistry, 59*, pp.50–58.
22. Marszałł, M.P., 2011. Application of magnetic nanoparticles in pharmaceutical sciences. *Pharmaceutical Research, 28*, pp.480–483.
23. Shen, Y.F., Tang, J., Nie, Z.H., Wang, Y.D., Ren, Y. and Zuo, L., 2009. Preparation and application of magnetic Fe3O4 nanoparticles for wastewater purification. *Separation and Purification Technology, 68*(3), pp.312–319.
24. Hozayn, M., Abdallha, M.M., AA, A.E.M., El-Saady, A.A. and Darwish, M.A., 2016. Applications of magnetic technology in agriculture: A novel tool for improving crop productivity (1): Canola. *African Journal of Agricultural Research, 11*(5), pp.441–449.
25. Huang, Y., Zhang, X., Kadic, M. and Liang, G., 2019. Stiffer, stronger and centrosymmetrical class of pentamodal mechanical metamaterials. *Materials, 12*(21), p.3470.
26. Zeng, J., Wang, X., Sun, J., Pandey, A., Cartwright, A.N. and Litchinitser, N.M., 2013. Manipulating complex light with metamaterials. *Scientific Reports, 3*(1), p.2826.
27. Aghighi, F., Morris, J. and Amirkhizi, A.V., 2019. Low-frequency micro-structured mechanical metamaterials. *Mechanics of Materials, 130*, pp.65–75.
28. Liu, L., Kang, L., Mayer, T.S. and Werner, D.H., 2016. Hybrid metamaterials for electrically triggered multifunctional control. *Nature Communications, 7*(1), p.13236.
29. Nefedov, I.S. and Tretyakov, S.A., 2005. On potential applications of metamaterials for the design of broadband phase shifters. *Microwave and Optical Technology Letters, 45*(2), pp.98–102.
30. Iyer, A.K., Alu, A. and Epstein, A., 2020. Metamaterials and metasurfaces—Historical context, recent advances, and future directions. *IEEE Transactions on Antennas and Propagation, 68*(3), pp.1223–1231.

31. Park, S.J., Hong, J.T., Choi, S.J., Kim, H.S., Park, W.K., Han, S.T., Park, J.Y., Lee, S., Kim, D.S. and Ahn, Y.H., 2014. Detection of microorganisms using terahertz metamaterials. *Scientific Reports*, *4*(1), pp.1–7.
32. Li, J., Zhou, X., Huang, G. and Hu, G., 2016. Acoustic metamaterials capable of both sound insulation and energy harvesting. *Smart Materials and Structures*, *25*(4), p.045013.
33. Lapine, M., Powell, D., Gorkunov, M., Shadrivov, I., Marqués, R. and Kivshar, Y., 2009. Structural tunability in metamaterials. *Applied Physics Letters*, *95*(8), p.084105.
34. Shadrivov, I.V., Kozyrev, A.B., van der Weide, D. and Kivshar, Y.S., 2008. Nonlinear magnetic metamaterials. *Optics Express*, *16*(25), pp.20266–20271.
35. Özbay, E., 2008. The magical world of photonic metamaterials. *Optics and Photonics News*, *19*(11), pp.22–27.
36. Lv, C., Krishnaraju, D., Konjevod, G., Yu, H. and Jiang, H., 2014. Origami based mechanical metamaterials. *Scientific Reports*, *4*(1), p.5979.
37. Lee, W., Kang, D.Y., Song, J., Moon, J.H. and Kim, D., 2016. Controlled unusual stiffness of mechanical metamaterials. *Scientific Reports*, *6*(1), pp.1–7.
38. Han, D., Ren, X., Zhang, Y., Zhang, X.Y., Zhang, X.G., Luo, C. and Xie, Y.M., 2022. Lightweight auxetic metamaterials: Design and characteristic study. *Composite Structures*, *293*, p.115706.
39. Chen, Y., Ye, W., Xu, R., Sun, Y., Feng, J. and Sareh, P., 2023. A programmable auxetic metamaterial with tunable crystal symmetry. *International Journal of Mechanical Sciences*, *249*, p.108249.
40. Mizzi, L., Azzopardi, K.M., Attard, D., Grima, J.N. and Gatt, R., 2015. Auxetic metamaterials exhibiting giant negative Poisson's ratios. *Physica Status Solidi (RRL)–Rapid Research Letters*, *9*(7), pp.425–430.
41. Kolken, H.M. and Zadpoor, A.A., 2017. Auxetic mechanical metamaterials. *RSC Advances*, *7*(9), pp.5111–5129.
42. Kshetrimayum, R.S., 2004. A brief intro to metamaterials. *IEEE Potentials*, *23*(5), pp.44–4.
43. Krzysztofik, W.J. and Cao, T.N., 2018. Metamaterials in application to improve antenna parameters. *Metamaterials and Metasurfaces*, *12*(2), pp.63–85.
44. Zheludev, N.I., 2011. A roadmap for metamaterials. *Optics and Photonics News*, *22*(3), pp.30–35.
45. Tadic, M., Kralj, S., Jagodic, M., Hanzel, D. and Makovec, D., 2014. Magnetic properties of novel superparamagnetic iron oxide nanoclusters and their peculiarity under annealing treatment. *Applied Surface Science*, *322*, pp.255–264.
46. Kralj, S., Makovec, D., Čampelj, S. and Drofenik, M., 2010. Producing ultra-thin silica coatings on iron-oxide nanoparticles to improve their surface reactivity. *Journal of Magnetism and Magnetic Materials*, *322*(13), pp.1847–1853.
47. Grass, R.N. and Stark, W.J., 2006. Gas phase synthesis of fcc-cobalt nanoparticles. *Journal of Materials Chemistry*, *16*(19), pp.1825–1830.
48. Akbarzadeh, A., Samiei, M. and Davaran, S., 2012. Magnetic nanoparticles: Preparation, physical properties, and applications in biomedicine. *Nanoscale Research Letters*, *7*, pp.1–13.
49. Kodama, R.H., 1999. Magnetic nanoparticles. *Journal of Magnetism and Magnetic Materials*, *200*(1–3), pp.359–372.
50. Majetich, S.A., Wen, T. and Mefford, O.T., 2013. Magnetic nanoparticles. *MRS Bulletin*, *38*(11), pp.899–903.
51. Gubin, S.P. ed., 2009. *Magnetic Nanoparticles*. John Wiley & Sons.
52. Faivre, D. and Bennet, M., 2016. Magnetic nanoparticles line up. *Nature*, *535*(7611), pp.235–236.

53. He, X. and Shi, H., 2012. Size and shape effects on magnetic properties of Ni nanoparticles. *Particuology*, *10*(4), pp.497–502.
54. Li, Q., Kartikowati, C.W., Horie, S., Ogi, T., Iwaki, T. and Okuyama, K., 2017. Correlation between particle size/domain structure and magnetic properties of highly crystalline Fe3O4 nanoparticles. *Scientific Reports*, *7*(1), p.9894.
55. Park, T.J., Papaefthymiou, G.C., Viescas, A.J., Moodenbaugh, A.R. and Wong, S.S., 2007. Size-dependent magnetic properties of single-crystalline multiferroic BiFeO3 nanoparticles. *Nano Letters*, *7*(3), pp.766–772.
56. Kemp, S.J., Ferguson, R.M., Khandhar, A.P. and Krishnan, K.M., 2016. Monodisperse magnetite nanoparticles with nearly ideal saturation magnetization. *RSC Advances*, *6*(81), pp.77452–77464.
57. Sosa, I.O., Noguez, C. and Barrera, R.G., 2003. Optical properties of metal nanoparticles with arbitrary shapes. *The Journal of Physical Chemistry B*, *107*(26), pp.6269–6275.
58. Wang, J., Sun, J., Sun, Q. and Chen, Q., 2003. One-step hydrothermal process to prepare highly crystalline Fe3O4 nanoparticles with improved magnetic properties. *Materials Research Bulletin*, *38*(7), pp.1113–1118.
59. Zipare, K., Dhumal, J., Bandgar, S., Mathe, V. and Shahane, G., 2015. Superparamagnetic manganese ferrite nanoparticles: Synthesis and magnetic properties. *J. Nanosci. Nanoeng*, *1*(3), pp.178–182.
60. Lin, X.M. and Samia, A.C., 2006. Synthesis, assembly and physical properties of magnetic nanoparticles. *Journal of Magnetism and Magnetic Materials*, *305*(1), pp.100–109.
61. Batlle, X., Moya, C., Escoda-Torroella, M., Iglesias, Ò, Rodríguez, A.F. and Labarta, A., 2022. Magnetic nanoparticles: From the nanostructure to the physical properties. *Journal of Magnetism and Magnetic Materials*, *543*, p.168594.
62. Vékás, L., Raşa, M. and Bica, D., 2000. Physical properties of magnetic fluids and nanoparticles from magnetic and magneto-rheological measurements. *Journal of Colloid and Interface Science*, *231*(2), pp.247–254.
63. Jönsson, P.E., 2003. Superparamagnetism and spin glass dynamics of interacting magnetic nanoparticle systems. *Advances in Chemical Physics*, *128*, pp.191–248.
64. Reddy, K.R., Lee, K.P., Kim, J.Y. and Lee, Y., 2008. Self-assembly and graft polymerization route to monodispersed Fe3O4@ SiO2—polyaniline core–shell composite nanoparticles: Physical properties. *Journal of Nanoscience and Nanotechnology*, *8*(11), pp.5632–5639.
65. Rocha-Santos, T.A., 2014. Sensors and biosensors based on magnetic nanoparticles. *TrAC Trends in Analytical Chemistry*, *62*, pp.28–36.
66. Eberbeck, D., Dennis, C.L., Huls, N.F., Krycka, K.L., Gruttner, C. and Westphal, F., 2012. Multicore magnetic nanoparticles for magnetic particle imaging. *IEEE Transactions on Magnetics*, *49*(1), pp.269–274.
67. Tripathy, A., Nine, M.J. and Silva, F.S., 2021. Biosensing platform on ferrite magnetic nanoparticles: Synthesis, functionalization, mechanism and applications. *Advances in Colloid and Interface Science*, *290*, p.102380.
68. Chen, H.J., Zhang, Z.H., Luo, L.J. and Yao, S.Z., 2012. Surface-imprinted chitosan-coated magnetic nanoparticles modified multi-walled carbon nanotubes biosensor for detection of bovine serum albumin. *Sensors and Actuators B: Chemical*, *163*(1), pp.76–83.k

4 Fabrication of Magnetic Polymer Composites

Ipsita Chinya

4.1 INTRODUCTION

Magnetic polymer composites have emerged as a fascinating class of materials that combine the unique properties of magnetic nanoparticles with the structural and functional characteristics of polymers. These composites possess tunable magnetism and magnetic responsiveness, while also offering the flexibility, processability, and diverse applications associated with polymers. The synthesis and fabrication of magnetic polymer composites involve the integration of magnetic nanoparticles within a polymer matrix, resulting in materials with tailored magnetic and mechanical characteristics.

This chapter aims to provide a comprehensive overview of the fabrication techniques employed in the creation of magnetic polymer composites. These techniques play a crucial role in determining the structure, properties, and performance of the composites, making them key factors in the successful development and application of these materials. By understanding and utilizing the appropriate fabrication techniques, researchers and engineers can effectively control the dispersion, alignment, and interfacial interactions of magnetic nanoparticles within the polymer matrix, thus achieving desired properties and functionalities.

The chapter will discuss both conventional and advanced fabrication techniques for magnetic polymer composites. Conventional techniques, such as solution casting, melt mixing, in situ polymerization, electrospinning, and sol-gel methods, will be explored in detail. These techniques offer simplicity, scalability, and compatibility with various polymer systems, making them widely used in the field. The advantages and limitations of each technique will be discussed, along with the considerations for achieving uniform nanoparticle dispersion and preventing agglomeration. In addition to conventional techniques, advanced fabrication methods that harness external fields or specialized processes will be covered. Magnetic field–assisted alignment techniques enable the controlled orientation of magnetic nanoparticles within the polymer matrix, leading to anisotropic structures with improved magnetic and mechanical properties. Layer-by-layer assembly techniques provide precise control over the deposition of alternating layers of polymers and nanoparticles, allowing the creation of multilayer films and coatings with customizable properties. 3D printing, microfluidic fabrication, and electrochemical deposition are other advanced techniques that offer unique advantages for fabricating magnetic polymer composites with complex geometries, controlled morphologies, and tailored functionalities.

Throughout the chapter, the influence of various fabrication parameters on the final characteristics of magnetic polymer composites will be emphasized. Factors such as nanoparticle loading, particle size and shape, surface modification, polymer

DOI: 10.1201/9781003454236-4

matrix selection, and processing conditions will be discussed in detail. Understanding and optimizing these parameters is essential for tailoring the properties of the composites according to specific application requirements.

In summary, this chapter provides a comprehensive overview of the fabrication techniques employed in the synthesis of magnetic polymer composites. By delving into both conventional and advanced fabrication methods, as well as the influence of fabrication parameters, this chapter aims to empower researchers and engineers to design and develop magnetic polymer composites with tailored properties and functionalities. The applications of these composites across diverse fields demonstrate their immense potential for advancing technology and addressing various societal challenges.

4.2 CONVENTIONAL FABRICATION TECHNIQUES

4.2.1 SOLUTION CASTING

Solution casting involves dissolving the polymer and dispersing the magnetic nanoparticles (using an ultrasonicator) in a solvent. The resulting solution is then cast into a mold or onto a substrate, followed by evaporation of the solvent to obtain the composite material. This technique offers simplicity, scalability, and the ability to control the nanoparticle dispersion. Thus, it is the most common technique for lab-scale fabrication of the nanoparticle. However, achieving uniform dispersion and preventing nanoparticle agglomeration are critical challenges.

The solution casting method is the most widely employed technique for the fabrication of magnetic nanocomposites. Chinya et al. prepared a zinc ferrite/polyvinyl fluoride–based nanocomposite for energy harvesting purpose [1, 2]. Ashjari et al. [3] created a magnetite/polyurethane composite using the solution mixing technique and investigated the properties. Solution mixing is a facile method for fabrication of a soluble polymer-based nanocomposite. Yet the evaporation of solvent (toxic) may cause an environmental issue. Solution casting technology has been widely used for the production of engineering plastics and optical films.

This approach (Figure 4.1) involves the dispersion of magnetic nanoparticles within a polymer solution, followed by the casting of the solution onto a substrate and subsequent solvent evaporation. The critical steps involved are presented below.

Preparation of Polymer Solution: The first step is to prepare a polymer solution by dissolving the desired polymer in a suitable solvent. The choice of solvent depends on the compatibility with the polymer and the desired properties of the nanocomposite.

Dispersion of Magnetic Nanoparticles: Magnetic nanoparticles, typically synthesized beforehand, are dispersed in the polymer solution. Various techniques, such as ultrasonication, mechanical agitation, or magnetic stirring (less preferred), can be employed to ensure the uniform distribution of nanoparticles within the solution.

Casting and Solvent Evaporation: The nanoparticle-dispersed polymer solution is then cast onto a substrate, such as a glass plate or a silicon wafer.

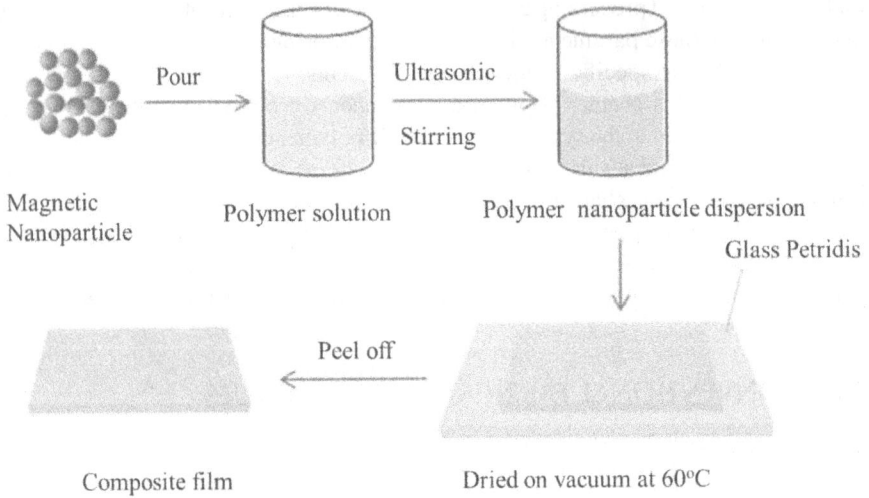

FIGURE 4.1 Schematic diagram of magnetic polymer nanocomposite fabrication via the solution casting technique.

The solution is spread evenly over the substrate using a doctor blade or spin coating technique. Subsequently, the solvent is allowed to evaporate, preferably in a vacuum, resulting in the formation of a thin film with embedded magnetic nanoparticles.

Advantages of the Solution Casting Method: The solution casting method offers several advantages for the fabrication of magnetic nanocomposites:

a. Simple and Cost-Effective: The method utilizes straightforward procedures and equipment, making it relatively simple and cost-effective compared to other fabrication techniques.
b. Control over Nanoparticle Dispersion: The solution casting method allows for excellent control over the dispersion of magnetic nanoparticles within the polymer matrix. This control is essential to ensure uniform distribution and prevent nanoparticle agglomeration.
c. Tailoring of Film Thickness: The thickness of the resulting film can be easily controlled by adjusting the concentration of the polymer solution or the casting conditions. This flexibility enables the fabrication of nanocomposite films with desired thicknesses for specific applications.

Limitations and Considerations: Despite its advantages, the solution casting method has certain limitations and issues for consideration:

a. Film Uniformity: Achieving a highly uniform film can be challenging, particularly for large-area coatings. Factors such as the evaporation rate of the solvent and the substrate surface characteristics can influence the final film uniformity.

b. Polymer-Solvent Compatibility: The selection of an appropriate solvent is critical to ensure compatibility with the polymer and prevent phase separation or other undesirable effects.

4.2.2 MELT MIXING

In melt mixing, the polymer and magnetic nanoparticles are combined in the molten state of the polymer using various mixing methods, such as twin-screw extrusion or high-shear mixing [4–7]. The molten mixture is subsequently solidified by cooling, resulting in a composite material. This method reduces the environmental hazard caused by evaporation of toxic solvent. Melt mixing is advantageous for large-scale production and provides good control over nanoparticle dispersion. However, it may cause thermal degradation of the polymer, leading to uneven distribution of nanoparticles. The melt mixing (Figure 4.2) method involves several key steps:

Selection of Polymer and Magnetic Nanoparticles: The first step is the selection of a suitable polymer matrix and magnetic nanoparticles. The choice of polymer depends on the desired properties and application requirements, while the magnetic nanoparticles should exhibit the desired magnetic properties and compatibility with the polymer. Melt mixing may provide broaden aspect of the constituent selection (although polymer selection will be limited to lower-melting polymers) for here no solution is present as in the solvent casting technique.

Preparation of Nanoparticle-Polymer Mixture: The magnetic nanoparticles are dispersed into the molten polymer matrix. This can be achieved by

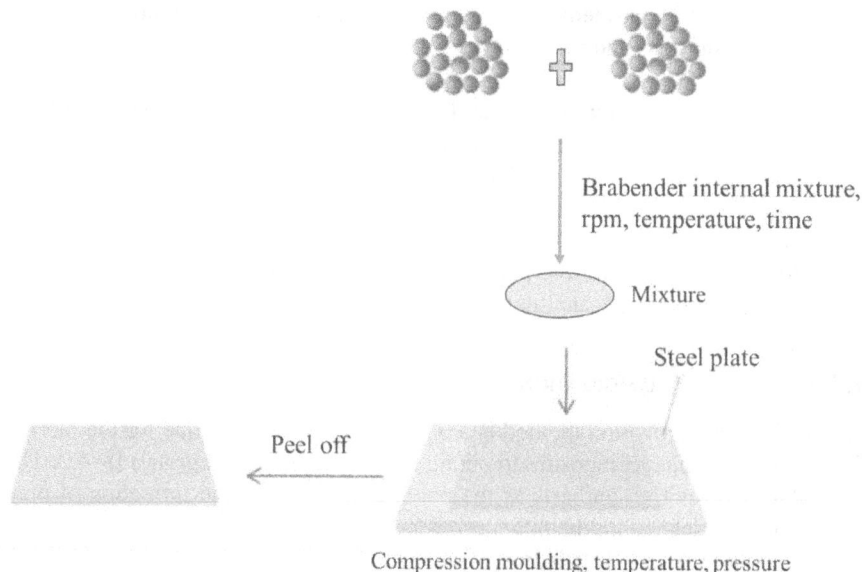

FIGURE 4.2 Schematic diagram of magnetic polymer nanocomposite fabrication via melt mixing technique.

various techniques such as twin-screw extrusion, internal mixers, or roll mills. The high shear forces generated during the mixing process facilitate the uniform dispersion of the nanoparticles within the molten polymer.

Solidification: After thorough mixing, the molten mixture is cooled and solidified to form the magnetic nanocomposite. This can be achieved by cooling the mixture to room temperature or using specific cooling methods such as quenching in water circulation or controlled air cooling to influence the material's microstructure and properties.

Advantages of the Melt Mixing Method: The melt mixing method offers several advantages for the fabrication of magnetic nanocomposites:

a. Homogeneous Dispersion: The high shear forces generated during melt mixing promote the uniform dispersion of magnetic nanoparticles within the polymer matrix, leading to improved mechanical and magnetic properties.

b. Scalability: The melt mixing method is highly scalable and can be easily adapted for large-scale production. This makes it suitable for industrial applications where bulk production is required.

c. Control over Nanoparticle Loading: The melt mixing method allows for precise control over the loading of magnetic nanoparticles in the nanocomposite by adjusting the composition and processing parameters. This enables the tailoring of the composite's magnetic and physical properties.

Limitations and Considerations: Despite its advantages, the melt mixing method has certain limitations and raises some issues for consideration:

a. High Processing Temperatures: The method requires the polymer to be melted at elevated temperatures, which limits the choice of polymers to those that can withstand such conditions without degradation.

b. Nanoparticle Aggl omeration: During melt mixing, there is a risk of nanoparticle agglomeration, especially at higher nanoparticle loadings. Proper control of mixing parameters and the use of surface modification techniques can mitigate this issue.

4.2.3 In Situ Polymerization

The in situ polymerization method is a widely employed technique for the fabrication of magnetic nanocomposites to obtain tunable magnetic properties [8–12]. This approach involves the synthesis of the polymer matrix and the formation of magnetic nanoparticles within the polymerization process, resulting in excellent control over the homogeneous distribution of nanoparticles and interfacial interactions in the final composite. It enables the formation of covalent bonds between the polymer matrix and nanoparticles, resulting in precisely controlled properties. However, careful selection of polymerization conditions and compatibility between monomers and nanoparticles are essential for successful synthesis. Elimination of unreacted monomer from the resulting dispersion is also an issue.

The in situ polymerization (Figure 4.3) method involves several key steps:

Selection of Monomers and Magnetic Precursors: The first step is the selection of suitable monomers for the polymer matrix and appropriate magnetic precursors for the nanoparticles. The choice of monomers and precursors depends on the desired polymer properties and the magnetic characteristics required for the nanocomposite.

Polymerization and Nanoparticle Formation: The monomers and magnetic precursors are combined in a reaction vessel under controlled conditions. The polymerization process, initiated by the addition of a suitable initiator or catalyst, simultaneously leads to the formation of the polymer matrix and the generation of magnetic nanoparticles. This allows for the incorporation of the nanoparticles within the growing polymer chains, resulting in a homogeneous distribution.

Post-treatment: After the polymerization reaction, the resulting magnetic nanocomposite can undergo additional post-treatment steps including removal of unreacted monomer, drying, grinding, or milling to obtain the desired particle size distribution and final composite morphology.

Advantages of In Situ Polymerization: The in situ polymerization method offers several advantages for the fabrication of magnetic nanocomposites:

Homogeneous Nanoparticle Dispersion: The simultaneous formation of the polymer matrix and nanoparticles within the polymerization process ensures a uniform distribution of magnetic nanoparticles throughout the composite. This leads to enhanced magnetic properties and improved performance.

Control over Composite Properties: By adjusting the monomer composition, reaction conditions, and precursor concentrations, the properties of the

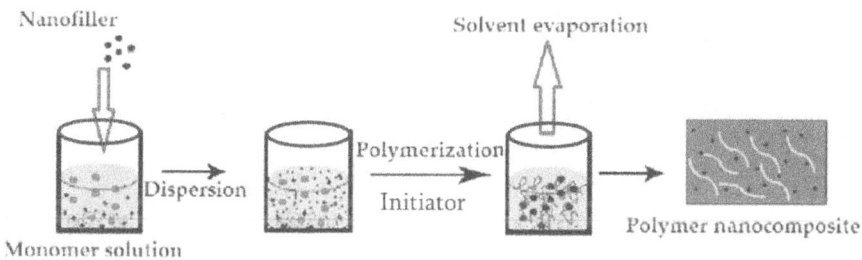

FIGURE 4.3 Schematic diagram of magnetic polymer nanocomposite fabrication via the in situ polymerization technique.

resulting nanocomposite can be finely tuned. This allows for control over parameters such as nanoparticle size, loading, and the overall composite structure.

Scalability: The in situ polymerization method can be easily scaled up for large-scale production, making it suitable for industrial applications where bulk production is required.

Limitations and Considerations : Despite its advantages, the in situ polymerization method has certain limitations and issues for consideration:

a. Monomer-Precursor Compatibility: The selection of compatible monomers and magnetic precursors is crucial to ensure a successful polymerization process and the formation of desired magnetic nanoparticles. Compatibility considerations include solubility, reactivity, and stability of the components.

b. Particle Size Control: Control over nanoparticle size and size distribution can be challenging in the in situ polymerization method. Additional techniques such as seed-mediated growth or the use of stabilizers may be necessary to achieve precise control over particle size.

4.2.4 ELECTROSPINNING

Electrospinning utilizes an electric field to produce polymer nanofibers with embedded magnetic nanoparticles [13–19]. The process involves the extrusion of a polymer solution or melt through a spinneret, followed by fiber elongation and solidification. Electrospinning offers a high surface area-to-volume ratio, tunable fiber morphology, and excellent nanoparticle dispersion. However, controlling fiber alignment and achieving uniform nanoparticle distribution across fibers can be challenging. The schematic for a typical electrospinning process is shown in Figure 4.4.

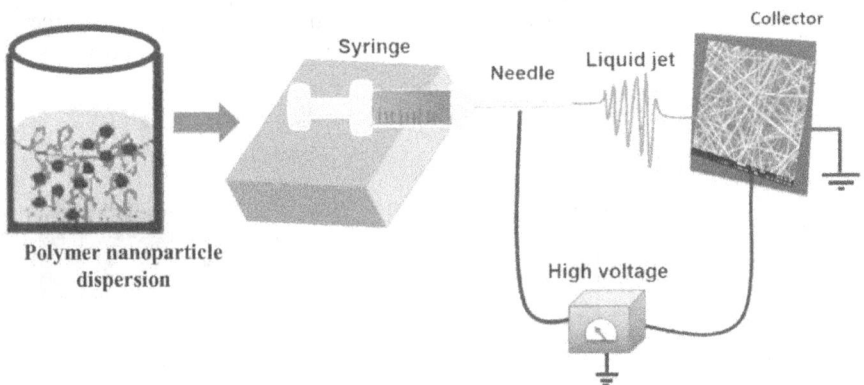

FIGURE 4.4 Schematic diagram of magnetic polymer nanocomposite fabrication via the electrospinning technique.

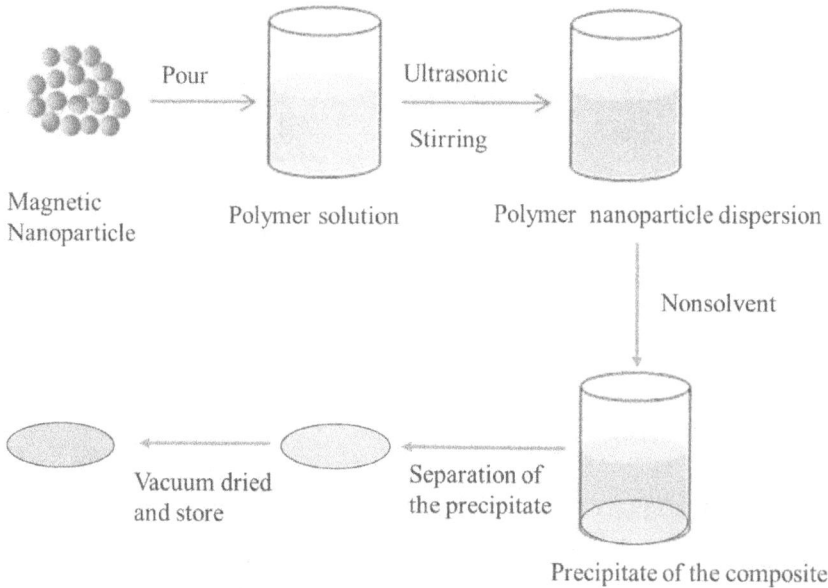

FIGURE 4.5 Schematic diagram of magnetic polymer nanocomposite fabrication via the electrospinning technique.

4.2.5 SOL-GEL METHOD

The sol-gel method involves the hydrolysis and condensation of precursor molecules to form a sol, which is subsequently transformed into a gel and then dried to obtain the composite material [20–23]. This technique allows the incorporation of magnetic nanoparticles during the sol formation or gelation stage, enabling homogenous dispersion. Sol-gel offers versatility in terms of compositional control and the ability to create thin films and coatings. However, it requires careful control of reaction conditions and long processing times.

4.2.6 CO-PRECIPITATE METHOD

The co-precipitate method is similar to solution casting, but instead of pouring the solution in the petri dish, another nonsolvent (with respect to the polymer) is added to the solution and the composite is precipitated out (Figure 4.5). It is then dried in a vacuum at a slightly elevated temperature and collected. Very good properties with excellent homogeneity dispersion are achieved with this technique [24].

4.3 ADVANCED FABRICATION TECHNIQUES

4.3.1 MAGNETIC FIELD–ASSISTED ALIGNMENT

Magnetic field–assisted alignment techniques exploit external magnetic fields to induce alignment of magnetic nanoparticles within the polymer matrix [23–29]. Techniques such as magnetic field alignment, magnetic field–assisted assembly, and

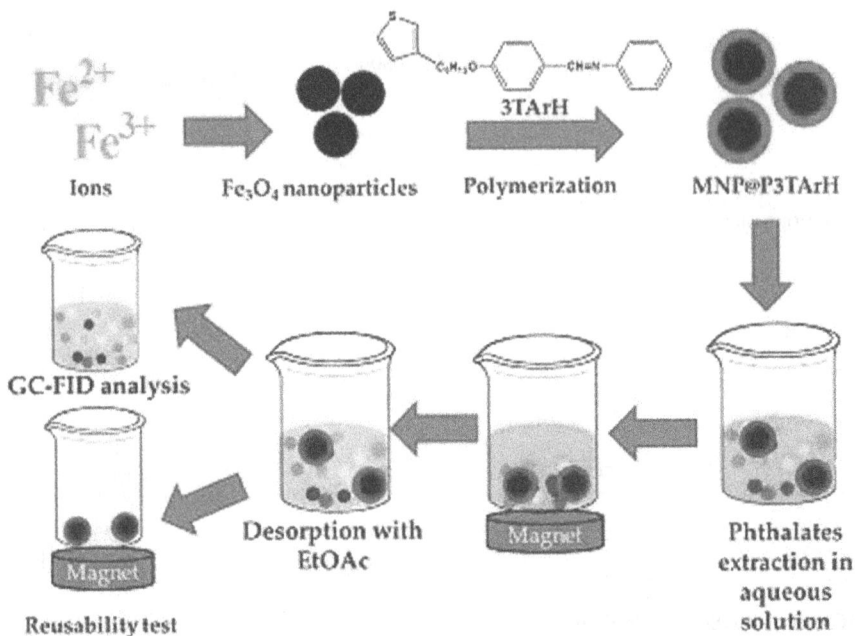

FIGURE 4.6   Magnetic field–assisted polymer nanocomposite fabrication (Fe$_3$O$_4$/GO-QPVA composite). [Copyright Polymers **2016**, *8*(5), 117.]

magnetic field–induced self-assembly can enhance the magnetic and mechanical properties of composites. They enable the creation of anisotropic structures with controlled orientation, leading to improved functionalities. The schematic of magnetic field–assisted polymer nanocomposite fabrication of an Fe$_3$O$_4$/GO-QPVA composite is depicted in Figure 4.6 [28].

4.3.2   LAYER-BY-LAYER ASSEMBLY

Layer-by-layer assembly involves the sequential deposition of alternating layers of oppositely charged polymers and magnetic nanoparticles. This technique (demonstrated in Figure 4.7) allows precise control over the film thickness, composition, and nanoparticle distribution [29–32]. Electrostatic interactions drive the layer-by-layer assembly, enabling the creation of multi-layer films, coatings, or capsules with customizable properties. However, achieving good interlayer adhesion and controlling film porosity are critical considerations. This is the common method for energy applications of the magnetic polymer nanocomposite, such as supercapacitors or harvesters or EMI shielding.

4.3.3   3D PRINTING

3D printing, or additive manufacturing, enables the fabrication of nanocomposites with complex geometries and intricate internal structures [33–35]. Various 3D

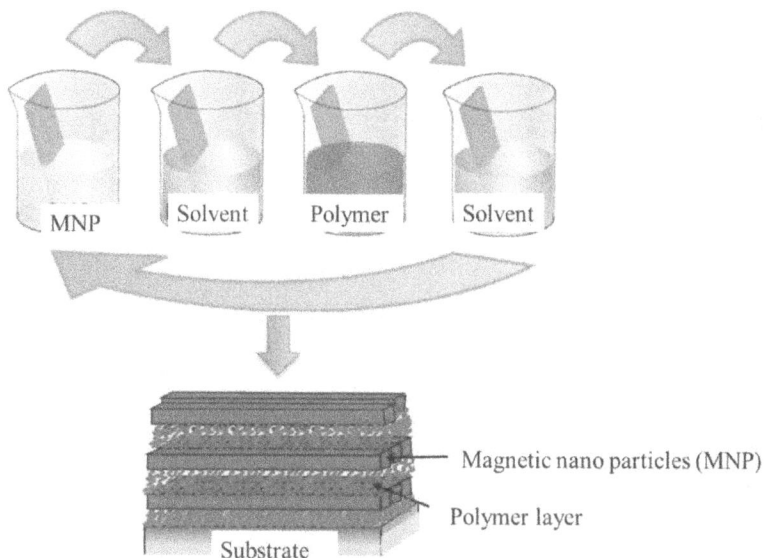

FIGURE 4.7 Schematic of layer-by-layer assembly of polymer magnetic nanocomposite fabrication.

printing techniques, such as fused deposition modeling, stereolithography, and selective laser sintering, can be employed. By incorporating magnetic nanoparticles into the printing materials or embedding them within the printed structures, functional magnetic composites can be created. However, optimization of printing parameters and material formulations is essential for achieving the desired properties. T. Hupfeld et al. prescribed a method for 3D printing an iron oxide nanoparticle functionalized polyamide composite [34]. This method is in general use in magnetic scaffold preparation. A schematic representation of the fabrication strategy is shown in Figure 4.8.

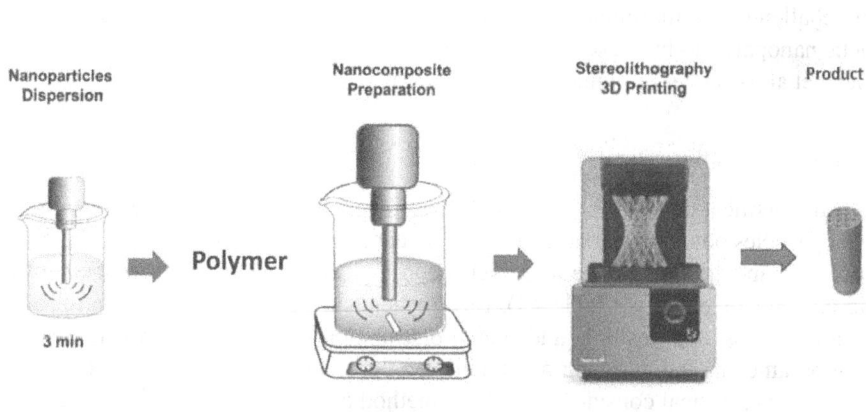

FIGURE 4.8 Schematic of 3D-printed polymer magnetic nanocomposite fabrication procedure.

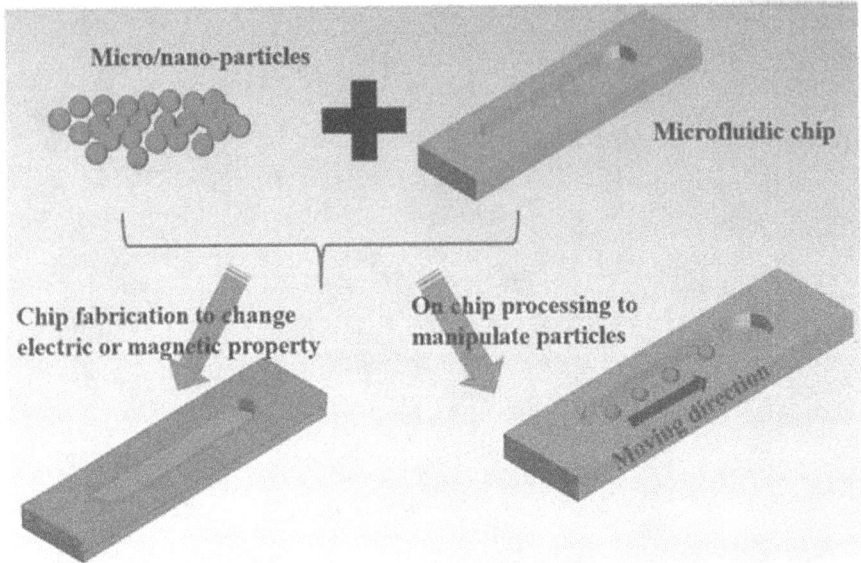

FIGURE 4.9   A typical schematic of microfluid-assisted magnetic nanoparticle polymer–based sensor fabrication procedure. (Copyright Multidisciplinary Digital Publishing Institute, 2014.[37])

4.3.4 MICROFLUIDIC FABRICATION

Microfluidic techniques [36, 37] involve the precise manipulation of fluids at the microscale to create magnetic polymer composites. Microfluidic devices enable controlled mixing of polymer and nanoparticle solutions, resulting in uniform nanoparticle dispersion. The continuous flow nature of microfluidics allows the fabrication of long fibers, microcapsules, or droplets with precise control over composition and morphology. However, scaling up the production and maintaining long-term stability are challenges in microfluidic fabrication. This is the most useful method for magnetic nanoparticle–based sensor fabrication [36, 37]. A simple schematic adopted by Jiang et al. is shown in Figure 4.9 [37].

4.3.5 ELECTROCHEMICAL DEPOSITION

Electrochemical deposition involves the electrodeposition of polymer and magnetic nanoparticles onto conductive substrates. By applying an electric potential, the polymer and nanoparticles can be selectively deposited, allowing the creation of composite coatings or patterns (Table 4.1). Electrochemical deposition offers control over film thickness, composition, and spatial distribution. However, optimizing deposition parameters and ensuring good adhesion between the deposited layers and the substrate are critical considerations. This method is extensively used for conducting polymer-magnetic nanocomposite fabrication [38–42]. The strategic schematic of the fabrication of the polypyrol-Co complex is shown in Figure 4.10 [42].

TABLE 4.1

Comparative Studies of Pros and Cons of Various Fabrication Technique

Fabrication Technique	Pros	Cons
Solution casting	• Simple and widely used • Suitable for large-scale production	• Limited control over nanoparticle dispersion • Potential for nanoparticle agglomeration
Melt mixing	• Good control over nanoparticle dispersion • Suitable for large-scale production	• Potential for thermal degradation of polymer • Uneven distribution of nanoparticles
In situ polymerization	• Excellent control over nanoparticle dispersion • Enhanced mechanical properties	• Requires careful selection of monomers and nanoparticles • Complex synthesis and compatibility considerations
Electrospinning	• High surface area-to-volume ratio • Excellent nanoparticle dispersion	• Challenging control of fiber alignment • Potential for uneven nanoparticle distribution
Sol-gel method	• Versatility in compositional control • Ability to create thin films and coatings	• Requires careful control of reaction conditions • Long processing times
Magnetic field–assisted alignment	• Enhanced magnetic and mechanical properties • Controlled orientation of nanoparticles	• Complex equipment and setup requirements • Limited to anisotropic structures
Layer-by-layer assembly	• Precise control over film thickness and composition • Customizable properties	• Adhesion between layers may be a challenge • Controlling film porosity can be difficult
3D printing	• Complex geometries and intricate internal structures • Tailored magnetic composites	• Requires optimization of printing parameters • Tailored magnetic composites • Need for optimization of material formulations
Microfluidic fabrication	• Controlled mixing for uniform nanoparticle dispersion • Scaling up production is challenging	• Creation of long fibers, capsules, or droplets • Long-term stability may be a concern
Electrochemical deposition	• Selective deposition onto conductive substrates • Control over film thickness and composition	• Adhesion to substrate may be challenging

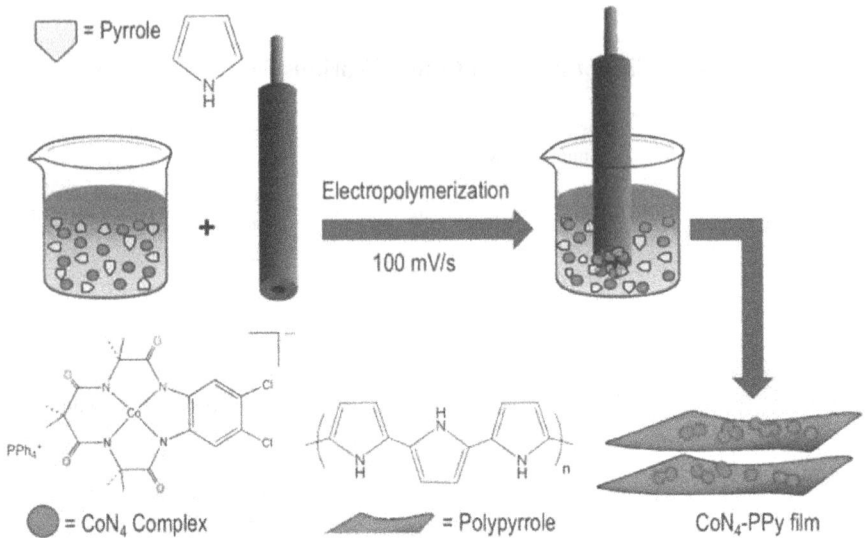

FIGURE 4.10 Schematic representation of preparation of PPy -CoN₄ composite via electro-chemical deposition. (Copyright Scientific reports, 9(1), p.5650, Springer Nature 2019.[42])

4.4   INFLUENCE OF FABRICATION PARAMETERS

4.4.1   Magnetic Nanoparticle Loading

The loading, or concentration, of magnetic nanoparticles within the polymer matrix significantly affects the composite's magnetic and mechanical properties. Optimization of the nanoparticle loading is crucial to achieve the desired balance between magnetic response and mechanical integrity.

4.4.2   Particle Size and Shape

The size and shape of magnetic nanoparticles impact the composite's magnetic and physical properties. Small nanoparticles offer enhanced magnetic response, while larger nanoparticles may provide improved mechanical strength. The control of particle size and shape during fabrication is essential for tailoring the composite properties.

4.4.3   Surface Modification of Nanoparticles

Surface modification of magnetic nanoparticles with functional groups or coatings can improve their compatibility with the polymer matrix and enhance interfacial interactions. Surface modification techniques, such as silanization or polymer graft-ing, can be employed during fabrication to promote efficient nanoparticle dispersion and interfacial adhesion.

4.4.4 POLYMER MATRIX SELECTION

The choice of polymer matrix plays a crucial role in determining the composite's properties and performance. Different polymers offer varying mechanical, thermal, and chemical properties. Compatibility between the polymer and magnetic nanoparticles, as well as the intended application requirements, guides the selection of the polymer matrix.

4.4.5 PROCESSING CONDITIONS

Processing conditions, such as temperature, pressure, mixing duration, and curing time, significantly influence the fabrication of magnetic polymer composites. Optimization of these parameters is vital to achieve uniform nanoparticle dispersion, minimize agglomeration, and promote interfacial interactions. Careful control of processing conditions ensures reproducibility and desired composite characteristics.

4.4.6 HOMOGENEOUS DISPERSION OF NANOPARTICLES IN POLYMER MATRIX

The most common problem arising in nanocomposite fabrication is dispersion of nanoparticles homogeneously over the matrix. And resolving the problem is more crucial for magnetic nanocomposites, as the magnetic nanoparticles are more prone to agglomeration. This also becomes a major factor in the reproducibility of the composite with the property.

4.5 CHALLENGES AND FUTURE PERSPECTIVES

4.5.1 SCALABILITY AND COST-EFFECTIVENESS

Scaling up the fabrication of magnetic polymer composites while maintaining consistent properties and minimizing costs remains a challenge. Its high surface energy and tendency to agglomerate are the main issues in reproducibility of the composite. Further advancements in processing techniques and materials are needed to enable large-scale production.

4.5.2 STABILITY AND DURABILITY

Ensuring the long-term stability and durability of magnetic polymer composites is crucial for their practical applications. Factors such as nanoparticle agglomeration, polymer degradation, and environmental conditions must be considered and addressed to enhance the composites' stability and reliability.

4.5.3 MULT-FUNCTIONALITY OF HYBRID COMPOSITES

Exploring the integration of additional functionalities, such as electrical conductivity, optical properties, or self-healing capabilities, into magnetic polymer composites could expand their range of applications. Developing hybrid composites that combine magnetic nanoparticles with other functional materials is a promising avenue for future research.

4.5.4   Sustainability and Environmental Impact

The environmental impact of magnetic polymer composites, including the disposal of nanoparticles and polymers, should be considered during the fabrication process. Strategies for sustainable synthesis, recycling, and environmentally friendly materials must be explored to mitigate potential ecological concerns.

4.6   CONCLUSION

The fabrication techniques employed in creating magnetic polymer composites significantly influence their properties and performance. The intention of all techniques is to produce a uniformly dispersed nanocomposite with minimal aggregation. This chapter provided an overview of both conventional and advanced fabrication techniques, highlighting their benefits and limitations. Melt-mixing, solution casting, in situ polymerization, electrospinning, and selective layer-by-layer deposition are commercially used techniques to produce polymer nanocomposites. The utilization of a nonsolvent may reduce the issue of nanoparticle aggregation in solvent mixing (co-precipitation). Also, ultrasonication with high frequencies plays the same role during mixing techniques. In situ polymerization delivers thermodynamically stable nanocomposites. Electrospinning offers porous microstructures. 3D printing is effectively used for scaffold preparation. A microfluidic–assisted fabrication offers selectively homogeneous dispersion over the matrix. A microfluidic-assisted fabrication strategy is in general use for sensor array preparation. In this chapter, the details of the methodology and application area of each technique, including its pros and cons, were addressed. Additionally, the influence of fabrication parameters on the final characteristics of the composites was discussed. Understanding these techniques and parameters is essential for designing and developing magnetic polymer composites tailored for specific applications across diverse fields.

REFERENCES

1. Chinya, I. and Sen, S. (2019), Influence of nanoparticle size on nucleation of electroactive phase and energy storage behavior of zinc ferrite/poly(vinylidene fluoride) nanocomposite, *J. Mater. Sci. Mater. Electron.*, 30, 5137–5148.
2. Chinya, I., Sasmal, A. and Sen, S. (2020), Conducting polyaniline decorated in-situ poled ferrite nanorod-PVDF based nanocomposite as piezoelectric energy harvester, *J. Alloys Compd.*, 815, 152312–152319.
3. Ashjari, M., Mahdavian, A. R., Ebrahimi, N. G. and Mosleh, Y. (2010), Efficient dispersion of magnetite nanoparticles in the polyurethane matrix through solution mixing and investigation of the nanocomposite properties, *J. Inorg. Organomet. Polym. Mater.*, 20, 213–219.
4. Kalia, S., Kango, S., Kumar, A., Haldorai, Y., Kumari, B. and Kumar, R. (2014), Magnetic polymer nanocomposites for environmental and biomedical applications, *Colloid Polym. Sci.*, 292, 2025–2052.

5. Wilson, J. L., Poddar, P., Frey, N. A., Srikantha, H., Mohomed, K., Harmon, J. P., Kotha S. and Wachsmuth, J. (2004), Synthesis and magnetic properties of polymer nanocomposites with embedded iron nanoparticles, *J. Appl. Phys.*, 95, 1439–1443.

6. Nisar, M., Bergmann, C., Geshev, J., Quijada, R. and Galland, G. B. (2016), An efficient approach to the preparation of polyethylene magnetic nanocomposites, *Polymer*, 97, 131–137.

7. Chow, W. S. and Mohd Ishak, Z. A. (2020), Smart polymer nanocomposites: A review, *eXPRESS Poly Lett.*, 14(5), 416–435.

8. Neves, J. S., Souza Jr., F. G. de, Suarez, P. A. Z., Umpierre, A. P. and Machado, F. (2011), In situ production of polystyrene magnetic nanocomposites through a batch suspension polymerization process, *Macromol. Mater. Eng.*, 296(12) 1107–1118.

9. Jiang, J., Li, L. C. and Zhu, M. (2008), Polyaniline/magnetic ferrite nanocomposites obtained by in situ polymerization, *React. Funct. Polym.*, 68(1), 57–62.

10. Yuvaraj, H., Woo, M. H., Park, E. J., Jeong, Y. T. and Lim, K. T. (2008), Polypyrrole/γ-Fe2O3 magnetic nanocomposites synthesized in supercritical fluid, *Eur. Polym. J.*, 44 (3), 637–644.

11. Reddy, K. R., Park, W., Sin, B. C., Noh, J. and Lee, Y. (2009), Synthesis of electrically conductive and superparamagnetic monodispersed iron oxide-conjugated polymer composite nanoparticles by in situ chemical oxidative polymerization, *J. Colloid Interface Sci.*, 335(1) 34–39.

12. Nathani, H. and Misra, R. D. K. (2004), Surface effects on the magnetic behaviour of nanocrystalline nickel ferrites and nickel ferrite-polymer nanocomposites, *Mater. Sci. Eng. B*, 113(3), 228–235.

13. Chen, X., Wei, S., Gunesoglu, C., Zhu, J., Southworth, C. S, Sun, L., Karki, A. B, Young, D. P. and Guo, Z. (2010), Electrospun magnetic fibrillar polystyrene nanocomposites reinforced with nickel nanoparticles, *Macromol. Chem. Phy.*, 211(16), 1775–1783.

14. Zhu, J., Wei, S., Rutman, D., Haldolaarachchige, N., Young, D. P. and Guo, Z. (2011), Magnetic polyacrylonitrile-Fe@FeO nanocomposite fibers - electrospinning, stabilization and carbonization, *Polymer*, 52(13), 2947–2955

15. Mincheva, R., Stoilova, O., Penchev, H., Ruskov, T., Spirov, I., Manolova, N. and Rashkov, I. (2008), Synthesis of polymer-stabilized magnetic nanoparticles and fabrication of nanocomposite fibers thereof using electrospinning, *Eur. Poly. J.*, 44(3), 615–627.

16. Chen, M., Qu, H., Zhu, J., Luo, Z., Khasanov, A., Kucknoor, A. S., Haldolaarachchige, N., Young, D. P., Wei, S. and Guo, Z. (2012), Magnetic electrospun fluorescent polyvinylpyrrolidone nanocomposite fibers, *Polymer*, 53(20), 4501–4511.

17. Munaweera, I., Aliev, A. and Balkus, Jr. K. J. (2014), Electrospun cellulose acetate-garnet nanocomposite magnetic fibers for bioseparations, *ACS Appl. Mater. Interfaces*, 6(1), 244–251.

18. Gradinaru, L. M., Bercea, M., Vlad, S., Mandru, M. B., Drobota, M., Aflori, M. and Ciobanu, R. C. (2022), Preparation and characterization of electrospun magnetic poly(ether urethane) nanocomposite mats: Relationships between the viscosity of the polymer solutions and the electrospinning ability, *Polymer*, 256, 125186.

19. Zhang, D., Karki, A. B., Rutman, D., Young, D. P., Wang, A., Cocke, D., Ho, T. H. and Guo, Z. (2009), Electrospun polyacrylonitrile nanocomposite fibers reinforced with Fe3O4 nanoparticles: Fabrication and property analysis, *Polymer*, 50(17), 4189–4198.

20. Topkaya, R., Kurtan, U., Baykal, A. and Toprak, M. S. (2013), Polyvinylpyrrolidone (PVP)/MnFe2O4 nanocomposite: Sol–Gel autocombustion synthesis and its magnetic characterization, *Ceram. Int.*, 39(5), 5651–5658.

21. Baykal, A., Günay, M., Toprak, M. S. and Sozeri, H. (2013), Effect of ionic liquids on the electrical and magnetic performance of polyaniline–nickel ferrite nanocomposite, *Mater. Res. Bull.*, 48(2), 378–382.

22. Baykal, A., Bıtrak, N., Ünal, B., Kavas, H., Durmus, Z., Özden, Ş. and Toprak, M.S. (2010), Polyol synthesis of (polyvinylpyrrolidone) PVP–Mn$_3$O$_4$ nanocomposite, *J. Alloys Comp.*, 502(1), 199–205.

23. Wu, S., Zhang, J., Ladani, R. B., Ghorbani, K., Mouritz, A.P., Kinloch, A. J. and Wang, C. H. (2016), A novel route for tethering graphene with iron oxide and its magnetic field alignment in polymer nanocomposites, *Polymer*, 97, 273–284.

24. Xu, C., Ouyang, C., Jia, R., Li, Y. and Wang, X. (2009), Magnetic and optical properties of poly(vinylidene difluoride)/Fe$_3$O$_4$ nanocomposite prepared by coprecipitation approach, *J. Appl. Poly. Sci.*, 111(4), 1763–1768.

25. Younes, H., Kuang, X., Lou, D., Vries, B. D, Rahman, Md M. and Hong, H. (2022), Magnetic-field-assisted DLP stereolithography for controlled production of highly aligned 3D printed polymer-Fe$_3$O$_4$@graphene nanocomposites, *Mater. Res. Bull.*, 154, 111938.

26. Younes, H., Kuang, X., Lou, D., Vries, B. D, Rahman, Md M., Hong, H, and Ramakrishna, S. (2023), "Recent advances in the additive manufacturing of stimuli-responsive soft polymers." *Adv. Eng. Mater.*, 25, no. 21: 2301074. https://doi.org/10.1002/adem.202301074

27. Ganguly, K., Jin, H., Dutta, S. D., Patel, D. K., Patil, T. V. and Lim, K.T. (2022), Magnetic field-assisted aligned patterning in an alginate-silk fibroin/nanocellulose composite for guided wound healing, *Carbohydr. Polym.*, 287, 119321.

28. Baharin, S. N. A., Sarih, N. M. and Mohamad, S. (2016), Novel functionalized polythiophene-coated Fe3O4 nanoparticles for magnetic solid-phase extraction of phthalates, *Polymers*, 8(5), 117.

29. Le, T. D., Tran, L.T., Dang, H. T. M., Tran, T. T. H. and Tran H. V. (2021), Graphene oxide/polyvinyl alcohol/Fe3O4 nanocomposite: An efficient adsorbent for co(II) ion removal, *J. Anal. Methods Chem.*, 2021, 6670913.

30. Maria A.G. (2018), Layer-by-layer assembled iron oxide based polymeric nanocomposites, *J. Magn. Magn. Mater.*, 467, 37–48.

31. Caruso, F., Spasova, M., Susha, A., Giersig, M. and Caruso, R. A. (2001), Magnetic nanocomposite particles and hollow spheres constructed by a sequential layering approach, *Chem. Mater.*, 13(1), 109–116.

32. Paterno, L. G., Soler, M. A. G., Fonseca, F. J. Sinnecker, J. P. Sinnecker, E. H. C. P., Lima, E. C. D., Báo, S. N., Novak, M. A. and Morais, P. C. (2010), Magnetic nanocomposites fabricated via the layer-by-layer approach, *J. Nanosci. Nanotechnol.*, 10(4), 2679–2685.

33. Manapat, J. Z., Chen, Q., Ye, P. and Advincula R. C. (2017), 3D printing of polymer nanocomposites via stereolithography, *Macromol Mater Eng.*, 302(9), 1600553.

34. Hupfeld, T., Salamon, S., Landers, J., Sommereyns, A., Doñate-Buendía, C., Schmidt, J., Wende, H., Schmidt, M, Barcikowski, S. and Gökce, B. (2020), 3D printing of magnetic parts by laser powder bed fusion of iron oxide nanoparticle functionalized polyamide powders, *J. Mater. Chem. C*, 8, 12204–12217.

35. Farahani, R.D., Dubé, M. and Therriault, D. (2016), Three-dimensional printing of multifunctional nanocomposites: Manufacturing techniques and applications, *Adv. Mater.*, 28(28), 5794–5821.

36. Lebel, L.L., Aissa, B., Monroy, O. A. P., Khakani, M. A. El. and Therriault, D. (2009), Three-dimensional micro structured nanocomposite beams by microfluidic infiltration, *J. Micromech. Microeng.*, 19(12), 125009.

37. Jiang, Y., Wang, H., Li, S. and Wen, W. (2014), Applications of micro/nanoparticles in microfluidic sensors: A review, *Sensors*, 14(4), 6952–6964.

38. Rao, H., Lu, Z., Ge, H., Liu, X., Chen, B., Zou, P., Wang, X., He, H., Zeng, X. and Wang, Y. (2017), Electrochemical creatinine sensor based on a glassy carbon electrode modified with a molecularly imprinted polymer and a Ni@polyaniline nanocomposite, *Microchimica Acta*, 184, 261–269.

39. Gangopadhya, R. and De, A. (2000), Conducting polymer nanocomposites: A brief overview, *Chem. Mater.*, 12, 608.

40. Xu, Z., Gao, M., Yu, L., Lu, L., Xu, X. and Jiang, Y. (2014), Co nanoparticles induced resistive switching and magnetism for the electrochemically deposited polypyrrole composite films, *ACS Appl. Mater. Interfaces*, 6(20) 17823–17830.

41. Naseri, M., Fotouhi, L. and Ehsani, A. (2018), Recent progress in the development of conducting polymer-based nanocomposites for electrochemical biosensors applications: A mini-review, *Chem. Rec.*, 18(6), 599–618.

42. Parnell, C. M., Chhetri, B. P., Mitchell, T. B., Watanabe, F., Kannarpady, G., Rangu Magar, A. B., Zhou, H., Alghazali, K. M., Biris, A. S. and Ghosh, A. (2019), Simultaneous electrochemical deposition of cobalt complex and poly(pyrrole) thin films for supercapacitor electrodes, *Sci. Rep.*, 9, 5650.

5 Characterization Techniques for Magnetic Polymer Composites

Anamika Pandey, Ankita Rawat, Monika Mishra,
Ajay Singh, Ritu Painuli, and Shivam Pandey

5.1 INTRODUCTION

Because of their biomimetic behavior and possible application in smart or sagely systems, materials that respond to stimulation have received a great deal of attention in recent decades. In fact, adaptability and the capacity to react to changes in the environment at all scales—from the molecular level to the macromolecular level—are interesting and essential for sustaining and controlling usual operations in any biological system. Therefore, stimuli-responsive or "smart" materials are frequently described as those that can adapt to changes in pH, light, temperature, ionic strength, and electric or magnetic field variations, among other triggers, by altering their own properties, such as size, surface area, solubility, permeability, shape, mechanical, and optical ones [1]. Materials made of polymers are among the extensively developed and researched materials in this field of study due to their great adaptability and capacity to drastically modify their inherent characteristics in response to minute environmental changes. A wide range of applications, including coatings (smart textiles and fibers), microelectronics (actuators, electromechanics), and biomedicine (drug delivery, tissue engineering, biosensors, active diagnostics), are increasingly relying on smart polymers. In addition, the enormous advancements in polymer chemistry and synthesis enable the right design of clearly specified macromolecules with stimuli-responsive building blocks for all of the aforementioned triggers. However, polymers that are innately magnetically sensitive are rare and often exhibit subpar effectiveness [2].

"Doping" of polymer materials with magnetic nanoparticles (MNPs) made of inorganic matter seemed to be the most appealing and effective solution for generating materials having extremely effective magnetic responses. In fact, these "small" magnets may respond to mild stimuli (static or alternating magnetic field) with a substantial impact as their magnetic moment is significantly bigger than that of molecular magnets. The resulting composite can be named a magnetic polymer composite [3].

DOI: 10.1201/9781003454236-5

A magnetic polymer composite refers to a material that is made up of a polymer matrix along with implanted magnetic particles or fillers that consists of iron oxide particles (commonly magnetite and maghemite). The combination of the polymer and magnetic components creates a material with unique properties and functionalities. The polymer matrix provides structural integrity and flexibility, while the magnetic particles add magnetic properties to the composite. The magnetic particles could be ferromagnetic, ferrimagnetic, or superparamagnetic, depending on their composition and size. Magnetic polymer composites have attracted significant interest in various fields as a result of their tunable magnetic properties, lightweight nature, and ease of processing. Such a smart inorganic/organic system has been proven to be applicable in many fields, including drug delivery, separation and/or purification of chemicals, development of shape-memory materials, and catalysis.

The key features and applications of magnetic polymer composites are the following.

1. **Magnetic Response**: A polymer composite constituted of magnetic particles is able to respond to external magnetic fields. This property can be exploited in applications such as magnetic sensors, actuators, and magnetic shielding.
2. **Biomedical Applications:** Magnetic polymer composites have been broadly studied for their potential use in biomedical applications. They can be used for targeted drug delivery systems, where the magnetic particles help in guiding the composite to specific locations in the body. Magnetic polymer composites are also investigated for utilization in magnetic hyperthermia, in tissue engineering, and as MRI (magnetic resonance imaging) contrast agents.
3. **Electromagnetic Interference (EMI) Shielding:** Magnetic characteristic of the composites make them suitable for EMI shielding applications. They can be used as coatings or components in electronic devices to attenuate electromagnetic radiation and prevent interference.
4. **Magnetic Separation:** Magnetic polymer composites can be employed for efficient separation of magnetic materials from mixtures or suspensions. These composites are used in applications such as waste treatment, mineral processing, and environmental remediation.
5. **Flexible Electronics:** The combination of polymer flexibility and magnetic properties makes these composites suitable for flexible electronics. They can be used in flexible displays, sensors, and actuators [4, 5].

The specific properties and applications of magnetic polymer composites can vary depending on the choice of polymer matrix, magnetic particles, and their concentration within the composite (Figure 5.1). Researchers continue to explore new compositions and fabrication techniques to broaden the uses of these materials and improve their performance [6].

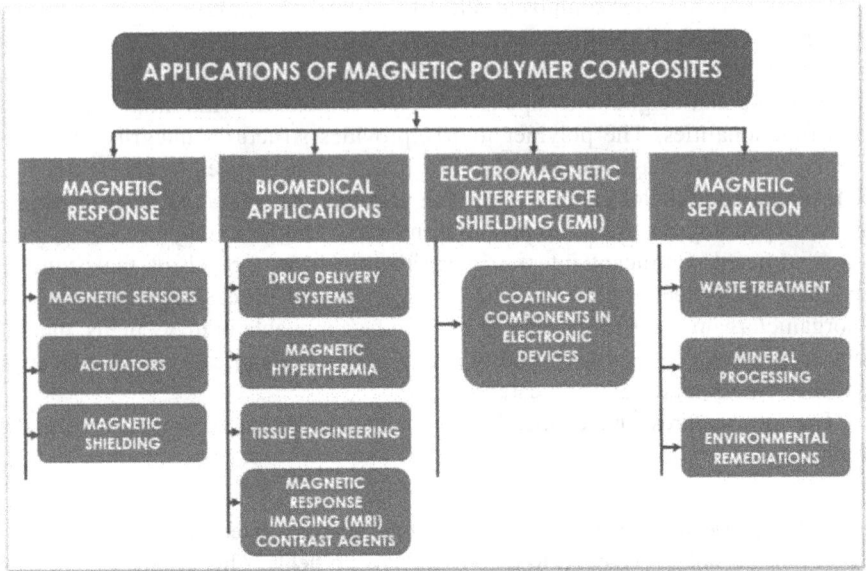

FIGURE 5.1   Applications of magnetic polymer composites.

5.2   CHARACTERIZATION TECHNIQUES

Modern society requires effective and reliable characterization procedures to meet the demands of upcoming research and development difficulties. For material analysis, academic and professional researchers require a variety of characterization techniques to support their practical work. The significance of these procedures might be attributed to the relationship between bulk characteristics and microscopic structure. These methods may be used to characterize various materials with a range of applications. The substance under research is affected by a variety of inputs, which then produce outputs (Figure 5.2). The material structure (micro, macro, nano, etc.) is then commented on by analyzing these outputs using a variety of detectors [7].

There are two common techniques used for material characterization study: -

1. Spectroscopy
2. Microscopy

5.2.1   MICROSCOPY

Microscopy is a versatile characterization technique that involves the use of microscopes to study the structure, composition, and properties of materials at the micro- and nanoscale. It provides detailed visual and analytical information about samples, allowing for the investigation of their morphology, crystal structure, chemical composition, and more [8]. Here are some commonly used microscopy techniques for characterization:

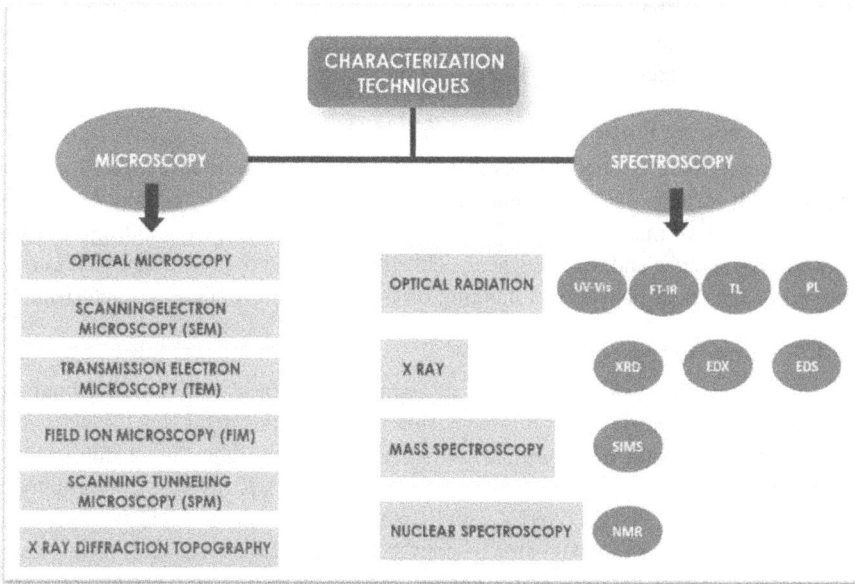

FIGURE 5.2 Characterization techniques.

5.2.1.1 Optical Microscopy

Light microscopy is another name for optical microscopy that uses visible light to observe and analyze samples. It can reveal details on the distribution, size, shape, and surface properties of particles or structures. Bright-field, dark-field, phase contrast, fluorescence, confocal microscopy, and differential interference contrast (DIC) microscopy are a few examples of different optical microscopy techniques.

5.2.1.2 Electron Microscopy

A beam of electrons rather than light is used in electron microscopy to analyse materials. It offers higher magnification and resolution compared to optical microscopy. There are two main forms of electron microscopy:

- **Scanning Electron Microscopy (SEM):** SEM scans a focused electron beam across the sample to provide detailed pictures of the sample surface. It offers details about the sample's topography, morphology, and elemental makeup.
- **Transmission Electron Microscopy (TEM):** TEM involves the transmission of electrons through a thin sample to create an image. It offers detailed pictures of the sample's internal organization, crystal lattice, and defects. Additionally, TEM can be utilized for elemental investigation using electron energy-loss spectroscopy (EELS) and selected area electron diffraction (SAED) to determine crystal orientations.

5.2.1.3 Atomic Force Microscopy (AFM)

Atomic force microscopy (AFM) uses a sharp probe to scan a sample's surface. The probe interacts with the sample's surface forces, and the resulting changes are detected, allowing the creation of a topographical image. AFM can achieve atomic-scale resolution and is capable of characterizing surface roughness, mechanical properties, and molecular interactions.

5.2.1.4   Scanning Probe Microscopy (SPM)

SPM encompasses several techniques that employ a sharp probe to investigate the sample surface. It includes AFM as mentioned above, as well as other techniques such as scanning tunneling microscopy (STM), which measures electron tunnelling between the probe and sample, and scanning near-field optical microscopy (SNOM), which combines AFM with near-field optical imaging for high-resolution optical characterization.

5.2.1.5   X-ray Microscopy

This microscopy uses X-rays instead of light or electrons to image and analyze samples. It offers high-resolution information in great detail regarding the chemical composition, elemental distribution, and structure of materials. X-ray microscopy techniques include X-ray fluorescence microscopy (XFM), X-ray absorption microscopy (XANES and EXAFS), and X-ray diffraction microscopy (XRD).

These are just a few examples of microscopy techniques used for characterization. Each technique has its own strengths and is applicable to different sample types and research goals. Researchers often combine multiple microscopy techniques to gain comprehensive insights into the properties and behavior of materials at the micro- and nanoscale [9].

5.2.2   SPECTROSCOPY

In material characterization investigations, spectroscopy is a commonly used method. It involves the interaction of radiation with material. The analysis or measurement of the resulting spectra (absorption/emission) yields valuable knowledge about the substance being investigated. Spectroscopy is utilized in astronomy, material characterization, and medicinal applications [10]. Spectroscopy methods are frequently divided into groups based on the wavelength range employed, the sort of interaction involved, or the type of substance under study, as shown in Table 5.1.

TABLE 5.1
Types of Spectroscopies Based on Electromagnetic Radiations

Type of Spectroscopy	Electromagnetic Radiation	Function
X-Ray Spectroscopy	X-rays	Determine the crystallographic structure of a material.
UV-Visible Spectroscopy	UV-visible	Determines the amount of absorbance and transmittance done by compounds.
IR Spectroscopy	IR	Determines functional groups.
Mass Spectroscopy	None (Uses beam of electrons and not photons)	Determines the molar mass and branches.
Nuclear Magnetic Resonance Spectroscopy	Radio waves	Determines the types of hydrogens present.

5.2.2.1 UV-Visible Spectroscopy

An essential physical instrument that uses light in the electromagnetic spectrum's ultra-violet, visible, and near infrared regions is ultraviolet-visible (UV-vis) spectroscopy. The Beer-Lambert law establishes a linear relationship between absorbance, the number of absorbers (or absorbing species) in the solution, and pathlength. It can therefore be used to determine the concentration of the absorbing species for a specific route length. This method is incredibly straightforward, adaptable, rapid, precise, and economical. It can be used to analyze liquids, gases, and solids by utilizing radiative energy related to the far and near ultraviolet (UV), visible (vis), and near infrared region (NIR) portions of the electromagnetic spectrum. The following wavelengths have been established for these regions: UV: 300–400 nm; vis: 400–765 nm; and NIR: 765–3200 nm [11].

5.2.2.1.1 Principle

The wavelength of the light that reaches the detector after traveling through an item in a light beam is measured. The number of molecules and chemical structure that are present canbe determined from the intensity of the observed signal at calculated wavelength. As a result, it is possible to gather data that are both quantitative and qualitative. Information can be obtained from the transmittance, absorbance, or reflectance of light with a wavelength between 160 and 3500 nm. The promotion of electrons to excited states or the anti-bonding orbitals results from the absorption of incoming energy. For this transfer to occur, the photon energy must be equal to the energy needed to promote an electron to the next higher energy state. This method is absorption spectroscopy's basic premise [12].

Three different types of ground state orbitals are generally involved: i.e., sigma (σ bonding), pi (π bonding), and n (non-bonding) atomic orbitals. In addition, the anti-bonding orbitals σ* (sigma star) orbital and π* (pi star) are also present.

Electronic transitions caused by UV and visible light absorption are as follows:

1. σ (sigma bonding orbital) to σ* (sigma anti-bonding orbital)
2. n (non-bonding orbital) to σ* (sigma anti-bonding orbital)
3. n (non-bonding orbital) to π* (pi anti-bonding orbital)
4. π (pi bonding orbital) to π* (pi anti-bonding orbital)

In UV spectroscopy, transitions occur when a molecule absorbs UV radiation and undergoes a change in its electronic energy levels. Based on the characteristics of the electronic states involved, these transitions can be divided into a number of different categories [13].

Some common types of transitions observed in UV spectroscopy are as follows:

σ → σ* Transitions: An electron is excited during these transitions from a filled σ bonding orbital to an empty σ* anti-bonding orbital. They are noticed in saturated organic compounds, where the electronic transitions occur within the σ framework of the molecule.

n → σ* Transitions: These transitions involve compounds having one hetero-atom, such as halogens (fluorine, chlorine, etc.), oxygen, and nitrogen. This type of transition is typically carried out by saturated halides, alcohols, ethers, aldehydes, ketones, and amines.

π → π* Transitions: The excitation of an electron occurs during these transitions from a filled π orbital to an empty π* anti-bonding orbital. They are most commonly observed in conjugated systems, such as aromatic compounds and conjugated double bonds.

n → π* Transitions: Here, an electron is excited from a non-bonding orbital to a π* anti-bonding orbital. These transitions happen in molecules like amines and carbonyl compounds (like aldehydes and ketones) that have heteroatoms with lone pairs of electrons [14].

5.2.2.1.2 Spectrophotometer and Its Instrumentation

The spectrophotometer is an analytical instrument used to calculate a substance's transmission or absorption of light as a function of its wavelength or frequency. The Beer-Lambert Law states that a sample's absorbance is directly proportional to its concentration and the length of the light's journey through it, which serves as the foundation for how it works.

$$\text{Beer Lambert's Law}: A = \varepsilon dc$$

The molar absorption coefficient, or molar absorptivity (ε), is a sample-dependent feature that measures the sample's absorption efficiency at a certain wavelength of light. The thickness of the cuvette used for the absorbance analysis is determined by the optical path length (d), which is typically 1 cm. The moles per liter (M) of the sample dissolved in the solution is referred to as the concentration (c).

Single- and double-beam spectrophotometer are the absorbance devices that can be used to record the UV-visible spectrum.

5.2.2.1.2.1 Single Beam Spectrophotometer
A single-beam spectrophotometer utilizes a single beam of light to work within a certain wavelength range (for example, 300–1150 nm). In the same cuvette, it successively examines the absorbance of the test solution and a solution used as a reference (Figure 5.3)[15, 16].

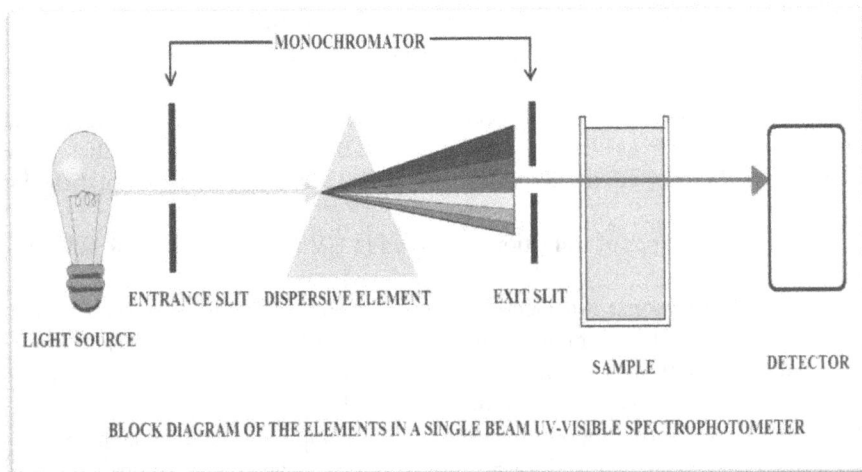

BLOCK DIAGRAM OF THE ELEMENTS IN A SINGLE BEAM UV-VISIBLE SPECTROPHOTOMETER

FIGURE 5.3 Diagram of single-beam spectrophotometer.

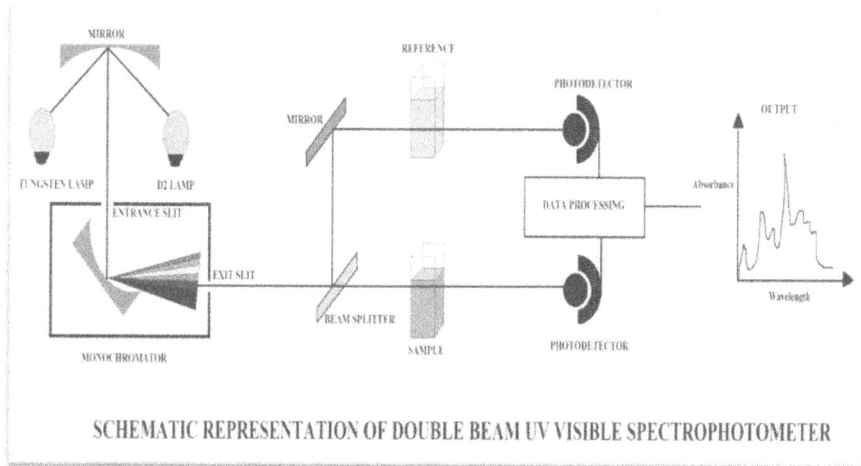

SCHEMATIC REPRESENTATION OF DOUBLE BEAM UV VISIBLE SPECTROPHOTOMETER

FIGURE 5.4 Representation of double-beam spectrophotometer.

5.2.2.1.2.2 Double Beam Spectrophotometer In a double-beam spectrophotometer, a monochromator's light beam divides into two beams. One beam performs
as a reference, while the other is utilized to measure the sample as it goes through it.
The reference and sample may be measured simultaneously with this configuration
(Figure 5.4), improving accuracy and usability [14–16].

The basic instrumentation of a spectrophotometer typically includes the following components [17]:

1. **Light Source:** The spectrophotometer incorporates a stable and controllable light source that emits light across a wide range of wavelengths.
Tungsten-halogen and deuterium lamps are typical sources of light for the
visible and ultraviolet (UV) spectra, respectively.
2. **Monochromator:** A monochromator is employed to separate a certain
wavelength of light from the source. It is made out of a prism or diffraction
grating that separates light into its constituent wavelengths and only enables
a small band of wavelengths to flow through.
3. **Sample Holder:** The sample holder, often known as a cuvette, is a
clear container used to hold samples for analysis. Typically, it is made
of quartz or glass, and its path length (the distance the light travels
through the sample) can be adjusted to control the sensitivity of the
measurement.
4. **Detector:** The detector calculates the intensity of light after it has passed
through the sample. Photodiodes, photomultiplier tubes (PMTs), and
charge-coupled devices (CCDs) are common detectors that transform the
light signal into an electrical signal.
5. **Electronics and Data Display:** The spectrophotometer includes electronic circuits to amplify and process the electrical signal from the

detector. It also has a display unit to show the measured absorbance or transmittance values.

6. **Control and Data Processing:** Spectrophotometers often have control buttons or a touchscreen interface to select measurement parameters, such as wavelength range and scanning speed, and data processing options. They may also have software or data analysis capabilities to perform calculations, generate spectra, and store or export data.

During operation, the spectrophotometer scans through a range of wavelengths or frequencies and measures the intensity of light transmitted through or absorbed by the sample at each point. The resulting spectrum provides information about the sample's absorption or transmission characteristics, which can be used to determine its concentration, identify substances, or study its molecular structure.

Different types of spectrophotometers exist, such as UV-visible spectrophotometers, which measure light in the visible and ultraviolet ranges of the electromagnetic spectrum, and infrared (IR) spectrophotometers, which measure absorption in the infrared range. The principles and basic instrumentation remain similar, with variations specific to the spectral range of interest [18].

5.2.2.2   Fourier Transform Infrared (FTIR) Spectroscopy

An effective analytical method for examining how infrared radiation interacts with materials is called Fourier transform infrared (FTIR) spectroscopy. It enables the identification and characterization of chemical compounds by providing details about the vibrational and rotational modes of molecules [19].

5.2.2.2.1   Principle and Instrumentation

The principle of FTIR spectroscopy is based on the measurement of the absorption or transmission of infrared light by a sample. The technique utilizes an interferometer to modulate the intensity of the infrared light, and a mathematical algorithm called the Fourier transform is employed to convert the interferogram into a spectrum [20].

1. **Light Source:** A broad-spectrum infrared light source is used, which can be a globar (a heated silicon carbide rod) or a nitrogen-cooled mercury-cadmium-telluride (MCT) detector for particular wavelength ranges.
2. **Interferometer:** The heart of the FTIR instrument is the interferometer, which splits the infrared beam into two paths—one is the sample beam, and the other is the reference beam. The Michelson interferometer is the most typical form of interferometer used in FTIR spectroscopy. In this device a beam splitter separates the infrared beam, mirrors reflect the beams back, and an interferometer arm recombines the beams.
3. **Sample Holder:** A sample holder or cell is used to hold the sample for analysis. It is often made of potassium bromide (KBr) or other suitable materials. The sample holder is transparent to infrared radiation to allow the passage of light.

4. **Detector:** A detector measures the intensity of the infrared light after it has interacted with the sample and the reference beams. The detector commonly used in FTIR spectrometers is a liquid nitrogen–cooled MCT detector or a room-temperature detector such as a deuterated triglycine sulfate (DTGS) detector.

5. **Data Acquisition and Fourier Transform:** The detector produces an interferogram, which is an intensity vs. time signal. This interferogram contains information about the constructive and destructive interference of the sample and reference beams. The interferogram is then digitized and processed using a computer. A mathematical algorithm called the Fourier transform is applied to convert the interferogram from the time domain to the frequency domain, generating an infrared spectrum.

6. **Data Analysis:** The resulting infrared spectrum represents the absorption or transmission of infrared light by the sample as a function of the wavenumber (inverse of the wavelength). The spectrum includes distinctive peaks that correspond to the molecules' vibrational modes in the sample. It can be compared to reference spectra or analyzed using spectral databases to identify and characterize the chemical compounds present in the sample.

Information regarding functional groups, molecular structure, and chemical composition can be obtained via FTIR spectroscopy (Figure 5.5). It is broadly used in various fields such as chemistry, pharmaceuticals, materials science, environmental analysis, and forensic science. The technique offers high sensitivity, rapid data acquisition, and the ability to analyze solid, liquid, and gas samples [21].

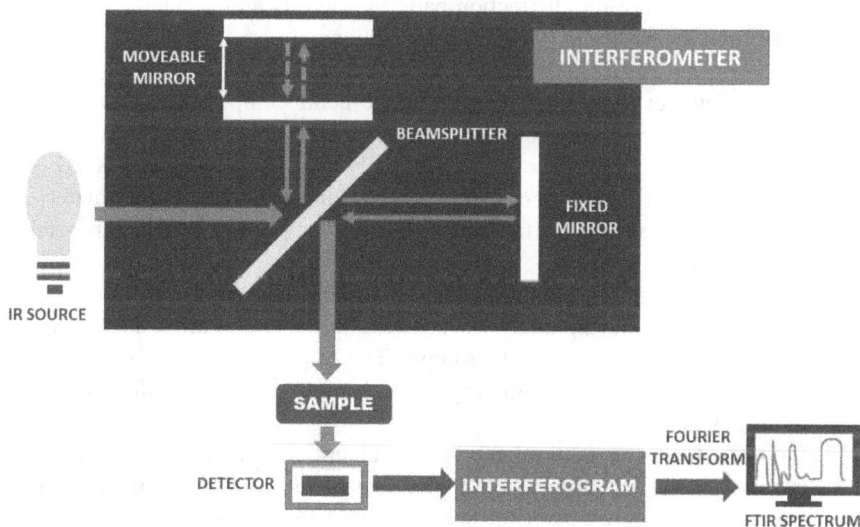

FIGURE 5.5   Schematic diagram of FTIR.

5.2.2.3   X-Ray Diffraction (XRD) Spectroscopy

X-ray diffraction (XRD) spectroscopy is an effective analytical method for examining the atomic and molecular structure of materials. It offers valuable information about the crystallographic properties, phase composition, and lattice parameters of crystalline substances. XRD is based on the principle of X-ray scattering by the regular arrangement of atoms in a crystal lattice, resulting in a characteristic diffraction pattern [22].

5.2.2.3.1   Principle and Instrumentation of XRD

The principle of XRD is rooted in Bragg's Law, which explains the connection between the X-ray wavelength (λ), angle of incidence (θ), and spacing of the crystal planes (d). According to Bragg's Law, X-rays diffract at particular angles when they come into contact with a crystal lattice that satisfies the condition for constructive interference between the X-rays reflected from the crystal planes [23].

$$\text{Bragg's Equation}: n\lambda = 2d \sin \theta$$

The diffracted X-rays form a diffraction pattern consisting of bright spots known as diffraction peaks. The diffraction pattern is a direct reflection of the crystal structure of the material. Each peak corresponds to a specific set of crystal planes and provides information about their spacing and orientation. The position, intensity, and shape of the diffraction peaks contain valuable information about the crystal symmetry, lattice parameters, and the arrangement of atoms within the crystal lattice [24].

The instrumentation of XRD involves several key components that work together to generate and analyze the diffraction patterns. Here is an overview of the main components and their roles in XRD instrumentation:

1. **X-ray Source:** The X-ray source is a crucial component of an XRD instrument. It typically consists of an X-ray tube that emits X-rays with a specific wavelength. The most commonly used X-ray source in XRD is copper (Cu) with a wavelength of 1.5406 Å (angstroms), but other materials such as cobalt (Co) or molybdenum (Mo) may be used for specialized applications.
2. **Sample Holder:** The sample holder or sample stage holds the sample in the X-ray beam during analysis. It should be stable and allow for precise positioning and rotation of the sample. The sample can be in the form of a powder, thin film, or single crystal, depending on the specific analysis requirements.
3. **Goniometer:** The goniometer is a mechanical assembly that controls the movement and orientation of the sample in the XRD instrument. It enables precise alignment of the sample and the X-ray beam at specific angles. The goniometer allows for angular scanning, where the sample is rotated, and diffraction patterns are recorded at different angles of incidence.

4. **X-ray Detector:** The X-ray detector records the intensity of diffracted X-rays as a function of the scattering angle. Several types of detectors are used in XRD instruments, including scintillation counters, proportional counters, or solid-state detectors. Scintillation detectors utilize scintillating crystals to convert X-ray photons into light, which is then detected and converted into an electrical signal. Solid-state detectors, such as CCD (charge-coupled device) or CMOS (complementary metal-oxide-semiconductor) detectors, offer high sensitivity, fast data acquisition, and improved resolution.

5. **Collimation System:** The collimation system consists of various components, including collimators and slits, which help control the size and divergence of the X-ray beam. Collimators are used to narrow down and focus the X-ray beam, while slits help define the beam path and remove unwanted scattering.

6. **Monochromator:** In some XRD instruments, a monochromator may be included to further refine the X-ray beam. A monochromator filters out unwanted X-ray wavelengths and produces a narrower and more monochromatic beam. This is particularly useful for reducing background noise and improving the resolution of diffraction patterns.

7. **Data Acquisition and Analysis System:** XRD instruments are equipped with data acquisition and analysis systems to capture and process the diffraction data. Specialized software is used to control the instrument, acquire the diffraction patterns, and perform data analysis. The software enables peak identification, peak fitting, and quantitative analysis of the diffraction data. It also provides tools for data visualization and the generation of crystallographic information, such as lattice parameters and crystal structure.

Modern XRD instruments may incorporate additional features and accessories, such as automated sample changers for high-throughput analysis, temperature-controlled stages for in situ measurements, and various sample preparation devices. These advancements enhance the capabilities and versatility of XRD instruments, allowing for a wide range of applications and research requirements. Overall, XRD instrumentation combines X-ray sources, sample holders, goniometers, detectors, collimation systems, and data analysis software to generate high-quality diffraction patterns and extract valuable structural information from materials (Figure 5.6). The instrumentation plays a critical role in enabling the accurate characterization and analysis of crystal structures, phase identification, and other important properties of materials using XRD techniques [25].

5.2.3 Thermal Methods

Thermal analysis techniques are effective tools for determining phase transitions and degradation processes of nanostructured polymeric materials and for improving knowledge of the "nano-effects" seen in polymeric nanocomposites [26]. Thermal analysis helps us to have better insight about structure and properties of various polymer nanocomposites especially at elevated temperatures. It comprises various investigation

X-RAYS

CRYSTAL

X-RAY TUBE

PHOTOGRAPHIC FILM

COLLIMATOR

SCHEMATIC DIAGRAM OF X-RAY DIFFRACTOMETER

INCIDENT BEAM   REFLECTED BEAM

θ θ

θ

d

d Sin θ

BRAGG'S LAW OF X-RAYS DIFFRACTION

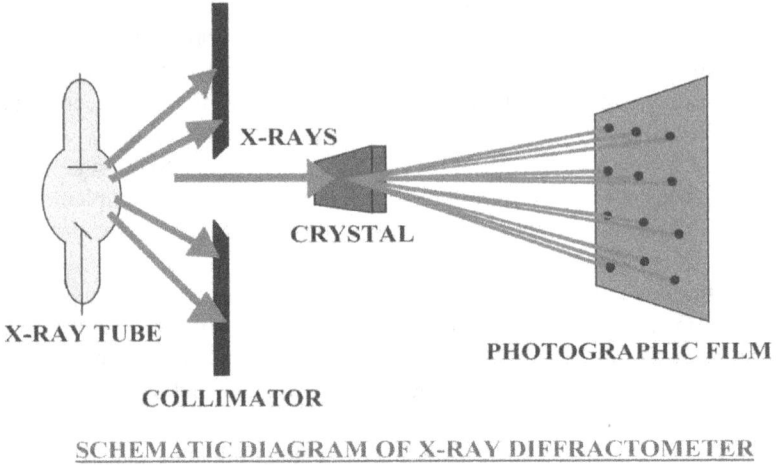

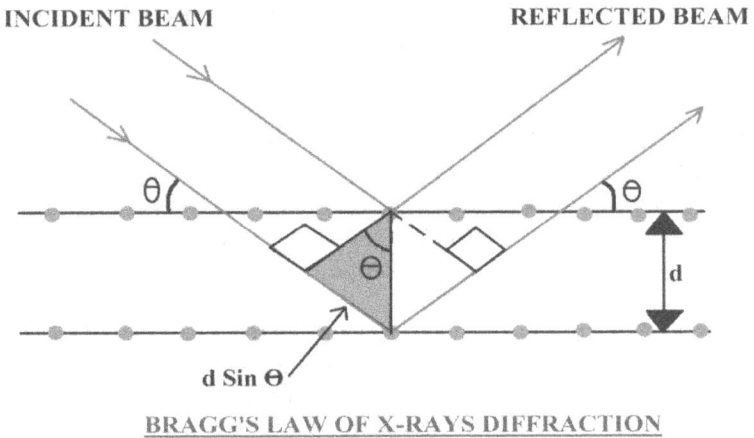

FIGURE 5.6   Schematic diagram of XRD.

methods like TGA, DSC, DTA, DMTA, and TMA for characterization of nanocomposite materials [27]. Various thermal analysis methods for polymers along with their instrumentation and applications may be understood from Chartoff and Sirkar [28].

5.2.3.1   Thermogravimetric Analysis (TGA)

Thermogravimetric analysis (TGA) is one of the chief thermal methods which is used to determine changes in weight of a substance as a function of elevating temperature. So, this method can be used to provide information in cases characterized by weight changes. A precise balance and a furnace with a linear temperature rise program are the basic instrumental requirements for TGA [29]. TGA is primarily used to examine thermal and oxidative stabilities as well as compositional properties of the sample (Tables 5.2 and 5.3).

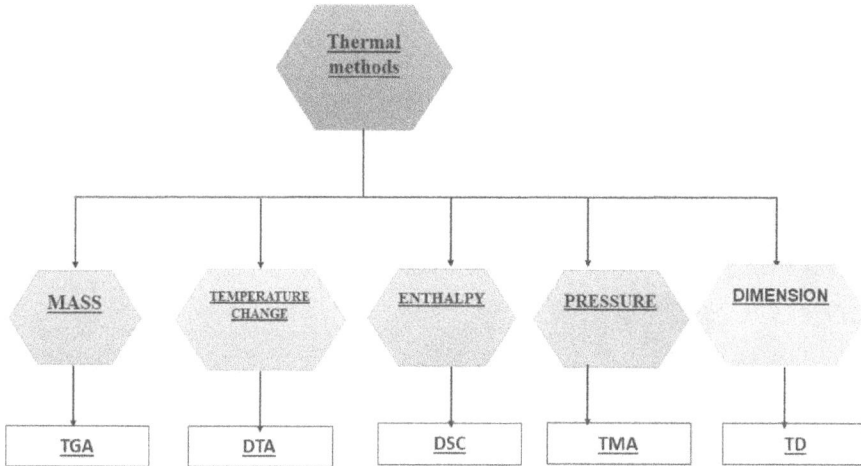

FIGURE 5.7 Classification of thermal analysis methods.

TABLE 5.2
Example of TGA Analysis for Nanocomposite

S. No.	Nanocomposite	Nanoparticle	Use of TGA	Reference
1	PVA-based nanocomposites	$CoFe_2O_4$	a. To determine ferro-fluid composition b. Characterization of mineralized sulfonated polystyrene c. To check thermal stability of PVC nanocomposite	[31]
2	Polyurethane foam nanocomposites	Fe_3O_4@APTS	Evaluation of role of MNPs on thermal properties of prepared magnetic nanocomposite	[32]
3	Aromatic polyamides/ Fe_3O_4 magnetic nanocomposites	Fe_3O_4	Estimation of adhered hyperbranched polymer to Fe_3O_4 surface by heating of nanocomposite	[33]
4	PMMA nanospheres	Fe_3O_4@$CaCO_3$	To estimate composition of composite nanospheres	[34]
5	PLA/PEI- functionalized Fe_3O_4 nanocomposites	Fe_3O_4@PEI	a. To examine thermal stability of PLA and its nanocomposites b. To examine decomposition of MNP@PEI nanoparticles	[35]
6	Magnetic hexacyanoferrate (II) polymeric nanocomposites	Potassium and zinc hexacyanoferrate	Through TG curves to determine the total weight loss percentage of the nanocomposite	[36]

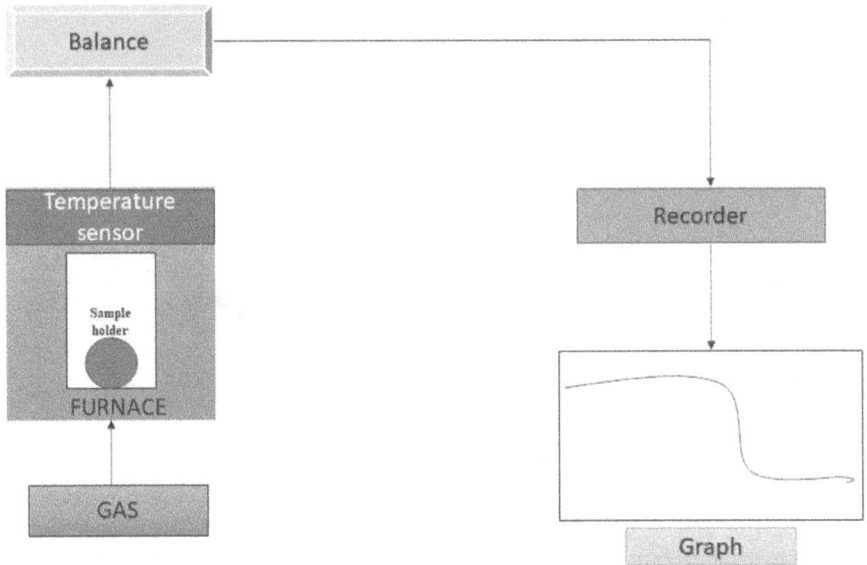

FIGURE 5.8 Block diagram for instrumentation of TGA.

This analysis is also important to know about the viscoelastic behavior of polymeric samples [30].

TG instruments comprise the following components (Figure 5.8):

a. sample holder
b. balance to weigh the sample
c. furnace to provide linear heating
d. temperature programmer for controlling temperature
e. recorder to record the changes in weight and temperature

5.2.3.2 Differential Thermal Analysis (DTA)

Differential thermal analysis (DTA) involves measuring the temperature difference between a sample and a reference material as a function of temperature. During this measurement a specified atmosphere with controlled temperature is maintained. Basically, this technique is implemented for determining the temperature changes in cases of phase transitions. A DTA thermogram may be used to study T_g and T_m of polymers. The method for evaluating T_g from a DTA thermogram is explained in S. Strella, 1963 [37].

5.2.3.3 Differential Scanning Calorimetry (DSC)

Differential scanning calorimetry (DSC) involves measurement of variation in the rate of heat flow between a sample and a reference material under controlled temperature conditions. The DSC measurement values help to determine various properties

TABLE 5.3

Some Polymer Nanocomposites Characterized by the DTA Method

S.No.	Nanocomposite	Nanoparticle	Method of Synthesis	Result and Use of DTA	Reference
1	Polyaniline-Co_3O_4 nanocomposite	Co_3O_4	Interfacial polymerization	An endothermic peak at 90°C and exothermic peak at 120°C	[38]
2	Polyaniline/CeO_2 nanocomposite	CeO_2	In situ polymerization	An endothermic peak at 102.4°C and an exothermic peak at 214.88°C	[39]
3	ZnSe/EVA nanocomposite	ZnSe	Direct sonicator method	Thermal stability up to 309°C for the endothermic peak and 315°C, the exothermic peak	[40]
4	ZnO/chitosan nanocomposite	ZnO	Sol-gel method	One endothermic and two exothermic peaks	[41]

of substances like heat capacity, transition enthalpy, purity, and glass transition. DSC curves help to establish phase diagrams and determine degree of crystallinity [42]. This technique for analysis is often confused with DTA, but the two are different from each other [43] (Figure 5.9).

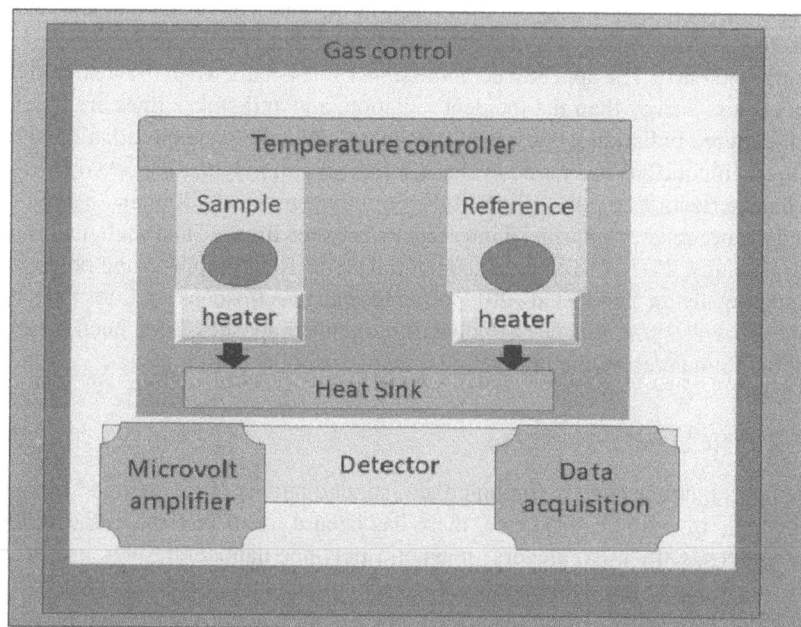

FIGURE 5.9   Diagram of scanning calorimeter.

TABLE 5.4

Some Polymer Nanocomposites Analyzed by the DSC Technique

S. No.	Nanocomposite	Nanoparticle	Reference
1	Polystyrene-based polymer nanocomposites	$CoFe_2O_4$	[31]
2	PLLA-ESO/surface-grafted silica nanocomposites	SiO_2	[44]
3	Poly (urethane-imide) nanocomposites	Nano graphene and nano-graphene oxide	[45]
4	poly(ε-caprolactone)/ Fe_3O_4 nanocomposites	Fe_3O_4	[46]
5	$Ni_0.5ZnO\cdot 5Fe_2O_4$ + polyurethane nanocomposites	$Ni_0\cdot 5ZnO\cdot 5Fe_2O_4$	[47]

This method has been used for characterizing a number of polymer nanocomposites, some of which are listed in Table 5.4.

5.2.4 RAMAN ANALYSIS

Raman micro spectroscopy helps us to understand the interactions that exist between polymers and nanoparticles that are in contact with them. Besides its contribution in identifying vibrations of specific functional groups, it also allows us to determine crystallinity parameters that help us to analyze the interactions within nanocomposites [48]. Raman analysis is a sensitive method that gives precise details about the chemical structure, phase, and polymorphism in the material. The Raman spectra are basically obtained due to the difference in the energies of the incident and the scattering photons. Two types of lines, called the stokes and the anti-stokes line, are basically seen in it. The appearance of the stokes line occurs when the scattered radiation has less energy than the incident radiation, and anti-stokes lines are observed when scattered radiation possesses higher energy than the incident radiation [49].

This technique has been used to characterize several polymer nanocomposites. In the characterization of polyaniline/$MgFe_2O_4$ nanocomposites, Raman analysis confirmed the presence of interfacial interactions between the core and shell materials in R. M. Khafagy, 2011. The composite material comprising graphene and polypyrrole was investigated in Yanik et al., 2017, using Raman spectroscopy. In Luo et al., 2016, graphene oxide/strontium ferrite/polyaniline nanocomposites have been analyzed using the Raman technique [50–52].

5.3 CONCLUSION

Since hybrid nanomaterials combine the distinct qualities of organic and inorganic components in a single substance, there has been a lot of interest in the study of these materials. In this category, magnetic polymer nanocomposites are of special interest due to the combination of excellent magnetic properties, stability, and good biocompatibility. Techniques such as co-precipitation, melt blending, microwave reflux, ceramic–glass processing, and plasma polymerization can be used to prepare organic–inorganic magnetic nanocomposites. These nanocomposites have been utilized as superparamagnetic or negative contrast agents, drug transporters, heavy metal adsorbents, and magnetically recoverable photocatalysts for the degradation of organic pollutants. In vivo imaging has also been performed utilizing these nanocomposites. Thus, characterization and identification of these composites is of utmost importance. Several methods have been employed in their investigation, but only a few are discussed in this chapter. Numerous techniques, such as UV-vis, XRD, FTIR, thermal, and Raman analysis, are used extensively for their characterization and further research. However, with the expanding field of nanocomposites, these methods have advanced, and more research can be conducted by using advanced instrumentation techniques.

REFERENCES

1. Medeiros, S. F., Santos, A. M., Fessi, H., & Elaissari, A. (2011). Stimuli-responsive magnetic particles for biomedical applications. *International journal of pharmaceutics*, *403*(1-2), 139–161.
2. Gong, B., Sanford, A. R., Ferguson, J. S., Filipcsei, G., Csetneki, I., Szilágyi, A., & Zrínyi, M. (2007). Magnetic field-responsive smart polymer composites. In *Oligomers-polymer composites-molecular imprinting*, (pp. 137–189). Springer.
3. Thevenot, J., Oliveira, H., Sandre, O., & Lecommandoux, S. (2013). Magnetic responsive polymer composite materials. *Chemical society reviews*, *42*(17), 7099–7116.
4. Choi, E. S., Brooks, J. S., Eaton, D. L., Al-Haik, M. S., Hussaini, M. Y., Garmestani, H., & Dahmen, K. (2003). Enhancement of thermal and electrical properties of carbon nanotube polymer composites by magnetic field processing. *Journal of applied physics*, *94*(9), 6034–6039.
5. Nguyen, V. Q., Ahmed, A. S., & Ramanujan, R. V. (2012). Morphing soft magnetic composites. *Advanced materials*, *24*(30), 4041–4054.
6. Faupel, F., Zaporojtchenko, V., Strunskus, T., & Elbahri, M. (2010). Metal-polymer nanocomposites for functional applications. *Advanced engineering materials*, *12*(12), 1177–1190.
7. Mourdikoudis, S., Pallares, R. M., & Thanh, N. T. (2018). Characterization techniques for nanoparticles: Comparison and complementarity upon studying nanoparticle properties. *Nanoscale*, *10*(27), 12871–12934.
8. Clarke, A., & Eberhardt, C. N. (2002). *Microscopy techniques for materials science*. Woodhead Publishing.
9. Schmolze, D. B., Standley, C., Fogarty, K. E., & Fischer, A. H. (2011). Advances in microscopy techniques. *Archives of pathology & laboratory medicine*, *135*(2), 255–263.
10. Pavia, D. L., Lampman, G. M., Kriz, G. S., & Vyvyan, J. A. (2014). *Introduction to spectroscopy*. Cengage learning.
11. Noda, I. (1993). Generalized two-dimensional correlation method applicable to infrared, raman, and other types of spectroscopy. *Applied spectroscopy*, *47*(9), 1329–1336.
12. Perkampus, H. H. (2013). *UV-VIS spectroscopy and its applications*. Springer Science & Business Media.
13. Clark, B. J., Frost, T., & Russell, M. A. (Eds.). (1993). *UV spectroscopy: Techniques, instrumentation and data handling* (Vol. 4). Springer Science & Business Media.
14. Chen, Z., Dinh, H. N., Miller, E., Chen, Z., Deutsch, T. G., Dinh, H. N., & Turner, J. (2013). UV-vis spectroscopy. *Photoelectrochemical water splitting: Standards, experimental methods, and protocols* (pp. 49–62). Springer.
15. Sommer, L. (2012). *Analytical absorption spectrophotometry in The visible and ultraviolet: The principles*. Elsevier.
16. Upstone, S. L. (2000). Ultraviolet/visible light absorption spectrophotometry in clinical chemistry. In *Encyclopedia of analytical chemistry*, (pp. 1699–1714). Springer.

17. Rocha, F. S., Gomes, A. J., Lunardi, C. N., Kaliaguine, S., & Patience, G. S. (2018). Experimental methods in chemical engineering: Ultraviolet visible spectroscopy—UV-Vis. *The Canadian journal of chemical engineering, 96*(12), 2512–2517.

18. Wellburn, A. R. (1994). The spectral determination of chlorophylls a and b, as well as total carotenoids, using various solvents with spectrophotometers of different resolution. *Journal of plant physiology, 144*(3), 307–313.

19. Faix, O. (1992). Fourier transform infrared spectroscopy. In *Methods in lignin chemistry* (pp. 83–109). Springer Berlin Heidelberg.

20. Ismail, A. A., van de Voort, F. R., & Sedman, J. (1997). Fourier transform infrared spectroscopy: Principles and applications. In *Techniques and instrumentation in analytical chemistry* (Vol. 18, pp. 93–139). Elsevier.

21. Bhargava, R., Wang, S. Q., & Koenig, J. L. (2003). FTIR microspectroscopy of polymeric systems. *Liquid chromatography/FTIR microspectroscopy/microwave assisted synthesis, 137*–191. https://doi.org/10.1007/b11052

22. Chauhan, A., & Chauhan, P. (2014). Powder XRD technique and its applications in science and technology. *Journal of Analytical & Bioanalytical Techniues, 5*(5), 1–5.

23. Feidenhans, R. (1989). Surface structure determination by x-ray diffraction. *Surface science reports, 10*(3), 105–188.

24. Bunaciu, A. A., UdriŞTioiu, E. G., & Aboul-Enein, H. Y. (2015). X-ray diffraction: Instrumentation and applications. *Critical reviews in analytical chemistry, 45*(4), 289–299.

25. Smith, F. (Ed.). (1999). *Industrial applications of x-ray diffraction*. CRC press.

26. Pielichowski, K., & Pielichowska, K. (2018). Polymer nanocomposites. In *Handbook of thermal analysis and calorimetry* (Vol. 6, pp. 431–485). Elsevier Science BV.

27. Corcione, C. E., & Frigione, M. (2012). Characterization of nanocomposites by thermal analysis. *Materials, 5*(12), 2960–2980.

28. Chartoff, R. P., & Sircar, A. K. (2002). Thermal analysis of polymers. *Encyclopedia of polymer science and technology.* https://doi.org/10.1002/0471440264.pst367.

29. Coats, A. W., & Redfern, J. P. (1963). Thermogravimetric analysis. A review. *Analyst, 88*(1053), 906–924.

30. Ng, C., Chen, H., Goh, S. G., Haller, L., Wu, Z., Charles, F. R., & Gin, K. (2018). Microbial water quality and the detection of multidrug resistant *E. Coli* and antibiotic resistance genes in aquaculture sites of Singapore. *Marine pollution bulletin, 135,* 475–480.

31. Lopez, J. A. (2001). Evaluating the predictive accuracy of volatility models. *Journal of forecasting, 20*(2), 87–109.

32. Alavi Nikje, M. M., Tamaddoni Moghaddam, S., Noruzian, M., Farahmand Nejad, M. A., Shabani, K., Haghshenas, M., & Shakhesi, S. (2014). Preparation and characterization of flexible polyurethane foam nanocomposites reinforced by magnetic core-shell Fe_3O_4@APTS nanoparticles. *Colloid and polymer science, 292,* 627–633.

33. Wei, A., Sun, X. W., Xu, C. X., Dong, Z. L., Yang, Y., Tan, S. T., & Huang, W. (2006). Growth mechanism of tubular ZnO formed in aqueous solution. *Nanotechnology, 17*(6), 1740.

34. Wang, J., Kumar, S., & Chang, S. F. (2010). Sequential projection learning for hashing with compact codes. In *Proceedings of the 27th International Conference on Machine Learning (ICML-10),* 1279–1286.

35. Alikhani, L., Rahmani, M. S., Shabanian, N., Badakhshan, H., & Khadivi-Khub, A. (2014). Genetic variability and structure of Quercus brantii assessed by ISSR, IRAP and SCoT markers. *Gene, 552*(1), 176–183.

36. Sheha, M. A., Mabrouk, M. S., & Sharawy, A. (2012). Automatic detection of melanoma skin cancer using texture analysis. *International journal of computer applications, 42*(20), 22–26.

37. Strella, S. (1963). Differential thermal analysis of polymers. I. The glass transition. *Journal of applied polymer science, 7*(2), 569–579.
38. Bedre, J. S., & Sapkal, S. (2012). Comparative study of face recognition techniques: A review. *Emerging Trends in Computer Science and Information Technology–2012 (ETCSIT2012) Proceedings published in International Journal of Computer Applications®(IJCA), 12.*
39. Kumar, V. V., Kumar, M. P., Thiruvenkadaravi, K. V., Baskaralingam, P., Kumar, P. S., & Sivanesan, S. (2012). Preparation and characterization of porous cross linked laccase aggregates for the decolorization of triphenyl methane and reactive dyes. *Bioresource technology, 119,* 28–34.
40. Sebastian, M. T., Ubic, R., & Jantunen, H. (2015). Low-loss dielectric ceramic materials and their properties. *International materials reviews, 60*(7), 392–412.
41. Dakroury, G. A., El-Shazly, E. A. A., & Hassan, H. S. (2021). Preparation and characterization of ZnO/Chitosan nanocomposite for cs (I) and sr (II) sorption from aqueous solutions. *Journal of radioanalytical and nuclear chemistry, 330*(1), 159–174.
42. Seliger, B., Höhne, A., Knuth, A., Bernhard, H., Meyer, T., Tampe, R., & Huber, C. (1996). Analysis of the major histocompatibility complex class I antigen presentation machinery in normal and malignant renal cells: Evidence for deficiencies associated with transformation and progression. *Cancer research, 56*(8), 1756–1760.
43. Klančnik, A., Piskernik, S., Jeršek, B., & Možina, S. S. (2010). Evaluation of diffusion and dilution methods to determine the antibacterial activity of plant extracts. *Journal of microbiological methods, 81*(2), 121–126.
44. Le, Q. V., Zou, W. Y., Yeung, S. Y., & Ng, A. Y. (2011). Learning hierarchical invariant spatio-temporal features for action recognition with independent subspace analysis. In *CVPR* (Vol. 2011, pp. 3361–3368). IEEE.
45. Pooladian, B., & Nikje, M. M. A. (2018). Preparation and characterization of novel poly (urethane-imide) nanocomposite based on graphene, graphene oxide and reduced graphene oxide. *Polymer-plastics technology and engineering. 57*(18), 1845–1857.
46. Liu, Y., Liu, J., Xia, H., Zhang, X., Fontes-Garfias, C. R., Swanson, K. A., & Shi, P. Y. (2021). Neutralizing activity of BNT162b2-elicited serum. *New England journal of medicine, 384*(15), 1466–1468.
47. Burakov, A. E., Galunin, E. V., Burakova, I. V., Kucherova, A. E., Agarwal, S., Tkachev, A. G., & Gupta, V. K. (2018). Adsorption of heavy metals on conventional and nano-structured materials for wastewater treatment purposes: A review. *Ecotoxicology and environmental safety, 148,* 702–712.
48. Wesełucha-Birczyńska, A., Świętek, M., Sołtysiak, E., Galiński, P., Piekara, K., & Błażewicz, M. (2015). Raman spectroscopy and the material study of nanocomposite membranes from poly (ε-caprolactone) with biocompatibility testing in osteoblast-like cells. *Analyst, 140*(7), 2311–2320.
49. Njuguna, J., & Pielichowski, K. J. J. O. M. S. (2004). Recent developments in polyurethane-based conducting composites. *Journal of materials science, 39,* 4081–4094.
50. Khafagy, R. M. (2011). Synthesis, characterization, magnetic and electrical properties of the novel conductive and magnetic Polyaniline/MgFe2O4 nanocomposite having the core–shell structure. *Journal of alloys and compounds, 509*(41), 9849–9857.
51. Catalkopru, A. K., Kantarli, I. C., & Yanik, J. (2017). Effects of spent liquor recirculation in hydrothermal carbonization. *Bioresource technology, 226,* 89–93.
52. Wen, S., Liu, H., He, H., Luo, L., Li, X., Zeng, G., & Yang, C. (2016). Treatment of anaerobically digested swine wastewater by *Rhodobacter blasticus* and *Rhodobacter capsulatus. Bioresource technology, 222,* 33–38.

6 Magnetic Nanoparticles for Drug Delivery

*Sandhya Vasanth, Alima Misiriya, Sindhu
Priya E S, Prajitha Biju, and Deekshai Rai*

6.1 INTRODUCTION

Nanotechnology-enabled drug delivery systems have significantly enhanced the medication delivery process by changing the drug's pharmacokinetics. This phenomenon results in an extended release of the drug in the bloodstream, decreases toxicity, and increases half-life. Magnetic nanoparticles (MNPs) play an important role in transferring medicines due to their distinctive characteristics, such as superparamagnetic behavior, magnetic resonance, and robust molecular and cellular interactions [1–3]. Magnetic powders have been employed in the medical and biomedical fields due to their compatibility with biological systems, surface modification capabilities, and magnetic properties. These powders have been used since ancient times, specifically in ancient Greece and Rome. MNPs have various applications, one of which is the targeted administration of drugs to specific areas of the body by using a magnetic field. However, this process is difficult due to numerous internal and external barriers.

Enhancing the surface of nanoparticles with biocompatible compounds enhances their ability to transport and encapsulate molecules such as medications, resulting in more precise drug administration within the body [4]. Nanotechnologies have broad applications in medicine and pharmaceuticals, specifically in diagnosis, accurate doses, biosensors, tissue engineering, and gene therapy. They can play a significant role at the nanoscale that revolutionized current medical diagnostic and therapeutic methods. These methods are carried out by using an externally magnetic field, magnetically assisted gene therapy, magnetically stimulated hyperthermia, and electrically driven tissue engineering [5–8]. MNPs have been under thorough investigation for over thirty years as a highly favorable method for targeted drug delivery. This chapter is due to their physicochemical properties and ability to function at the cellular and molecular level in biological interactions [9, 10]. The most notable therapeutic potential is associated with applications that utilize "intelligent" particles consisting of a magnetic core, a recognition layer, and a therapeutic payload. The main disadvantage of most chemotherapeutic methods for cancer treatment is their limited specificity. These MNPs can tackle cancer, and they can be used for the treatment of tumors through three distinct methods: attaching specific antibodies to selectively bind to corresponding receptors, employing targeted for hyperthermia in tumor therapy, and loading pharmaceuticals onto the substrate for precise therapeutic targeting. An extensive study has been conducted on the disciplines of antisense and gene therapy, as they promise to transform the world of medicine for cancer treatment [11].

DOI: 10.1201/9781003454236-6

Magnetofection has the potential to offer several advantages, including reducing the required number of vectors, shortening the time needed for effective transfection and transduction, and enabling gene transfer to non-receptive cells. MNP drug delivery is a novel technique in nanomedicine that employs nanoparticles with magnetic characteristics to deliver drugs to specific targets [12]. This technique aims to optimize the accuracy of medication administration to targeted tissues or cells, reduce adverse effects, and enhance the therapeutic effectiveness of pharmaceuticals. Below is a concise explanation of the mechanism behind magnetic nanoparticle drug delivery.

Because of their biocompatibility and magnetic characteristics, MNPs are commonly made from minerals such as iron oxide, specifically magnetite or maghemite. The size of these nanoparticles typically falls between 10 and 100 nm, and something of this size can easily bypass the biological barriers [13]. These nanoparticles undergo a process where they are enveloped with biocompatible substances, such as polymers or lipids, to enhance their stability and compatibility with living organisms. Surface functionalization entails the attachment of targeted ligands or molecules that increase the nanoparticles' affinity for cells or tissue. A magnetic field originating from an external source directs the MNPs toward the desired location; it can be conducted through various methods, such as utilizing permanent magnets or electromagnetic fields. The magnetic field's intensity and orientation affect the motion and aggregation of the nanoparticles. Drug release can be initiated by many means, such as alterations in temperature, administration of an external stimulus (e.g., alternating magnetic fields), or pH, once the MNPs have reached the desired location. MNPs can also be modified into a precise medication and release drugs at the targeted site of their activity.

6.2 THE ADVANTAGE OF MAGNETIC POLYMER NANOPARTICLES IN DRUG DELIVERY

MNPs can be used with an external magnetic field to target diseases. Creating superparamagnetic iron oxide nanoparticles (SPIONs) has also shown promise as a novel drug delivery vehicle.

MNPs also act in multiple therapies; for example, multi-layer synthesized water-dispersible SPIONs have applications in hyperthermia, magnetic resonance imaging (MRI), and drug delivery for cancer treatment and diagnosis.

The biocompatible drug delivery vehicle can easily synthesize and characterize MNPs coated with a polylactic glycolic acid (PLGA) polymer. The polymeric shells of the designed nanoparticles can be loaded with the prospective anti-cancer medication quercetin. These nanoparticles are suitable for targeting lung cancer cells through nebulization.

MNPs are synthesized as micro and nanoparticle formulations for targeted drug delivery. MNPs can be synthesized easily in the required particle size and shape using different techniques. There is a requirement for intensive research and a lack of clinical data.

This study focuses on developing MNPs and conducting tests in laboratory settings and on animals to evaluate the effectiveness of using MNPs for drug and gene delivery. Additionally, clinical trials are being conducted to investigate the potential of using MNPs for targeted medication delivery.

6.3 TYPES OF MNPS

6.3.1 MAGNETIC IRON OXIDE NANOPARTICLES (MINPS)

Magnetic iron oxide nanoparticles (MINPs) are highly desirable nano sizes due to their exceptional magnetic characteristics, making them well-suited for biomedical applications. They can be covered with a range of inorganic substances, including metallic compounds, silica, sulfide, metal oxide, and non-metallic materials, or with organic materials such as dextran, starch, polyethylene glycol (PEG), and polyethyleneimine, either directly or indirectly. MINPs can undergo added modifications using specifically engineered molecules following the coating process. The three prominent types of MINPs include hematite, magnetite, and maghemite [14, 15]. Although it is challenging to regulate the sizes and shapes of MINPs, this can be achieved using sonochemical synthesis, thermal breakdown, coprecipitation, and hydrothermal synthesis. Aggregation plays a significant role in the synthesis of MINPs, and effectively managing electrostatic and steric repulsions is essential to addressing this issue. Surface coatings can enhance the stability of MINPs. It is also necessary to consider the impact of the coating material used for the magnetic properties and their biocompatibility.

6.3.2 FERRITES-BASED NANOPARTICLES

MNPs made of maghemite or magnetite do not react much on their surface because functionalization molecules are bonded with covalent bonds, making them hard. The reactive properties of MNPs can be enhanced by applying a coating with silica onto their surface. Being around organo-silane molecules makes it possible for covalent bonds to form with the silica shell, making adding different surface functional groups easy. The functionalized silica shell can be chemically bonded to other fluorescent dye molecules. Ferrite nanoparticle clusters, composed of eighty maghemite superpara MNPs and superparamagnetic oxide nanoparticles, have several benefits compared to metallic nanoparticles. These benefits include a narrow size distribution and enhanced magnetic properties. Due to their lack of magnetic agglomeration, they show improved colloidal stability. The magnetic moment can be adjusted by altering the nanoparticle cluster's size and maintaining superparamagnetic properties regardless of the size of the nanoparticle cluster. The silica surface enables straightforward covalent functionalization [14].

6.3.3 POLYMERS-BASED NANOPARTICLES

The high surface area-to-volume ratio makes MNPs highly reactive and prone to corrosion. Magnetic particles are coated with polymer (Table 6.1) to protect them from corrosion. This coating also prevents the leaching of potentially toxic components into the body during in vivo applications.

TABLE 6.1

Classification Based on the Type of Polymer

Classification	Examples
Natural polymer	i. Carbohydrate: Starch, dextran, chitosan
	ii. Protein: Albumin, arginyl glycyl aspartic acid, lipids
Synthetic polymer	Polyethyleneglycol, polyvinylalcohol, poly-L-lactic acid, silica, gold

Classification is based on the nature of the polymers:

a. Hydrophilic: Starch, dextran, albumin, arginyl glycyl aspartic acid, poly-ethylene glycol, polyvinyl alcohol
b. Hydrophobic: Lipids
c. Amphiphilic: Polystyrene-polyacrylic acid block copolymer, tetradecy-lphosphonate and polyethylene glycol-2- tetradecyl ether [16].

6.3.4 ORGANIC LINKERS-BASED MNPS

The surface modifications in MNPs lead the particles to bind with biomolecules. If the binders and MNPs bind weakly, the drug may be exposed to the environment and released. As a result, surface modifications are necessary to create strong interactions to enhance biomolecules' binding process and control the release mechanism.

Examples include amine, carboxylic acids, thiol, aldehyde, carbonyl, and biotin. Active biomolecules are drug molecules, proteins, DNA molecules, etc. [16].

6.3.5 NANOPARTICLES OF COBALT

Research has been conducted on the magnetic characteristics of nanoparticles with cobalt. Nevertheless, the use of cobalt gives rise to apprehensions owing to its toxicity and incompatibility with biological systems like in vivo and in vitro.

6.3.6 NANOPARTICLES OF NICKEL

Nanoparticles composed of nickel show magnetic characteristics and have been extensively studied for their potential use in medication delivery applications. Like cobalt, nickel has toxicity problems, so endeavors are undertaken to apply a coating or change these particles to improve their biocompatibility in the nano-formulation for therapeutic action.

6.3.7 MULTIPLE MNPS

A few of the MNPs are engineered to fulfill multiple purposes. For example, nanoparticles may have magnetic properties that enable them to be used for precise targeting and administration of drugs and diagnostic imaging. These versatile nanoparticles often include components such as polymers or silica in addition to the magnetic core [17].

6.3.8 HYBRID MNPs

Hybrid nanoparticles are formed by combining magnetic and non-magnetic materials to reach certain qualities. One instance of this is magnetic core-shell nanoparticles, which include a magnetic core enveloped by a non-magnetic shell [18].

6.3.9 METAL-ORGANIC FRAMEWORKS (MOFs)-BASED MNPs

Metal-organic frameworks (MOFs) are a class of substances made up of metal ions or clusters joined together by organic linkers. MOFs that include magnetic components have the potential for drug delivery applications because of their adjustable structure and porosity [19].

6.3.10 NANOPARTICLES COMPOSED OF CARBON

Carbon-based nanoparticles, such as graphene oxide or carbon nanotubes, can be changed with magnetic components to enable medication delivery. These materials have distinctive characteristics, such as a substantial surface area and exceptional biocompatibility.

6.4 MECHANISM ACTION MNPs IN THE BIOLOGICAL SYSTEM

The matrix or encapsulated drug complex is introduced into the biological system by arterial injection or intravenously. This formulation will directly load the drug in concentrated form to the tumor site without affecting target tissues using an external magnetic field generated by permanent magnets and a gradient beyond the field.

6.4.1 TARGETING

Targeting is often targeted via non-specific approaches, such as tissue-specific pore size or the benefits of the increased permeability and retention effect in tumor tissues. An alternative technique involves utilizing electrostatic forces by introducing a positively charged peptide to radioactive agents and superparamagnetic iron oxide.

6.4.1.1 PASSIVE TARGETING

Passive targeting is a technique that enhances the resilience of blood nanoparticles in scenarios such as aberrant structures, vascular damage, tumors, inflammation, and infection. This approach relies on the augmented permeation and retention phenomenon, which enhances the leaking of macromolecules and nanoparticles from blood vessels and their accumulation in tumor tissue. Capillary insufficiency, blood pressure, and lymphatic drainage are key factors in this aggregation. The reticuloendothelial system (RES) has a natural way of getting rid of it so that contrast agents or drug delivery vectors can reach where they need to go.

6.4.1.2 ACTIVE TARGETING

To target tissues specifically, superparamagnetic iron nanoparticles are engineered with different catalysts, such as minor molecules, peptides, proteins, and antibodies.

These particles could attach to molecules on the surface of the patient's cells, which helps in ligand-induced endocytosis and cellular internalization. Monoclonal antibodies, like Herceptin, are employed for drug administration and molecular identification. The arginylglycylaspartic acid part of the RGD peptide works as a ligand to bring MNPs to cancerous cells in conditions like squamous cell carcinoma, breast cancer, and malignant melanoma; the utilization of short peptides and tiny molecules enhances the polyvalent binding pathway, wherein numerous ligands bind to receptors simultaneously [20].

6.5 METHODS OF PREPARATION OF MNPS

The current research has reported different methods for synthesizing MNPs and novel methods for synthesizing Fe_3O_4 nanomaterials with different shapes.

6.5.1 GREEN SYNTHESIS

Green nanotechnology is an emerging discipline that specifically targets reducing or eliminating harmful compounds to restore the environment. One method involves the manufacture of metal nanoparticles using inactivated plant tissue, plant extracts, and other components of living plants. This approach is ecologically sustainable, devoid of toxicity, and secure, making it a compelling substitute for chemical and physical methodologies. Biological approaches for producing nanoparticles involve the utilization of microbes, enzymes, fungi, plants, or plant extracts as environmentally friendly alternatives to chemical and physical processes. Plants are ideal for biosynthesis because they have reducing agents such as citric acid, ascorbic acids, flavonoids, reductases, dehydrogenases, and extracellular electron shuttles. The research has suggested a quick, non-toxic, and environmentally friendly approach for synthesizing MNPs. The procedure uses ferric chloride hexahydrate, ferrous chloride tetrahydrate, carob leaf extract, and sodium hydroxide. Magnetite nanoparticles may be synthesized at a comparatively low temperature range of 80–85°C. The green synthesis of MNPs has numerous benefits, such as cost-effectiveness, eco-friendliness, non-toxicity, and the ability to regulate the product's treatment and size in a single-vessel reaction under mild conditions [21].

6.5.2 PRECIPITATION FROM SOLUTION

Precipitation reactions are an ancient technique for preparing MNPs. The magnetic particles are in dissolved metal precursors in a solvent; a precipitating agent is added to create an insoluble solid. The main advantage is large particle quantities, typically synthesized through homogeneous precipitation reactions involving nucleation separation and growth [22].

6.5.3 CO-PRECIPITATION

Co-precipitation is a technique commonly employed for producing magnetic MNPs with precise sizes and characteristics. Its widespread use in biomedical applications is due to its simplicity of implementation and reduced reliance on hazardous substances and procedures. MNPs are synthesized by introducing a base into aqueous salt solutions in an inert environment at ambient or elevated temperatures. There are two primary methods for synthesis. One involves partially oxidizing suspensions of ferrous hydroxide using various oxidizing agents. In contrast, the other involves aging mixtures of stoichiometric amounts of ferrous and ferric hydroxides in water-based solutions. Both methods result in the formation of magnetite particles that are spherical and uniform in size. The size and shape of iron oxide nanoparticles depend on the salts, the amount of ferric and ferrous ions present, the reaction temperature, the concentration of ions in the medium, and other important factors in the reaction. In an oxygen-free environment, full precipitation should happen between 8 and 14 on the pH scale, with a stoichiometric ratio of 2/1 (Fe^{3+}/F^{2+}). By manipulating the pH and ionic strength of the precipitation medium, it is possible to regulate the average size of the particles within a range of 2 to 15 nm, spanning ten times the difference in size. The size of MNPs decreases as the medium's pH value and ionic strength increase. This change in size affects the chemical structure of the surface, which in turn affects the electrostatic surface charge of the particles. Adding nitrogen gases to the solution protects the magnetite nanoparticles from major oxidation. Adding polyvinyl alcohol (PVA) to the iron salts makes the tiniest particles. Chemical co-precipitation is used to formulate Fe_3O_4 MNPs, but the process is not without problems. For example, the particles stick together because they are small, have a large specific surface area, and have high surface energy. In both the synthesis and purification steps, it is crucial to control the high pH value of the reaction mixture. However, this adjustment has resulted in minimal success in creating uniform and monodisperse nanoparticles [22].

6.5.4 MICROEMULSION

A microemulsion is a stable and uniform mixture of two phases, water and oil, which are normally unable to mix. The presence of a surfactant makes this mixture possible. Surfactant molecules could create a single layer at the boundary between oil and water. The hydrophilic head groups of the surfactant molecules are in the water phase, while the hydrophobic tails are dissolved in the oil phase. This process has several benefits, including using uncomplicated equipment, elevated control over particle size and content, manipulating the crystalline structure, and achieving a high specific surface area. The microemulsion process yields smaller particles with increased saturation magnetization. The characteristics of nanoparticles produced via the microemulsion technique are contingent upon the particular surfactant type and structure. The size of the droplets in the inverse microemulsion increases linearly when more water is supplied to the system. Microemulsions and inverse micelles can precisely control the shape and size of MNPs. The sequential utilization of reverse micelles is employed to synthesize an iron core initially, followed by enlarging the micelles to accommodate the shell. The main problems with the microemulsion method are that it is hard to make the process bigger, and any leftover surfactants can change the properties of the particles. Nevertheless, the technique has been encouraging in generating MNPs characterized by a substantial specific surface area and a well-defined crystalline structure [22].

6.5.5 POLYOL TECHNIQUE

The polyol technique is a highly promising approach for producing MNPs with consistent size and shape, specifically for biomedical purposes such as magnetic resonance imaging. Small metal particles can be made by reducing dissolved metal salts and letting metals precipitate from a polyol solution. Numerous variables, such as pH, temperature, adsorbent quantity, and contact time, can affect the adsorption of metal ions, which in turn can affect the size of the metal particles. The polyol approach is also a preparative procedure for synthesizing nanocrystalline alloys and bimetallic clusters. The liquid polyol is a solvent for the metallic precursor, a reducing agent, and a complexing agent for the metallic cations. The reaction mixture undergoes reflux at temperatures ranging from 180°C to 199°C, forming an intermediate compound. Subsequently, the intermediate compound is reduced to produce metal nuclei. Submicrometric-sized particles can be achieved by raising the reaction temperature or promoting heterogeneous nucleation. Iron particles of around 100 nm can be acquired by disproportionating ferrous hydroxide in an organic medium. Iron alloys can be produced using the coprecipitation of Fe, Ni, and Co in either ethylene glycol (EG) or PEG. Metallic particles that are uniform in size, spherical in shape, and not clumped together have been synthesized without using a seed material (homogeneous nucleation). On the other hand, particles ranging from 50 to 100 nm have been synthesized by using Pt as a nucleating agent (heterogeneous nucleation). The polyol technique creates metallic nanoparticles protected from oxidation by adsorbing glycol molecules on their surface. Using a non-watery solvent such as polyol mitigates the issue of hydrolysis in tiny metal particles, a common occurrence in aqueous environments [22].

6.5.6 THERMAL DECOMPOSITION OF ORGANIC PRECURSORS

The iron precursors undergo decomposition in the presence of heated organic surfactants, resulting in enhanced samples characterized by precise size control, a restricted size distribution, and excellent crystallinity of individual and dispersible magnetic iron oxide nanoparticles. The breakdown of organometallic precursors, such as $[M^{n+}(acac)_n]$, (M = Fe, Mn, Co, Ni, Cr; n = 2 or 3, acac = acetylacetonate), $M^x(cup)_x$ (cup = N-nitroso phenyl hydroxylamine) or carbonyls [such as $Fe(CO)_5$], at high temperatures can result in the production of nanoparticles that exhibit a high degree of uniformity and precise control over their size. This approach can potentially be utilized for biological applications such as MRI, which heavily rely on the size of particles. Accurate control of size and morphology in the reaction requires careful consideration of the temperature, duration, aging period, and annealing period. It is possible to make uniform magnetite nanoparticles 3 to 20 nm in size by reacting iron (III) acetylacetonate, phenyl ether, alcohol, oleic acid, and oleylamine at a high temperature (265°C). However, this strategy has many disadvantages, such as producing organic-soluble nanoparticles, the requirement for post-synthesis surface treatment, and the subsequent dissolution of the nanoparticles in nonpolar solvents [22].

6.5.7 Chemical Vapor Deposition (CVD)

Generating MNPs in the vapor phase creates more thermodynamically unstable conditions compared to forming solid materials. Chemical vapor deposition (CVD) allows versatility in manufacturing diverse materials and can use the extensive repository of precursor chemicals for CVD procedures. Precursors may exist in solid, liquid, or gaseous states under normal conditions and are introduced into the reactor as vapor. Wegner et al. investigated the use of this technique to synthesize bismuth nanoparticles by manipulating the flow field and combining cold and hot gases to regulate the distribution of particle sizes. Gas phase techniques for synthesizing nanomaterials include thermal decomposition, reduction, hydrolysis, disproportionation, oxidation, and other chemical processes [22].

6.5.8 Spray Pyrolysis

Spray pyrolysis is a technique wherein a solution is sprayed into reactors, causing the aerosol droplets to undergo evaporation, condensation, drying, and thermolysis. Scientists have used this method to make colloidal clusters of superparamagnetic MNPs that can be solid or hollow spheres. The surface of these MNPs may have more silica. The significant production rate suggests a favorable outlook for using MNPs in both in vivo and in vitro applications. A Fe3+ salt and an organic part that acts as a reducing agent are usually used to start pyrolysis-based processes. Various particle morphologies and sizes of homogenous g-Fe_2O_3 particles can be achieved in alcoholic solutions. Fe (III) nitrate and Fe (III) chloride solutions create compact clusters with spherical morphology [22].

6.5.9 Laser Pyrolysis

Laser pyrolysis is a technique used to heat precursors to induce reactions and homogeneous nucleation. This technology enables targeted heating and quick cooling. This process entails subjecting a combination of gases to a continuous-wave CO_2 laser, which initiates and maintains a chemical reaction. Biocompatible magnetic dispersions have been created utilizing Fe_2O_3 NPs. This process has also been used to synthesize nanoparticles of varied materials, such as Si NPs [22].

6.5.10 Sonochemical Reaction

The sonochemical approach is a widely used alternative for synthesizing new materials with unique features. The process entails using ultrasound to induce acoustic cavitation, which produces, enlarges, and then collapses bubbles inside a liquid medium. The procedure involves using volatile precursors in solvents with low vapor pressure to enhance the production of particles. The process of acoustic irradiation consists of using an ultrasound probe, specifically a titanium horn, which is used at a frequency of 20 kHz. This technique has been used to produce various nanocomposites, such as MNPs. A sonochemistry method was described for synthesizing highly pure Fe_3O_4 powder in the nanometer size range with a particle size of 10 nm.

The produced Fe_3O_4 nanoparticles show superparamagnetic properties and display a low magnetism level at ambient temperature. The powders typically show porosity, lack a defined crystalline structure, and are formed into clusters. A study was also conducted on differences between the sonochemical and co-precipitation products regarding their crystallinity and magnetic properties. The researchers discovered that the sonochemical method showed superior crystallinity and saturation magnetization compared to the co-precipitation method [22].

6.5.11 SOL-GEL METHOD

Metallo-organic precursors have been effectively utilized in synthesizing magnetite nanoparticles by the sol-gel method. Shakeel Akbar and his colleagues effectively created nanoparticles with a modified sol-gel method, obtaining optimal outcomes for alpha-phase particles under two specific circumstances. Sara Shaker and her colleagues effectively synthesized magnetite nanoparticles (Fe_3O_4) using a soul-gel approach and an annealing process. The synthesis involved using ferric nitrate and ethylene glycol at different temperatures. The sol-gel process offers several benefits, including producing pure amorphous phases, giving precise control over particle size, and creating materials with a specified structure. Additionally, it mitigates pollution resulting from byproducts and the post-treatment of items [22].

6.6 APPLICATION

6.6.1 GENE DELIVERY

In therapeutic settings, MNPs have long been employed as agents to increase contrast in magnetic resonance imaging. Nevertheless, for them to function as efficient conveyors for DNA or pharmacological substances, it is imperative to change the surface of the particles to ease the binding of specific molecules. Cleavable linkers, electrostatic interactions between the particle surface and the therapeutic agent, and encasing the target molecule(s) in a biodegradable outer shell are all methods for attaching molecules to the particles' surfaces. In the first investigation illustrating the specific transportation of DNA using magnetic particles, Cathryn Mah, Barry Byrne, and team applied a layer of adeno-associated virus (AAV) having green fluorescent protein (GFP) onto the surface of magnetic particles by using a detachable heparin sulfate linker. It was easier for the transduction to work in both C12 cells grown in a lab and those in living mice after they were injected intramuscularly. An alternative method for binding DNA to particle surfaces involves electrostatic interactions between DNA's negatively charged phosphate backbone and positively charged molecules attached to the particles' surface [23–26]. Polyethyleneimine (PEI) is commonly used for this method because it binds and compacts DNA due to its many secondary amine groups distributed along its chain length. The discovery of polyethyleneimine-coated magnetic particles marked the first instance of using MNPs to spread genes in a non-viral manner in laboratory settings. This method has been employed to introduce genetic material into several types of cells, such as primary lung epithelial cells and blood vessel endothelial cells [27]. The group's recent research has concentrated on enhancing the overall transfection effectiveness of this method with dynamic magnetic fields generated by oscillating arrays of permanent rare earth magnets. A revolutionary method for delivering genes via nanoparticles, called nanotube spearing, involves carbon nanotubes embedded with nickel and coated with DNA. When exposed to cells in a precisely aligned magnetic field, the nanotubes position themselves along the magnetic flux lines, enabling them to penetrate cells, traverse the membrane, and transport the desired DNA [28–30].

6.6.2   Drug Delivery

The concept of using magnetic forces to direct therapeutic substances has been a subject of fascination since the late 1970s. Widder et al.'s initial research proved the efficiency of magnetically sensitive microspheres for delivering anti-tumor drugs. Nevertheless, the number of clinical trials conducted thus far is limited. In 1996, the first Phase I clinical trial was conducted, wherein nanoparticles were used to form complexes with epirubicin through electrostatic interactions. The study did not find an LD 50 (lethal dose) for the particles in these experiments; however, the particles did not show an affinity for the tumor location. In 2002, research was conducted on 32 patients diagnosed with hepatocellular carcinoma. Doxorubicin hydrochloride mixed with MNPs was administrated through a sub-selective hepatic artery catheterization. The particle-drug complex was directed toward the tumor location with an external magnetic field, and the localization of the particles was assessed using MRI. The findings showed that the particle/drug complex targeted the tumor sites, resulting in drug impact on around 64 to 91% of the tumor volume [23–30].

6.6.3   Cancer Treatment

6.6.3.1   Hyperthermia

Studies used alternating magnetic fields (AMF) as a targeted therapy method to make it more prevalent to heat only cancer cells. Hyperthermia is a medical procedure that involves increasing the temperature of a particular tissue or the entire body above the usual physiological levels. Cancer cells show sensitivity to hyperthermia because of the reduction in pH levels inside the malignant microenvironment, leading to a loss in thermotolerance. The common hyperthermia therapies can be produced using ultrasonic waves, radio-frequency waves, microwaves, infrared radiation, and hot water. Nevertheless, these treatments are associated with many adverse effects, such as blisters, burns, discomfort, and uncontrolled tissue proliferation. Localized hyperthermia is being studied as a complementary treatment to standard cancer therapies, including chemotherapy and radiation therapy, with the advantage of fewer side effects. The progress in nanotechnology has enhanced the safety, practicality, and efficacy of local hyperthermia therapy for treating malignant tissues. High magnetism has proved effective in generating localized heat within malignant tissue areas. Particles can be introduced into malignant tissue through direct injection or concentrated at the tumor site using a magnetic field. By subjecting the particles to an AMF, intense heat is produced specifically at the malignant location, resulting in the demise of cancer cells. The studies conducted on living organisms have proved the favorable impact of hyperthermia therapy on decreasing the size of cancerous tissue by directly or intravenously administering particles into the tumor, which is then exposed to an AMF. By creating heat in response to AMF, magnetic hyperthermia has been shown to effectively stop the growth of cancer cells and make it easier for drugs to get into cancer cells [23–30].

MNPs coated in gold are used in photothermal ablation therapy to kill cancer cells by heating them by absorbing electromagnetic photons. Cells expressing the A33 antigen showed enhanced accumulation of NP and increased cell death upon exposure to NIR light for 6 minutes. Gold-coated magnetite nanoparticles have diverse possibilities in targeted photothermal therapy and multimodal imaging. Photodynamic therapy (PDT) uses a chemical that makes light more sensitive to cytotoxic singlet oxygen (1O2), which hurts cancer cells by creating free radicals. The therapeutic efficacy is enhanced by conjugating these medicines. In one of the research studies using pheophorbide-A conjugated SPIONs to prove photodynamic therapy (PDT), the results showed that these SPIONs can supply bimodal MRI contrast and fluorescence. In vivo fluorescence and MRI imaging of Ce6-coated SPIONs in mice with tumors also confirmed the enhanced cellular absorption of chlorin-e6-conjugated PEGylated SPIONs [23–30].

6.6.3.2   Drug Delivery

SPIONs have garnered considerable interest due to their potential applications in tumor therapy, imaging, and drug delivery. Nanoparticles are well-suited for drug administration because of their varied sizes and functionalization. Nevertheless, it is crucial to consider factors such as colloidal stability and biocompatibility, which can potentially make it possible to treat neurological illnesses by transporting medications over the blood-brain barrier. However, obstacles such as embolization and significant distances between diseased locations and external magnetic fields still need to be overcome. Utilizing controlled drug delivery enables the maintenance of drug levels within the necessary concentration range, thereby preventing overdose and minimizing adverse effects [23–30].

6.6.3.3   Biotherapeutics and Chemotherapeutics

Biotherapeutics and chemotherapeutics target tumor growth by interfering with cellular processes, including DNA replication, protein expression, cell division, and anti-apoptotic pathways. Biotherapeutics involves administering biologically active substances such as peptides, proteins, DNA, or small interfering RNA (siRNA). At the same time, chemotherapeutics consists of the administration of small-molecule medications, including paclitaxel, 5-fluorouracil, temozolomide, and doxorubicin. MNPs have been used for biotherapeutic and chemotherapeutic purposes. Scientists have created formulations of iron oxide nanoparticles coated with chitosan-PEG-PEI and functionalized with chlorotoxin (CTX) and green fluorescent protein (GFP)–encoded DNA. They have led to a higher level of absorption in specific tumors. Researchers also created elongated magnetic mesoporous silica nanoparticles (M-MSNs) for suicide gene therapy. Compared to spherical M-MSNs, these nanoparticles were better at carrying drugs, releasing drugs fast, and delivering genes.

The synthesized MNPs can be engineered to regulate drug release by different stimuli, although accurately measuring the individual tumor microenvironment can present challenges [23–30].

6.6.3.4 Radiotherapeutics

Research is now being conducted on nanoparticles for their application in radiotherapy. These nanoparticles can carry radionuclides, emitters, and radiosensitizers to target tumor cells and cause DNA damage specifically. When compared to current radiation techniques, nanoparticles offer specific advantages because they reduce damage to non-targeted tissues and make it possible to combine treatments like gene therapy or chemotherapy. The application of MNPs containing neutron-activated holmium-166 and platinum-based chemotherapy drugs has demonstrated potential in managing non-small cell lung cancer. When holmium and platinum are mixed, they make the cytotoxicity much higher. It has been proved that platinum-based medicines are better at blocking radiation. The investigators attempted to treat the hypoxic tumor microenvironment by manipulating myeloid-derived suppressor cells (MDSCs) in gliomas, which produce arginases that reduce the efficiency of immune cells. To induce the inflammatory repolarization of MDSCs toward target tumor cells, zinc-doped iron oxide nanoparticles were modified to act as radiosensitizers and ROS generators, leading to a rise in the median survival rates of mice with glioma [23–30].

6.6.3.5 Cancer Immunotherapy

Cancer immunotherapy aims to target solid tumors and use the patient's immune system to identify and suppress cancer cells. Different approaches have been used, including adoptive cell transfer protocols, dendritic cell vaccines, inhibitory checkpoint molecules, and combinations of these approaches. One advantage of combining cancer immunotherapy techniques with nanoparticles is a targeted transport mechanism that can be precisely adjusted to maximize effectiveness. An improved method of precisely localizing the drug to the desired target site combines targeting agent conjugation with magnetic navigation. Combining MNPs can improve the therapeutic efficacy of hyperthermia therapies. Nanoparticle immunotherapeutic formulations can be changed to work as fluorescent probes and MRI contrast agents. It is possible to track how they move through living things. Dendritic cell vaccines employ nanoparticles as carriers for delivering antigens to dendritic cells, which then proceed to lymph nodes to activate antigen-specific cytotoxic T lymphocytes, hence impeding tumor progression. Recent studies have examined nanoparticles that combine checkpoint inhibitors with chemotherapeutic drugs or cancer cell antigens [23–30].

6.6.4 Nanothernostics

Theranostic nanoplatforms with MNPs are often used as T2 (negative) MRI contrast agents because they are more magnetic and can hold water protons inside a polymeric shell.\Positive T1 magnetic resonance imaging (MRI) contrast is created when polymer-nanoparticle (NP) structures have Gd3+ complexes or Fe3+-terpyridine complexes.

Magnetic field induced tomography (MPI) is a novel imaging technique that measures the arrangement of MNPs in tissue by analyzing their magnetization response to a magnetic field. MPI offers superior contrast compared to MRI. A polymer-nanoparticle system with highly favorable magnetic particle imaging (MPI) contrast qualities has been successfully created. Fluorescence imaging is commonly conducted with MRI using a polymer labeled with a fluorescent marker. The MNPs are changed with a polymer labeled with a fluorescent dye activated by two photons. Creating a nanoprobe that can accurately measure the mass of pancreatic beta cells would provide a reliable sign of the start of type 2 diabetes. Magneto-fluorescent nanogels exhibit promise for the use of fluorescence imaging in the detection of cancer cells. Photoacoustic imaging (PAI) is an innovative imaging technique that combines optical excitation and ultrasonic imaging to detect and analyze sound waves [23–30].

6.6.5 BIOSENSORS AND DETECTION

Traditional biosensors have difficulties with the fixation of biorecognition elements onto the transducer surface, resulting in problems such as decreased sensitivity and selectivity. Nanomaterials, specifically magnetic beads (MBs), have been introduced to tackle these issues. They possess magnetic capabilities and other notable attributes, including compact dimensions, a high surface area to volume ratio, and exceptional biocompatibility. They can be customized for specific purposes and have been widely used in many domains, such as biosensing and clinical diagnostics. In addition, the process of preparing samples can be streamlined by isolating the desired substances using magnetic separation. Using MBs enhances sensitivity, reduces the time needed for analysis, and minimizes interference. The synthesis and alteration are essential in fine-tuning their characteristics for diverse applications [23–30].

6.6.6 CELL LABELLING AND TRACKING

A study looked at the prospects for using ultra-small iron oxide nanoparticles (USPIO-PAA), superparamagnetic iron oxide nanoparticles (SPIO-PAA), and glucosamine-modified iron oxide nanoparticles (USPIO-PAA-GlcN) as markers for mesenchymal stem cells (MSCs) in magnetic resonance imaging to help track cells. Significant disparities were seen in cellular dosage and labeling efficacy of the three samples. When used with polylysine, SPIO-PAA exhibited uneven cell internalization, while no uptake was seen for USPIO-PAA. On the other hand, USPIO-PAA-GlcN, which is made up of ridiculously small iron oxide nanoparticles functionalized with glucosamine, showed an important level of cellular uptake and biocompatibility. MRI experiments in vitro and in vivo showed sensitive detection, suggesting that adding glucosamine can make it easier for MSCs to accept these nanoparticles [23–30].

6.6.7 WOUND HEALING

Endothelial progenitor cells (EPCs) are used for wound healing research. Researchers have developed a hydrogel sheet with groove patterns for in situ cell delivery to optimize their therapeutic effectiveness. The hydrogel sheet's structured surface enables endothelial cell shape and structure alterations, enhanced communication with neighboring cells, and the release of growth factors like PDGF-BB. MNPs were added to the hydrogel sheet to ease magnetic field movement. This approach improved wound healing by introducing EPCs into a microphysiological system, improving blood vessel formation and skin wound healing [23–30].

6.7 CONCLUSION

The integration of composite magnetic ions (IONPs) with inorganic and organic materials for biomedical applications has been examined in this chapter. The main goal of the research has been to combine organic dyes and iron oxide MNPs into a single platform for agents of bimodal imaging, such as photodynamic treatment and dual-mode imaging for biomolecule identification. Because of their dual-diagnostic MRI/FI modalities, these devices offer more complementary datasets and improved diagnostic information.

The distinctive optical characteristics of carbon dots (CDs) and quantum dots (QDs), such as their high molar extinction coefficient and resistance to photobleaching, are well known. When noble metal nanoparticles and magnetic nanoparticles are combined into a single platform, numerous therapeutic and diagnostic modalities can be used concurrently, increasing diagnostic picture quality and accuracy while cutting costs and time.

Mixing polymers with magnetic nanoparticles allows you to make multi-functional systems that improve stability, blood flow, pH-sensitive magnetic resonance imaging, and compatibility with living things. Due to their capacity to catalyze fenton processes and transfer certain medications, gene segments, or magnetothermal heating to targeted regions of interest, magnetic nanoparticles (specifically, SPIONs) hold great promise for revolutionizing cancer research's diagnostic and therapeutic applications.

Incorporating magnetic particles into diagnostic environments expands the possibilities for multiple analyte detection while also satisfying the practical demand for faster and less expensive analysis. Nanoparticle-based cancer treatment is a multi-million-dollar industry, and in the next five years, a few investigations will advance to clinical trials.

REFERENCES

1. Kralj S, Makovec D. Magnetic assembly of superparamagnetic iron oxide nanoparticle clusters into nanochains and nanobundles. ACS Nano. 2015;9(10):9700–9707. doi: 10.1021/acsnano.5b02328.
2. Mohammadi Ziarani G, Malmir M, Lashgari N, Badiei A. The role of hollow magnetic nanoparticles in drug delivery. RSC Adv. 2019;9(43):25094–25106. doi: 10.1039/c9ra01589b.
3. Sakellari D, Brintakis K, Kostopoulou A, Myrovali E, Simeonidis K, Lappas A, Angelakeris M. Ferrimagnetic nanocrystal assemblies as versatile magnetic particle hyperthermia mediators. Mater Sci Eng C Mater Biol Appl. 2016;58:187–193. doi: 10.1016/j.msec.2015.08.023.

4. Öztürk Er E, Dalgıç Bozyiğit G, Büyükpınar Ç, Bakırdere S. Magnetic nanoparticles based solid phase extraction methods for the determination of trace elements. Crit Rev Anal Chem. 2022;52(2):231–249. doi: 10.1080/10408347.2020.1797465.
5. Fu H, Xu W, Wang H, Liao S, Chen G. Preparation of magnetic molecularly imprinted polymer for selective identification of patulin in juice. J Chromatogr B Analyt Technol Biomed Life Sci. 2020;1145:122101. doi: 10.1016/j.jchromb.2020.122101.
6. McCarthy JR, Kelly KA, Sun EY, Weissleder R. Targeted delivery of multifunctional magnetic nanoparticles. Nanomedicine (Lond). 2007;2(2):153–167. doi: 10.2217/17435889.2.2.153.
7. Yoo D, Lee JH, Shin TH, Cheon J. Theranostic magnetic nanoparticles. Acc Chem Res. 2011;44(10):863–874. doi: 10.1021/ar200085c. Erratum in: Acc Chem Res. 2012 Sep 18;45(9):1622.
8. Li X, Li W, Wang M, Liao Z. Magnetic nanoparticles for cancer theranostics: Advances and prospects. J Control Release. 2021;335:437–448. doi: 10.1016/j.jconrel.2021.05.042.
9. El-Boubbou K. Magnetic iron oxide nanoparticles as drug carriers: Clinical relevance. Nanomedicine (Lond). 2018;13(8):953–971. doi: 10.2217/nnm-2017-0336.
10. El-Boubbou K. Magnetic iron oxide nanoparticles as drug carriers: Preparation, conjugation and delivery. Nanomedicine (Lond). 2018;13(8):929–952. doi: 10.2217/nnm-2017-0320.
11. Vallabani NVS, Singh S, Karakoti AS. Magnetic nanoparticles: Current trends and future aspects in diagnostics and nanomedicine. Curr Drug Metab. 2019;20(6):457–472. doi: 10.2174/1389200220666181122124458.
12. Ko MJ, Min S, Hong H, Yoo W, Joo J, Zhang YS, Kang H, Kim DH. Magnetic nanoparticles for ferroptosis cancer therapy with diagnostic imaging. Bioact Mater. 2023;32:66–97. doi: 10.1016/j.bioactmat.2023.09.015.
13. Lee CW, Liu JF, Wei WC, Chiang MH, Chen TY, Liao SH, Chiang YC, Kuo WC, Chen KL, Peng KT, Liu YB, Chieh JJ. Synthesised conductive/magnetic composite particles for magnetic ablations of tumours. Micromachines (Basel). 2022;13(10):1605. doi: 10.3390/mi13101605.
14. Thakur P, Chahar D, Taneja S, Bhalla N, Thakur A. A review on MnZn ferrites: Synthesis, characterization and applications. Ceram Int. 2020;46(10):15740–15763. doi: 10.1016/j.ceramint.2020.03.287.
15. Albinali KE, Zagho MM, Deng Y, Elzatahry AA. A perspective on magnetic core-shell carriers for responsive and targeted drug delivery systems. Int J Nanomedicine. 2019;14:1707–1723. doi: 10.2147/IJN.S193981.
16. Manousi N, Rosenberg E, Deliyanni E, Zachariadis GA, Samanidou V. Magnetic solid-phase extraction of organic compounds based on graphene oxide nanocomposites. Molecules. 2020;25(5):1148. doi: 10.3390/molecules25051148.
17. Healy S, Bakuzis AF, Goodwill PW, Attaluri A, Bulte JWM, Ivkov R. Clinical magnetic hyperthermia requires integrated magnetic particle imaging. Wiley Interdiscip Rev Nanomed Nanobiotechnol. 2022;14(3):e1779. doi: 10.1002/wnan.1779.
18. Shams SF, Ghazanfari MR, Schmitz-Antoniak C. Magnetic-plasmonic heterodimer nanoparticles: Designing contemporarily features for emerging biomedical diagnosis and treatments. Nanomaterials (Basel). 2019;9(1):97. doi: 10.3390/nano9010097.
19. Safdar Ali R, Meng H, Li Z. Zinc-based metal-organic frameworks in drug delivery, cell imaging, and sensing. Molecules. 2021;27(1):100. doi: 10.3390/molecules27010100.
20. Cerdan K, Moya C, Van Puyvelde P, Bruylants G, Brancart J. Magnetic self-healing composites: Synthesis and applications. Molecules. 2022;27(12):3796. doi: 10.3390/molecules27123796.
21. Hossain N, Mahlia TMI, Saidur R. Latest development in microalgae-biofuel production with nano-additives. Biotechnol Biofuels. 2019;12:125. doi: 10.1186/s13068-019-1465-0.

22. Kappe D, Bondzio L, Swager J, Becker A, Büker B, Ennen I, Schröder C, Hütten A. Reviewing magnetic particle preparation: Exploring the viability in biosensing. Sensors (Basel). 2020;20(16):4596. doi: 10.3390/s20164596.

23. Zimina TM, Sitkov NO, Gareev KG, Fedorov V, Grouzdev D, Koziaeva V, Gao H, Combs SE, Shevtsov M. Biosensors and drug delivery in oncotheranostics using inorganic synthetic and biogenic magnetic nanoparticles. Biosensors (Basel). 2022;12(10):789. doi: 10.3390/bios12100789.

24. Nowak-Jary J, Machnicka B. Pharmacokinetics of magnetic iron oxide nanoparticles for medical applications. J Nanobiotechnology. 2022;20(1):305. doi: 10.1186/s12951-022-01510-w.

25. Andrade RGD, Veloso SRS, Castanheira EMS. Shape anisotropic iron oxide-based magnetic nanoparticles: Synthesis and biomedical applications. Int J Mol Sci. 2020;21(7):2455. doi: 10.3390/ijms21072455.

26. Mondal S, Manivasagan P, Bharathiraja S, Santha Moorthy M, Kim HH, Seo H, Lee KD, Oh J. Magnetic hydroxyapatite: A promising multifunctional platform for nanomedicine application. Int J Nanomedicine. 2017;12:8389–8410. doi: 10.2147/IJN.S147355.

27. Kubelick KP, Mehrmohammadi M. Magnetic particles in motion: Magneto-motive imaging and sensing. Theranostics. 2022;12(4):1783–1799. doi: 10.7150/thno.54056.

28. Tran HV, Ngo NM, Medhi R, Srinoi P, Liu T, Rittikulsittichai S, Lee TR. Multifunctional iron oxide magnetic nanoparticles for biomedical applications: A review. Materials (Basel). 2022;15(2):503. doi: 10.3390/ma15020503.

29. Maldonado-Camargo L, Unni M, Rinaldi C. Magnetic characterization of iron oxide nanoparticles for biomedical applications. Methods Mol Biol. 2017;1570:47–71. doi: 10.1007/978-1-4939-6840-4_4.

30. Alromi DA, Madani SY, Seifalian A. Emerging application of magnetic nanoparticles for diagnosis and treatment of cancer. Polymers (Basel). 2021;13(23):4146. doi: 10.3390/polym13234146.

7 Magnetic Polymer Composites for Bioimaging

*Bina Gidwani, Manisha Verma, Varsha Sahu,
Amber Vyas, Ravindra Kumar Pandey,
and Shiv Shankar Shukla*

7.1 INTRODUCTION

7.1.1 BIOIMAGING

A method called bioimaging enables scientists to view biological processes in real time without physically penetrating the skin or entering the body. Bioimaging seeks to interfere as little as possible with biological functions. It is widely used to learn more about the three-dimensional structure of the thing being seen without directly touching it [1].

Multi-scale and multi-modal imaging is frequently needed for complex bioimaging applications (such as those involving molecular, cellular, and organ levels of organization). Imaging that combines ultrasound and light is an illustration of this. Imaging makes it possible to comprehend intricate structures and dynamically interacting processes deep inside the body. Many imaging techniques make use of the entire energy spectrum. Ultrasound, optical coherence tomography (OCT), MRI, and CT employing X-rays are a few examples of clinical modalities [2].

In a broader sense, the word "bioimaging" refers to methods for visualizing biological substances that have been fixed for monitoring. The basic and medical sciences can employ bioimaging to analyze common anatomy and physiology and gather data. The multimodal nature of bioimaging research necessitates interdisciplinary teams with expertise in electrical engineering, mechanical engineering, biomedical engineering, and other fields [3].

The science of biomedical imaging has advanced since Roentgen first discovered the X-ray 100 years ago, culminating in a novel imaging approach with MRI, CT, and PET. A brief explanation of several bioimaging techniques is presented. A few of the methods under investigation right now are molecular imaging, functional MRI, diffusion-weighted MRI, and magnetic resonance spectroscopy (MRS). PET, SPECT, and optical imaging are a few examples of molecular imaging techniques [4].

7.1.2 MAGNETIC POLYMERS

Because polyaniline and tetracyanoquinodi methane are ferromagnetic compounds with the ability to attract metals, which is highly dependent on the presence of electrically uncharged components, they are used in magnetic polymers. Beads made of magnetic polymer are made to react to environmental factors like magnetic field, temperature, pH, and solvent composition. The magnetic field has the added benefits of immediate action and contactless control among these stimuli. The conductive, semi-flexible, air-stable polymer polyaniline is based on emerald. Because it is a stable electrical conductor that resembles metal, it can be regarded as a magnetic polymer. From aniline monomers, it can be polymerized. A metallic magnet is produced when polyaniline and a free radical combine to create tetracyanoquinodi methane as an acceptor molecule. Where, 1, 4-cyclohexanedione is combined with malononitrile to form tetracyanoquinodimethane, which is then produced by dehydrating the resultant diene with bromine.

7.1.2.1 Magnetic Polymer Materials

In general, ferromagnetic materials, alloys, oxides, or composite structures based on iron, cobalt, and nickel are referred to as magnetic particles. Among the several kinds of magnetic materials, paramagnetic materials have positive susceptibility and a magnetization proportional to the external H, whereas antimagnetic materials have the opposite property—a magnetization proportional to the external H—but a negative susceptibility. When H reaches a particular value, the magnetization of ferromagnetic materials begins to saturate after initially increasing considerably with H. Furthermore, due to their special significance in energy applications like motors, transformers, and sensors, soft magnetic particles are a significant category of magnetic materials. In relation to intrinsic coercivity (Hco), the term "soft" is used. The cutoff used to classify a substance as soft is arbitrary; soft magnetic materials are those with an Hco between 400 and 1000 A mL [5, 6].

The majority of the time, soft magnetic particles are found in cubic crystalline minerals, such as Fe, Ni, Co, Fe-Ni, and Fe-Co, and metal oxides like MFe_2O_4 (where M is a divalent metal). Depending on how the magnetic polymer composite particles will be used, consumers can choose between commercial plastics like polystyrene (PS) and polymethyl methacrylate (PMMA) and conducting polymers like polyaniline (PANI) and polypyrrole (PPy) [7].

7.1.2.2 Types of Magnetic Polymer

Three categories of magnetic polymers are distinguished:

- Magnetic polyradical molecules that are entirely organic in origin
- Association of metal ions in polymers
- Ferromagnetic particles based on polymer composites and metal oxides.

7.1.2.3 Preparation of Magnetic Polymer Particles

The preparation of magnetic polymer particles can be generally classified into three categories:

Three general categories in which magnetic polymer particle preparation falls:
- **Magnetic core–polymer shelled particles:** One approach to fabricating these structures is to prepare magnetic nanoparticles and polymer

microspheres separately and then combine the two elements by exploiting their physical interaction.

- **Particles with a polymer shell and a magnetic core:** The second method is the in situ precipitation of iron oxide nanoparticles on the polymer surface in the presence of polymeric microspheres.
- **Polymeric materials with incorporated magnetic particles:** Using magnetic nanoparticles to help the polymerization of monomers is a third technique [8].

Heterogeneous polymerization includes mini-emulsion polymerization, inverse emulsion/micro-emulsion polymerization, conventional emulsion polymerization, and emulsifier-free mini-emulsion polymerization [1]. The preparation methods for various magnetic polymer composites and their magnetic properties are listed in Table 7.1.

7.1.2.3.1   Iron Oxide

The magnetic community has made considerable use of iron oxide nanoparticles among other magnetic particles. Iron in particular has the highest saturation magnetization at ambient temperature and a suitably high Curie temperature. As a result, iron has gotten a lot of attention in the majority of practical applications. Iron's reactivity, particularly with oxygen and moisture, is perhaps its worst disadvantage. Fortunately, the utility of zero-valent iron nanoparticles is greatly enhanced by the employment of polymers as protective agents. Iron oxides come in a variety of forms, including gamma iron oxide (Fe_2O_3), magnetite (Fe_3O_4), and hematite [8].

7.1.2.3.2   Carbonyl Iron

Soft-magnetic carbonyl iron (CI) microbeads with high purity (>98%) and characteristic spherical form have found widespread use in MR systems due to their potential for commercial application. But because the dispersion medium and the magnetic particles have very different densities, CI particles typically have serious dispersion problems. Furthermore, over time, bare CI particles employed in engineering applications may undergo an oxidation process [9].

7.1.2.3.3   Other Magnetic Polymer Composites

Composites with other magnetic metallic polymers such as cobalt-based magnetic nanoparticles, which have high saturation magnetization values and high magnetic susceptibility, are recognized to be the most often utilized materials after iron-based ones. Their weakness, which stems from their potent magnetic and van der Waals force, is their instability [10]. They frequently have a limited range of practical applications since they are prone to forming agglomerates. The long-term colloidal stability of these particles' magnetic dispersions in solution is enhanced by polymer coating, which is a feasible and efficient way to prevent their oxidation and agglomeration [11].

TABLE 7.1

Preparation Methods for Various Magnetic Polymer Composites and Their Magnetic Properties

S. No.	Magnetic Properties	Polymer	Method	Density (g/cm³)	M_s Value (emu/g)
1	Fe_3O_4	Polydivinyl benzene	Distillation precipitation polymerization	1.83	41
2	Fe_3O_4	Poly(diphenyl amine)	Co-precipitation	2.44	77.1
3	Fe_3O_4	Poly(orthoanisidine)	Chemical oxidation polymerization	2.52	36
4	Fe_3O_4	Polystyrene	Mini emulsion polymerization	2.8	27
5	Fe_3O_4	Polystyrene	Pickering emulsion polymerization	1.69	59
6	Fe_3O_4	Polystyrene	Shirasu porous glass membrane technique	2.29	31.7
7	Fe_2O_3	Poly(methylCrylate)	Pickering emulsion polymerization	1.68	20.05
8	Cl	Polyindole	Chemical oxidation polymerization	7.3	180
9	Cl	Poly(diphenyl amine)	Oxidative dispersion polymerization	7.42	191
10	Cl	Polyaniline	Chemical oxidation polymerization	4.21	136
11	Cl	Polydopamine	In situ self-oxidation polymerization	6.17	135
12	Cl	Poly(methyl methacrylate)	In situ dispersion polymerization	4.5	151
13	$ZnFe_2O_4$	Polyaniline	Pickering emulsion polymerization	5.4	73.7

7.2 MAGNETIC NANOPARTICLES

Nanocomposites are multi-phasic materials that have at least one unit with a dimension in the size range below 100 m in the matrix material. In order to create a material with distinct characteristics relative to its individual constituent parts, magnetic nanocomposite materials are composed of two or more components with various physical or chemical properties. At least one of these substances must be magnetic, and at least one must be in the nanoscale size range. Due to their exceptional qualities, magnetic nanocomposites have received the most attention. There are several different ways to combine magnetic nanoparticles with an additional organic or inorganic component to create nanocomposites. Nanocomposites can exist in a variety of states, including colloidal, powder, fibers, membranes, and films. When utilized as anode materials in Li-ion batteries, for instance, carbon shells and graphene sheets mixed with Fe_2O_3 nanoparticles have been found to form electrical networks that enable quick and effective electron transport [12].

7.3 SYNTHESIS OF MAGNETIC NANOCOMPOSITES

Magnetic nanocomposites can be synthesized by following six methods listed in Figure 7.1. These methods are discussed in brief below.

- **Co-precipitation method:** Co-precipitation is the most efficient and straightforward wet chemical method for producing iron-based nanocomposites, and it typically begins with the synthesis of Fe2+ and Fe3+ salt species in alkali solution. The precipitation of Fe^{2+}/Fe^{3+} salt is pH-dependent

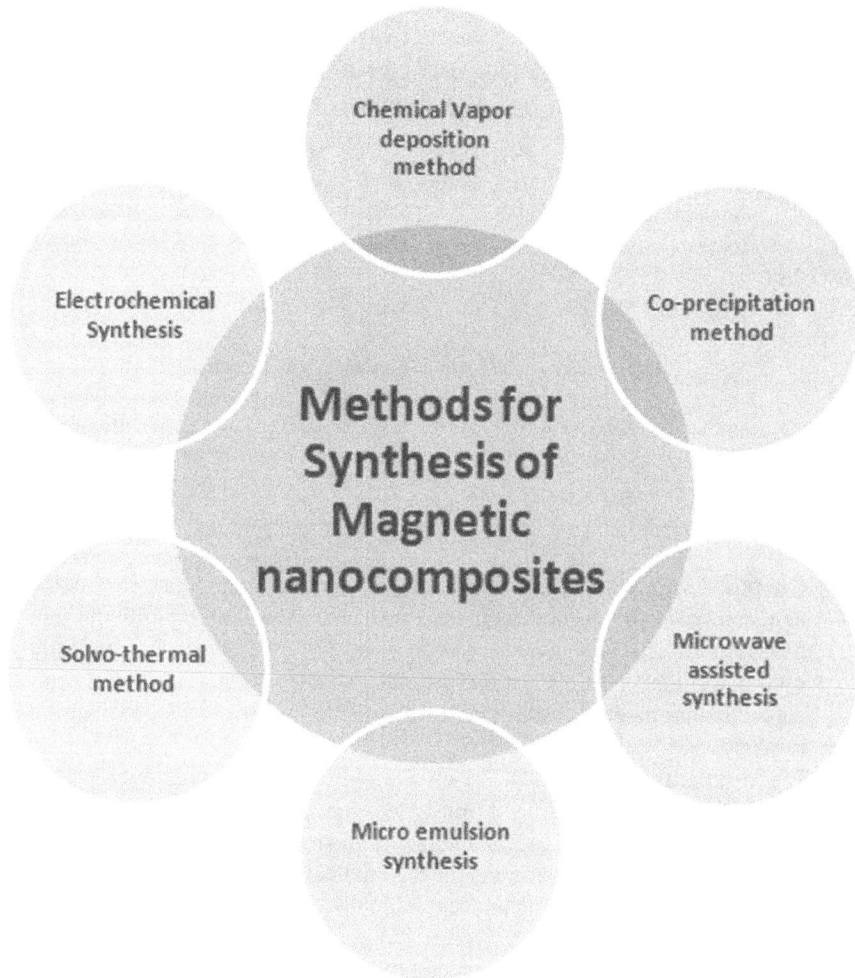

FIGURE 7.1 Methods for synthesis of magnetic nanocomposites.

and normally occurs at a pH between 8 and 14 with a 2:1 $Fe3^1/F^{2+}$ ratio under non-oxidizing circumstances. The equations below depict the reaction for the wet chemical production of magnetite nanoparticles [13, 14]. A list of some commonly used magnetic sources for producing iron-based nanocomposites by the co-precipitation method is shown in Table 7.2.

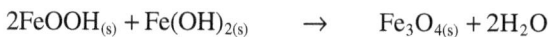

$$Fe^{3+} + 3OH^- \rightarrow Fe(OH)_{3(s)}$$

$$Fe(OH)_3(s) \rightarrow FeOOH(s) + H_2O$$

$$Fe^{2+} + 2OH^- \rightarrow Fe(OH)_{2(s)}$$

$$2FeOOH_{(s)} + Fe(OH)_{2(s)} \rightarrow Fe_3O_{4(s)} + 2H_2O$$

TABLE 7.2

Some Commonly Used Magnetic Sources for Producing Iron-Based Nanocomposites by the Co-precipitation Method

Magnetic Source	Modifier	Type of Iron Oxide	Reaction Temperature (°C)	Solvent	Magnetic Saturation Value	Reaction Time	Application
$FeCl_3$, $FeSO_4.7H_2O$	Zn-Mn	Magnetite	80	Water	74	20 min.	Hyperthermia
$FeCl_3$, $FeSO_4.7H_2O$	Eggshell	Magnetite	40	Water	14.46	2 hours	Cr Removal
$FeCl_3$	$SnCl_2.2H_2O$	Magnetite	100	Ethylene glycol	49.7	30 min.	-
$FeCl_3.6H_2O$	$ZnCl_2$	Ferrite	80	Water	12	40-60 min.	Hyperthermia
$FeCl_2.4H_2O$, $FeCl_3.6H_2O$	Cellulose	Hematite	30	Water	13.2	2.5 hour	AS removal

- **Chemical vapor deposition method:** This process can create iron oxide nanocomposites with good solid-state performance and high purity through one-dimensional nanomaterial production. This technique is frequently used in a chamber with a reactive gas that can deposit on a substrate to create a coating material,. Additionally, this technique offers a lot of material production versatility [15, 16].

- **Electrochemical synthesis:** The electrochemical method has been widely employed for the production of iron-based nanocomposites including different phases of iron oxide, such as magnetite and maghemite. Electrochemical synthesis is the process of transmitting an electric current between two or more electrodes, known as anodes and cathodes, on the electrolyte solution medium. This process involves oxidizing the anode metal to create its ion species, which is then reduced to metal by the cathode while being stabilized. When compared to chemical synthesis methods, this method can create a product with a higher percent yield [13, 17].

- **Solvo-thermal method:** Solvo-thermal synthesis is one of the extensively utilized methods for producing magnetite minerals. With the exception of the use of an organic solvent in place of water as the solvent, this technique of synthesis is nearly comparable to the hydrothermal approach. The solvo-thermal production of the most common magnetic material is carried out by hydrolyzing and oxidizing iron salt solution as a precursor or neutralizing mixed metal hydroxides in aqueous media, which can result in the development of ferrite. The solvo-thermal approach is also known as the alcohol-thermal method and the glycol thermal method when alcohol and glycerol are utilized as the reaction's solvents. Through the dissolution and dispersion of iron salts in the product under pressure and temperature conditions, the product crystallization process is aided. This is the one-step supercritical synthesis of magnetic iron nanoparticles. Once the iron salt aqueous solution reaches its subcritical temperature, the low dielectric constant will instantly drive the nanoparticles to precipitate, causing instantaneous nucleation [18, 19].

- **Micro emulsion synthesis:** With the help of surfactants (such as Tween 80, sodium dodecyl sulphate, brij, etc.), two phases that do not dissolve each other combine to form a persistent colloidal suspension known as a micro emulsion. The micro emulsion solvent may be a water/oil or oil/water system. Magnetic nanoparticles precursors, with sizes ranging from 1 to 100 nm, are dispersed in aqueous phase. The initial stage of the micro emulsion process is the micelle system, which is a nano-reactor made of water droplets containing magnetic nanoparticles and encased by surfactant molecules. Micelle layers control the nucleation, development, and agglomeration of magnetic nanoparticles. The subsequent addition of the second emulsion causes the precipitation of magnetic nanoparticles. A few advantages of the micro emulsion synthesis process include simple preparation, effective nucleation control, prevention of agglomeration, low viscosity, low energy consumption, thermodynamic stability, and the reversibility of micro emulsion production [8, 20].

- **Microwave assisted synthesis:** In order to speed up the reaction and effectively control the creation of magnetic nanoparticles, microwave-aided heating techniques have been used widely in the synthesis of multipurpose materials. In this synthesis, ionic conduction and molecular mobility of the precursor fluids and reducing agents cause electromagnetic radiation to arise. With a quick reaction time, the sample irradiation process exerts temperatures between 100 and 2000°C [21, 22].

7.4   ROLE OF MAGNETIC PARTICLES IN BIOMEDICAL APPLICATIONS

Iron-based magnetic nanoparticles demonstrate particularly good biosafety. There is growing evidence that applying magnetic nanoparticles alone or in combination with a sufficient magnetic field can have significant positive impacts on healing living things. Magnetic nanoparticles research and applications in cancer treatment, tissue engineering, bioseparation, and bioimaging have advanced dramatically. The diagnosis of human illnesses and the assessment of medication efficacy have both benefited considerably from the use of magnetic nanoparticles as contrast agents. In addition to creating magnetic scaffolds, magnetic nanoparticles have also gained the attention of researchers as biomaterials for repair and regeneration due to their magneto biology effect [23].

Magnetic materials find application in a wide range of areas in the biomedical field. Some of them include cell fate research, bioinspired fabrication, magnetic hyperthermia, and magnetic resonance imaging. These are briefly discussed below.

Biomolecules and biopolymers offer specialized binding, targeted delivery, high biocompatibility, biodegradation, and other benefits. Magnetic nanocomposites are expensive, have a low synthesis yield, and may not be suitable for scale manufacturing. They are also difficult to reproduce structurally. The biomedical applications of magnetic particles are shown in Figure 7.2.

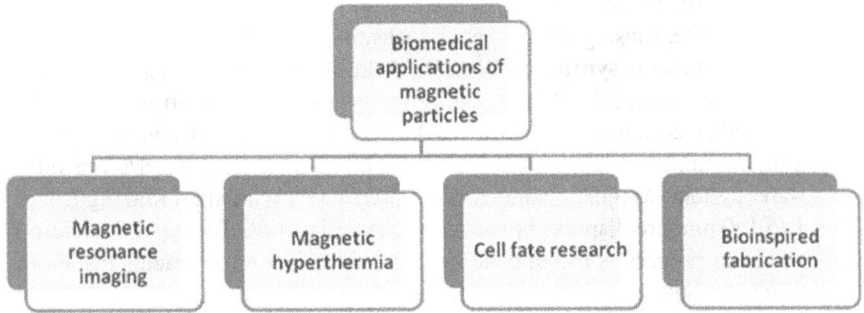

FIGURE 7.2 Biomedical applications of magnetic particles/materials in various fields.

Magnetic Resonance Imaging

Compared to the other imaging techniques like computed tomography scan and X-rays, MRI is a very superior imaging technique because of its high contrast. This technique helps diagnose various tissues and organs such as bones, tissue, and joints. It can accurately analyze and diagnose the abnormalities.

MRI utilizes a magnetic field to align the magnetization of the protons or atomic nuclei. Then a radio wave is applied which changes the aligned nuclei. The energy from the frequency is absorbed through nuclei and excites them to a higher energy level. The excited nuclei then fall back to the lower energy level or the ground state in what is called the relaxation time. For MRI there are two relaxation times: T1 and T2. T1 is the amount of time required for the excited nuclei to return to their previous position. This causes the loss of some energy surrounding the nuclei. T2 is known as the spin-spin relaxation time. In this, an energy exchange occurs between high- and low-energy nuclei without loss of surrounding nuclei. Some other relaxation time or T2, losses the transverse magnetization at faster rate compared to the T2. The relaxation time can be measured by second for biomedical applications.

In MRI, the contrast medium consists of a metal which is functionalized with biocompatible coating. A paramagnetic metal works well because of its unpaired electrons that produce a magnetic field. This causes a decreasing of the relaxation time of surrounding protons and produces an MR image with high contrast.

For MRI, magnetic nanoparticles are studied as one of the greatest contrast agents due to their chemical and physical properties. Magnetic nanoparticles have magnetic properties and also have the ability to target the particular cell through the magnetic field. For biomedical studies MNPs are useful because they have magnetic properties at room temperature. They do not change with variations in physiological conditions like pH, temperature, etc. MNPs are also stable in aqueous environments.

For their biological stability, sometimes biocompatible materials like hydrophilic materials are also conjugated to the magnetic core. For in vivo studies, it is essential to control their movement in blood, evade the reticuloendothelial system (RES), and maintain their toxicity in biological systems with high magnetization properties. Therefore, iron-based magnetic nanoparticles with further coating are used [24].

It was reported that Huang and co-workers demonstrated the synthesis of negatively charged MNPs, which were conjugated with poly(DL-lactic acid-co-malic acid) through covalent bonding of NH2-PEG. The result showed that PLMA-PEG-MNPs were efficient T2 contrast agents with 162.7mM^{-1} s^{-1} high r^2 value. Furthermore, their uptake could be increased by altering the PEG content. It was reported that the MNP uptake was decreased by increasing the PEG surface density [25].

Another study reported the development of water-dispersible superparamagnetic iron oxide nanoparticles. In this technique iron oxide cores were synthesized through precipitation of iron salt in the presence of ammonia and then coated with the b-cyclodextrin and pluronic polymer (F127). This formulation was used for hyperthermia, magnetic resonance imaging, and drug delivery. The water dispersibility allows the encapsulation of the drug in the polymer and provides for a superior hyperthermia property compared to the bare MNP. The nanoparticles' MRI property was observed in agar gel and cisplatin–resistant ovarian cancer cells (A12780CP). This formulation increased the imaging contrast many-fold. A curcumin-containing formulation showed equivalent inhibitory effects on ovarian (A2780CP), breast (MDA-MB-231), and prostate (PC-3) cancer cells in 40% drug release. The improved therapeutic effect was evaluated through western blotting and TEM results [26].

It was reported by Murali and co-workers that synthesized polyethylene glycol (PEG)–coated MNPs, which were used as the drug carrier, have the ability to conjugate with the antibody, as well as the MRI agent. In this study, the iron oxide core was coated with oleic acid and further conjugated with PEG. This formed water-dispersible MNPs with 8 nm of iron oxide core and 184 nm hydrodynamic diameter. In this formulation doxorubicin was loaded, which provided a sustained release drug delivery system. For MRI, T1 and T2 were evaluated in vitro and in vivo circulation was determined in mouse carotid arteries. The MNPs provided enhanced T2 MRI contrast, long blood circulation, and 30% relative concentration 50 min after injection. Using the N-hydroxysuccinimide group of oleic acid and PEG for the conjugation of the amine group on the transferrin antibody provides active targeting of the human MCF-7 breast cancer cell line [27].

MAGNETIC HYPERTHERMIA

MNPs are also used in magnetic hyperthermia–based biomedical applications due to their bioavailability, solubility, permeation, metabolism, and surface properties. MNPs are used in magnetic hyperthermia for cancer treatment. In studies of magnetic hyperthermia, small and deep-seated tumors treated with MNP in the presence of a magnetic field produced localized hyperthermia on cancer cells. In general, the cancer treatment required the exposure of tumor tissue to high temperatures of up to 41–47°C, which damaged the DNA and killed the cancer cells [28]. Magnetic hyperthermia provides targeted and superior cancer therapy. Magnetic hyperthermia takes advantage of the magnetic properties of MNPs in fluid and transforms electromagnetic energy into heat [29].

It was reported that Rabia et al. prepared and investigated the various properties of a ferrite filler and polypyrrole matrix composite. They took six samples of composites that contained polypyrrole and different concentrations of ferrite like 25%,

50%, 75%, and pure ferrite. They reported that the addition of ferrite in polypyrrole decreased the lattice constant, while the magnetization and core activity were increased by increasing ferrite content. It resulted that the composites with 25% and 50% ferrite showed low dielectric loss and may be suitable for hyperthermia in cancer treatment [30].

In another report, Sudip and co-workers synthesized iron oxide, hydroxyapatite, and hydroxyapatite-coated iron oxide through the solvo-thermal and wet chemical precipitation method. These formulations were used for the treatment of cancer through magnetic hyperthermia. They were evaluated through X-ray diffraction, Fourier transform infrared spectroscopy, field emission transmission electron microscopy, and energy dispersive X-ray spectroscopy. The non-toxic property of the formulations was inspected through MTT and trypan blue cytotoxicity assay. Their magnetic saturation at room temperature in the presence of a 15,000 Oe (1.5 T) magnetic field was found to be 83.2 emu/g for iron oxide and 40.6 emu/g for hydroxyapatite-coated iron oxide. The hydroxyapatite-coated iron oxide composite showed excellent hyperthermia effect on MG-63 osteosarcoma cells. Their SAR value was found to be 85 W/g. The in vitro study showed that the hyperthermia temperature was reached within 3 min and within 30 min approximately all Mg-63 osteosarcoma cells had been killed [31].

Hervault and team developed a magnetic nanocomposite with an iron oxide core and pH and thermosensitive polymer: P(DEGMA-co-PEGMA-b-[TMSPMA co-VBA]). This formulation could be used as a drug delivery carrier and hyperthermic agent. The drug and polymer are conjugated through an acid-cleavable linker which also provides targeted drug delivery. In this formulation, the core of iron oxide has a saturation magnetization of 70 emu/g. The conjugation contained 8.1 Wt % of polymer and had good heating properties in different magnetic fields. Drug release studies showed that a small amount of the drug was released at room temperature. But after 48 hr at an acidic pH in a tumor with hyperthermia conditions (50°C), 25.2% of the drug was released [32].

Jingmiao et al. developed Fe_3O_4 chitosan nanoparticles through the cross-linking method. Firstly, oleic acid–modified Fe_3O_4 nanoparticles were synthesized through the co-precipitation method and further chitosan was added through physical adsorption on the surface of the Fe_3O_4. The formulation was prepared through cross-linking of amino groups on the chitosan by using glutaraldehyde. The researchers demonstrated that the prepared formulation had a mean diameter of 10.5 nm with a quasi-spherical structure. The saturation magnetization was found to be 30.7 emu/g and the formulation showed a superparamagnetic property at room temperature. By altering the current magnetic field the inductive heating effect was significantly changed. The overall research work demonstrated that the formulation has potential in hyperthermia [33].

G. R. Iglesias and co-workers developed multi-functional nancomposites devised for the treatment of cancer in vitro by delivering the drug gemcitabine and producing magnetic hyperthermia. In this magnetite, particles were loaded onto a polymer shell of polyethylene glycol. Adsorption of the drug on nanoparticles was confirmed through IR spectroscopy and electrophoresis. The drug released was studied through heating the surroundings via magnetic hyperthermia. The drug released showed first-order release kinetics, and it could be faster when the studies are carried out on the thermostatted bath at 43°C instead of 37°C. The major finding of this investigation was that at 43°C the MNPs retained their hyperthermia response with high heating power and they released the drug at a faster rate at zero-order kinetic [34].

BIOINSPIRED FABRICATION

Bioinspired nanocomposites are used specially for the target site. The system contains natural polymers or biopolymers which are further conjugated with active molecules to improve their efficacy. It is a combination of two or more componenta which has particular properties. Therefore, the composite which has a biopolymer is referred as a bioinspired nanocomposite [35]. Magnetic nanoparticles are gaining much attention for biomedical applications. They can be widely used as the contrast agent for improving MRI and clinical diagnosis. They are also used as drug carriers for the delivery of a drug to a particular location and as magnetic hyperthermia agents for the treatment of cancer. Several researchers reported that Fe_3O_4 magnetic nanoparticles enhanced their in vitro activity via bio coating. In magnetic nanoparticles various biopolymers are used for conjugation. Hence the magnetic polymer composite helps in the fabrication of bioinspired formulations. Magnetic nanoparticles conjugated with bioactive compounds are used for various biomedical applications. It was reported that MNPs simply conjugated with organic or inorganic materials were used in the process of bone repair [36]. It was also reported that various types of biochemical stimulation like physical, mechanical, magnetic and electrical have been used for bone tissue repair or regeneration. Thus, several research works were done using this stimulation or combination of stimulations. Recently magnetic nanoparticles with polymer conjugation were widely used for the development of bioinspired fabrication for bone tissue repair [37]. It was demonstrated that MNPs mixed with chitosan and silk formed silk/fibroin/chitosan magnetite scaffolds via the freeze casting method. The results showed that their degradation decreased in phosphate buffer and influenced the cellular nature. In another report microporous ferrogel was incorporated into Fe_3O_4 and formed conjugation, which promoted the adhesion and cell viability of resident cells and which is controlled by the magnetic field. For bioinspired fabrication various natural polymers such as collagen, chitosan, and hydroxyapatite are used with magnetic nanoparticles. The composites have good biocompatibility and biological properties that help in biomedical applications. Magnetic responsive materials have great application in biomedicine. Particular superparamagnetic nanoparticles are used as magnetic systems for bone tissue engineering due to their biocompatibility and stability [36].

It was reported that Fe_3O_4 nanoparticles coated with hydroxyapatite were used for bioinspired fabrication. In this study, an appropriate amount of precursor solution of Fe_3O_4 nanoparticles and hydroxyapatite was added to the chitosan/collagen organic matrix and conjugation scaffolds were prepared through in situ technology. This formulation was evaluated for its magnetic properties and its microstructure. Researchers also studied the effect of Fe_3O_4 on the scaffold properties; osteoblast and osteogenic properties were investigated. Bone generation and biocompatibility were studied in vitro and in vivo. Hence this scaffold has the potential for regeneration of bone tissue [36].

Margarida Fernandes et al. developed magneto active 3D porous scaffolds through magneto strictive particles ($CoFe_2O_4$) and piezoelectricpolymer (poly vinylidene fluoride) by using the solvent casting method via overlapping of a nylon template with three different diameter fibers (60, 80, and 120 µm) and formed 3D scaffolds with different pore sizes. The formulation has 5–20 µm pore size similar structures with trabecular bone. The formulation was suitable for the proliferation of osteoblasts through magnetic stimuli [37].

CELL FATE RESEARCH

The cell is the integral part of the organism in which the cell membrane plays a crucial role in regulation of cell cycle, migration, differentiation, and proliferation. Its action potential helps in excitability of the neurons and cardiomyocytes. For this, electro-active biomaterials gain much attention because they not only support cell adhesion but also modulate or regulate the cell and its function, particularly for electrically excitable cells and tissue. Recently these materials have been used in tissue generation, research, and modulation of cell fate [38].

Magnetic nanoparticles are the nanoparticles most widely used in the biomedical field because of their biocompatibility, surface functionalization, and magnetism property. Recently biomaterials or polymers such as hydroxyapatite, calcium phosphate, and biopolymer have been widely used for the functionalized magnetic nanoparticles. They are then used for bone cell growth, differentiation, and proliferation. Magnetic nanoparticles activate the mechanical transduction receptor, mechano-sensitive signaling pathway, and signal transduction receptor [38, 39]. In magnetic nanoparticles, the magnetic field can be used for the transportation of biological molecules. Magnetic nanoparticles have the ability to use mechanical force for movement within cells. Therefore, they are used in controlling cell fate and cell signaling at the molecular level [40]. Magnetic nanoparticles in combination with the particular ligand can easily modulate cellular function or determine cells' fate in biological system by controlling them through external magnetic field [39, 41].

It was reported that iron oxide magnetic nanoparticles were used for the investigation of bone cells' behavior. Research shows that, in adjacent cells when the amount of nanoparticles was increased, osteogenic cell growth was reduced. Nanoparticles have various effects on cells and their fate when they are added to magnetic scaffolds. Cellular fate is controlled by several factors in nanoparticles like stiffness, chemical structure of the surface, wettability, hydrophilicity, matrix interaction, and micro environment. It was reported that the incorporation of magnetic nanoparticles into a polyurethane matrix changed their crystallinity, surface roughness, hydrophilicity, and cell fate. The hydrolytic degradation of polyurethane is based on the hydrolysis of ester and urethane group [39].

Organic and inorganic polyurethane nanocomposites with Fe_2O_3 nanoparticles were developed for understanding cell signaling and the effect of magnetite nanoparticles on cell proliferation and cell responses. They were used for the determination of cells' behavior and their communication with stem cells and the microenvironment. The properties of the formulation were evaluated through FTIR, AFM, SEM, and electrochemical impedance spectroscopy. It was also reported that the presence of Fe_2O_3 affects the morphology and mechanical, electrochemical, and biological properties of polyurethane nanocomposite. Their hydrophilicity and conductivity were increased by addition of magnetic nanoparticles, as were their water absorption, biodegradation, and cell viability. Biocompatibility of the formulation was evaluated through MTT assay and cell staining [39].

7.5 CONCLUSION

Magnetic nanoparticles have great potential in the biomedical field. However, magnetic nanoparticles show some toxicity, low magnetization, and aggregation in the biological systems; these are the major challenges for the uses of magnetic nanoparticles in bioimaging applications. Recently magnetic polymer composites have been widely used in bioimaging applications in place of magnetic nanoparticles, due to low toxicity, biocompatibility, and high magnetization. They also provide large-scale synthesis and surface functionalization. They are a versatile platform which can be easily functionalized for different applications with respect to external magnetic fields. They can be widely used in several bioimaging applications such as MRI, magnetic hyperthermia, bioinspired fabrication, and cell fate research. This chapter is based on the magnetic polymer composite and its bioimaging applications.

ACKNOWLEDGMENTS

The authors are grateful to the Department of Science and Technology (DST-FIST) Letter no. SR/FST/COLLEGE/2018/418, New Delhi, for providing financial assistance.

REFERENCES

1. H.S. Lahoti, S.D. Jogdand, Bioimaging: Evolution, significance and deficit, Cureus, 2022, 14(9), 1–8.
2. M.L. James, S.S. Gambhir, A molecular imaging primer: Modalities, imaging agents, and applications, Physiol Rev, 2012, 92, 897–965.
3. M. Moseley, G. Donnan, Multimodality imaging: Introduction, Stroke, 2004, 35, 2632–2634.
4. R. Weissleder, M. Nahrendorf, Advancing biomedical imaging, Proc Natl Acad Sci U S A, 2015, 112,14424–14428.
5. J. Prasad Rao, P. Gruenberg, K.E. Geckeler, Magnetic zero-valent metal polymer nanoparticles: Current trends, scope, and perspectives, Prog Polym Sci, 2015, 40,138–147.
6. I. Bloom, L.K. Walker, J.K. Basco, T. Malkow, A. Saturnio, G. De Marco, G. Tsotridis, A comparison of fuel cell testing protocols – a case study: Protocols used by the U.S. Department of Energy, European Union, International Electrotechnical Commission/Fuel Cell Testing and Standardization Network, and Fuel Cell Technical Team, J Power Sources, 2013, 243, 451–457.
7. Y. Deng, L. Wang, W. Yang, S. Fu, A. Elaïssari, Preparation of magnetic polymeric particles via inverse microemulsion polymerization process, J Magn Magn Mater, 2003, 257, 69–78.
8. Y. Mori, H. Kawaguchi, Impact of initiators in preparing magnetic polymer particles by miniemulsion polymerization, Colloids Surf B, 2007, 56, 246–254.

9. J. Choi, S. Han, H. Kim, E.-H. Sohn, H.J. Choi, Y. Seo, Suspensions of hollow polydivi-nylbenzene nanoparticles decorated with Fe_3O_4 nanoparticles as magnetorheological fluids for microfluidics applications, ACS Appl Nano Mater, 2019, 2, 6939–6947.

10. M. Mrlik, M. Sedlacik, V. Pavlinek, P. Bazant, P. Saha, P. Peer, P. Filip, Synthesis and magnetorheological characteristics of ribbon-like, polypyrrole-coated carbonyl iron suspensions under oscillatory shear, J Appl Polym Sci, 2013, 128, 2977–2982.

11. Q. Lu, K. Choi, J.D. Nam, H.J. Choi, Magnetic polymer composite particles: Design and magnetorheology, Polymers, 2021, 13, 512.

12. K. Cerdan, C. Moya, P.V. Puyvelde, G. Bruylands, J. Brancart, Magnetic self healing composites: Synthesis and applications, Molecules, 2022, 27, 3796.

13. I. Fatimah, G. Fadillah, S.P. Yudha, Synthesis of iron based nanocomposites: A review, Arab J Chem, 2021, 14, 103301.

14. A.V. Rane, K. Kanny, V.K. Abitha, S. Thomas, Methods for synthesis of nanoparticles and fabrication of nanocomposites, In: Nanoparticles in Drug Delivery, Imaging, and Sensing. 2018: 121–139.

15. J.O. Carllson, P.M. Martin, Chemical vapour deposition; Handbook of deposition technologies for films and coatings; William Andrew Publishing, 3rd edition; 2010: 314–363.

16. S. Majidi, F. Zeinali Sehrig, S.M. Farkhani, M. Soleymani Goloujeh, A. Akbarzadeh, Current methods for synthesis of magnetic nanoparticles, Artif Cells Nanomed Biotechnol, 2016, 44, 722–734.

17. X. Ye, L. Chen, L. Liu, Y. Bai, Electrochemical synthesis of selenium nanoparticles and formation of sea urchin like selenium nanoparticles by electrostatic assembly, Mater Lett, 2017, 196, 381–384.

18. J. Li, Q. Wu, j. Wu, Synthesis of Nanoparticles via Solvothermal and Hydrothermal Methods. Springer International Publishing; 2016:295–328.

19. S.H. Feng, G.H. Li, Hydrothermal and Solvothermal Synthesis, Modern Inorganic Synthetic Chemistry, 2nd Edition. Elsevier; 2017:73–104.

20. K. Qiao, W. Tian, J. Bai, L. Wang, J. Zhao, Z. Du, X. Gong, Application of mag-netic adsorbents based on iron oxide nanoparticles for oil spill remediation: A review, J Taiwan Inst Chem Eng, 2019, 97, 227–236.

21. D.C. Onwudive, Microwave-assisted synthesis of PbS nanostructures, Heliyon, 2019, 5(3), e01413.

22. K.K. Kefeni, T.A.M. Msagati, B.B. Mamba, Ferrite nanoparticles: Synthesis, charac-terisation and applications in electronic device, Mater Sci Eng, 2017, 215, 37–55.

23. K. Li, J. Xu, P. Li, Y. Fan, A review of magnetic ordered materials in biomedical field: Construction, application and prospects, Compos Part B Eng, 2022, 228, 109401.

24. Charlette Felton, Alokita Karmakar, Yashraj Gartia, Punnamchandar Ramidi, Alexandru S. Biris, Anindya Ghosh, Magnetic nanoparticles as contrast agents in bio-medical imaging: Recent advances in iron-and manganese-based magnetic nanopar-ticles, Drug Metab Rev, 2014, 46(2), 142–154.

25. Hira Fatima, Kyo-Seon Kim, Iron-based magnetic nanoparticles for magnetic reso-nance imaging, Adv Powder Technol, 2018, 29(11), 2678–2685.

26. Murali M. Yallapu, Shadi F. Othman, Evan T. Curtis, Brij K. Gupta, Meena Jaggi, Subhash C. Chauhan, Multi-functional magnetic nanoparticles for magnetic resonance imaging and cancer therapy, Biomaterials, 2011, 32, 1890–1905.

27. Murali Mohan Yallapu, Susan P. Foy, Tapan K. Jain, Vinod Labhasetwar, PEG-functionalized magnetic nanoparticles for drug delivery and magnetic resonance imag-ing applications, Pharm Res, 2010, 27, 2283–2295. doi: 10.1007/s11095-010-0260-1.

28. Atta Ullah Khan, Lan Chen, Guanglu Ge, Recent development for biomedical applica-tions of magnetic nanoparticles, Inorg Chem Commun, 2021, 134, 1–25. doi: 10.1016/j.inoche.2021.108995

29. Vanessa Fernandes Cardoso, António Francesko, Clarisse Ribeiro, Manuel Bañobre-López, Pedro Martins, Senentxu Lanceros-Mendez, Advances in magnetic nanoparticles for biomedical applications, Adv Healthcare Mater, 2017, 7(5), 1–35.

30. Rabia Qindeel, Norah H. Alonizan, Leda G. Bousiakou, Magnetic behavior of ferrite-polymer composites for hyperthermia applications, J Mater Sci: Mater Electron, 2020, 31, 19672–19679.

31. Sudip Mondal, Panchanathan Manivasagan, Subramaniyan Bharathiraja, Madhappan Santha Moorthy, Van Tu Nguyen, Hye Hyun Kim, Seung Yun Nam, Kang Dae Lee, Junghwan Oh, Hydroxyapatite coated iron oxide nanoparticles: A promising nanomaterial for magnetic hyperthermia cancer treatment, Nanomaterials, 2017, 7, 426. doi: 10.3390/nano7120426

32. A. Hervault, N. T. K. Thanh, S. Maenosono, M. Lim, C. Boyer, A. Dunn, D. M. Mott, Doxorubicin loaded dual pH- and thermo-responsive magnetic nanocarrier for combined magnetic hyperthermia and targeted controlled drug delivery applications, Nanoscale, 2016, 8, 12152–12161. doi: 10.1039/C5NR07773G

33. Jingmiao Qu, Guang Liu, Yiming Wang, Ruoyu Hong, Preparation of Fe_3O_4–chitosan nanoparticles used for hyperthermia, Adv Powder Technol, 2010, 21, 461–467.

34. G. R. Iglesias, Felisa Reyes-Ortega, B. L. Checa Fernandez, Ángel V. Delgado, Hyperthermia-triggered gemcitabine release from polymer-coated magnetite nanoparticles, Polymers, 2018, 10, 269. doi: 10.3390/polym10030269

35. Supriya Mishra, Shrestha Sharma, Md. Noushad Javed, Faheem Hyder Pottoo, Md. Abul Barkat, Harshita, Md. Sabir Alam, Md. Amir, Md. Sarafroz, Bioinspired nanocomposites: Applications in disease diagnosis and treatment, Pharm Nanotechnol, 2019, 7, 1–15.

36. Yao Zhao, Tiantang Fan, Jingdi Chen, Xin JiacanSu, Panpan Zhi, Lin Pan, Qiqing Zou, Zhang, Magnetic bioinspired micro/nanostructured composite scaffold for bone regeneration, Colloids Surf B Biointerfaces, 2018, 174, 70–79. doi: 10.1016/j.colsurfb.2018.11.003

37. Margarida Fernandes, Daniela M. Correia, Clarisse Ribeiro, Nelson Castro, Vitor Correia, Senentxu Lanceros-Mendez, Bioinspired three-dimensional magneto active scaffolds for bone tissue engineering, ACS Appl Mater Interfaces, 2019, 11(48), 45265–45275. doi: 10.1021/acsami.9b14001

38. Alberto Pardo, Manuel Gomez-Florit, Silvia Barbosa, Pablo Taboada, A. Domingues, Manuela E. Gomes, Magnetic nanocomposite hydrogels for tissue engineering: Design concepts and remote actuation strategies to control cell fate, ACS Nano, 2021, 15, 175–209. https://dx.doi.org/10.1021/acsnano.0c08253

39. Mohsen Shahrousvand, Monireh Sadat Hoseinian, Marzieh Ghollasi, Ali Karbalaeimahdi, Ali Salimi, Fatemeh Ahmadi Tabar, Flexible magnetic polyurethane/Fe_2O_3 nanoparticles as organic-inorganic nanocomposites for biomedical applications: Properties and cell behavior, Mater Sci Eng C, 2016, 74, 556–567. doi: 10.1016/j.msec.2016.12.117

40. Jiri Kudr, Yazan Haddad, Lukas Richtera, Vojtech Adam, Ondrej Zitka, Zbynek Heger, Mirko Cernak, Magnetic nanoparticles: From design and synthesis to real world applications, Nanomaterials, 2017, 7, 243. doi: 10.3390/nano7090243

41. Zhirong Liu, Xingyi Wan, Zhong Lin Wang, Linlin Li, Electroactive biomaterials and systems for cell fate determination and tissue regeneration: Design and applications, Adv Mater, 2021, 33(32), e2007429.

8 Magnetic Hydrogel and Its Biomedical Applications

Sanjay Kumar Gupta and Astha Verma

8.1 MAGNETIC HYDROGEL: INTRODUCTION

Despite the fact that the term "hydrogel" was first used in literature in 1894 (Van Bemmelen, 1894), the hydrogels described at the time were actually a type of colloidal gel produced from inorganic salts. Later, the term "hydrogel" was used to describe a 3D network of naturally occurring hydrophilic polymers and gums combined by physical or chemical cross-linking techniques. The use of this term was strongly dependent on the availability of water in the environment (Lee et al., 2013; Liu et al., 2020). The high water content, porosity, and ease with which their properties (mechanical, chemical, microstructure, etc.) can be changed make hydrogels ideal for biomedical applications. Hydrogels are the materials that are closest in appearance to the extracellular matrix of mammals (González-Díaz and Varghese, 2016).

Generation of the current version of hydrogel in the biological field was first performed by Liu et al. Magnetic hydrogel was fabricated by Wichterle and Lím (1960), indicating that glycol-dimethacrylate–based hydrophilic gels exhibited adjustable mechanical properties and water content. From then on, more and more hydrogels have been developed, and the smart hydrogels were then introduced in different fields of biological science, such as drug delivery, bioseparation, biosensors, and tissue engineering. Smart hydrogels are so described because they respond directly to changes in environmental conditions (Wichterle and Lím, 1960), and numerous studies of smart hydrogels in the applications of nanotechnology, drug delivery, and tissue engineering have been carried out in the last few decades (Li X. et al., 2019; Li Z. et al., 2019; Zhang Y. et al., 2019).

One type of smart hydrogel, magnetically responsive hydrogel, has recently been used in biomedical applications to enhance the biological activities of cells, tissues, or organs (Abdeen et al., 2016). This is primarily due to the magnetic responsiveness of the system to the external magnetic field and the development of functional structures for remote control of the mechanical, biochemical, and physical characteristics of the surroundings around the tissues, cells, or organs (Wang et al., 2009). In biomedical applications like tissue engineering, drug delivery, biosensors, and cancer therapy, magnetic hydrogels are particularly promising due to their distinctive properties. The development of smart hydrogels that respond to environmental factors like temperature, pH, electric field, specific analytes, and enzymes has received a lot of attention over the past 20 years in an effort to find hydrogels

DOI: 10.1201/9781003454236-8

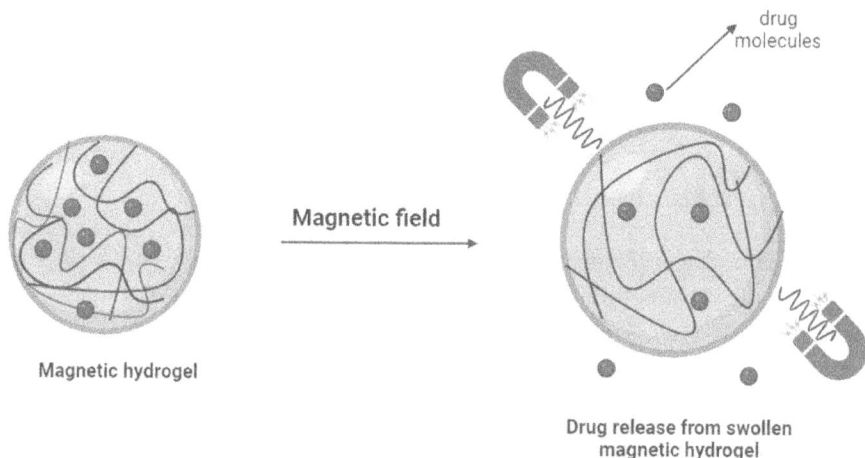

drug
molecules

Magnetic field

Magnetic hydrogel

Drug release from swollen
magnetic hydrogel

FIGURE 8.1 Schematic representation of drug release from magnetic-responsive hydrogel under the influence of an external magnetic field.

with better environmental-specific controllability, release kinetics, actuation, and response qualities. A important type of stimuli-responsive hydrogels are magnetic hydrogels, sometimes referred to as ferrogels (a hydrogel mixture containing magnetic micro- and/or nanoparticles), which are capable of changing their mechanical behavior and microstructure in response to an electromagnetic field from the outside (Figure 8.1) (Ebara et al., 2014).

Magnetic hydrogels (or ferrogels) are recognized as smart materials because they can respond to the surrounding magnetic field (da Silva Fernandes et al., 2021). Typically, magnetic hydrogels are formed of a matrix hydrogel and a magnetic substances that was embedded into the matrix (Liu et al., 2020). Recent research suggests that magnetic hydrogel may provide a great drug delivery and targeting method. For instance, Gao et al. (2019) constructed a magnetic hydrogel using ferromagnetic vortex-domain iron oxide, and they hypothesized that this distinctive magnetic hydrogel might considerably decrease the recurrence of local breast tumors.

The development of magnetic responsive poly(vinyl alcohol) (PVA) hydrogels by Manjua et al. (2019) suggests prospective use for tissue engineering, drug administration, or a biosensor system. These hydrogels may be motivated by an ON/OFF magnetic field and non-invasively controlled protein sorption and motility.

In contrast to magnetic hydrogels, a variety of smart biomaterials, such as scaffolds, biofilms, and other smart hydrogels, are triggered by external stimuli which involve light, pH, glucose, temperature, stress, or charge and have lots of potential for biomedical applications (Amani et al. 2019; Chen H. et al., 2019; Cui et al., 2019; Zhao et al., 2019; Yang et al., 2020; Ali et al., 2023). However, these stimuli-responsive competent polymers' two primary disadvantages are their prolonged response times and less precisely regulated structures.

8.2   PROPERTIES OF MAGNETIC HYDROGELS

Due to the intricacy of their molecular makeup at the nano- and micrometric dimensions, magnetic hydrogels are one of the most adaptable magnetic colloids because their macroscopic properties, especially the mechanical features, can be altered in a variety of ways. These systems can be developed by adjusting the following parameters (Abrougui Mariem et al., 2019):

  i. the size, shape, component, and concentration of the magnetic particles;
  ii. the polymer chains' length, chemistry, composition, and concentration;
  iii. the degree of cross-linking in the polymer network;
  iv. the density of the bonds between the polymer chains and the magnetic particles, if any are present; and
  v. lastly, the degree of hydration that can attain higher than 90% v/v.

The majority of the magnetic-responsive hydrogels described in the literature feature a spherical shape, being made of ferrites (magnetite) or ferromagnetic metals (iron). The high density of the particles in these situations may facilitate gravitational settling during the synthesis of the magnetic hydrogels, which leads to a poor homogeneity of the final gel. Core-shell composites (such as iron-silica, magnetite-polymer) provide an alternative to facing the above challenge (Bonhome-Espinosa et al., 2017; Gila-Vilchez et al., 2018; Abrougui Mariem et al., 2019).

Baron et al. (2021) developed a cost-effective magnetic hydrogel with a blending technique using biocompatible and biodegradable materials. Without the use of any cross-linking agents, the magnetic hydrogel was created utilizing a double layer of oleic acid-coated magnetite and PVA, a renewable water-soluble cellulose derivative.

Conventional hydrogels have an array of benefits, yet their use is frequently constrained by their inadequate controllability, responsiveness, and actuation properties. Traditional hydrogels, for instance, are unable to control release merely via passive mechanisms, such as molecular diffusion and matrix breakdown, when used as therapeutic carriers, and that results in poor safety and low efficacy. Therefore, there is a rising focus on the creation of hydrogels with personalized, controllable, and responsive physicochemical properties when required (Maitra and Shukla, 2014; Gaharwar et al., 2014).

8.3   METHODS OF MAGNETIC HYDROGELS PREPARATION

Magnetic hydrogels are commonly fabricated by the interaction among magnetic ingredients (γ-Fe_2O_3, Fe_3O_4, etc.) and hydrogel polymer matrix through covalent or non-covalent bonds. A number of preparation methods of magnetic hydrogels have been reported (Table 8.1), primarily categorized as follows:

  i. Blending method;
  ii. Co-precipitation method;

iii. Cross-linking method;
iv. Incorporation method; and
v. Other methods.

There are differences in raw material selection, design strategies, and application fields of magnetic hydrogels with specific performance.

8.3.1 BLENDING METHOD

The most commonly used method by far for preparing magnetic hydrogels is to blend an MNP suspension with hydrogel precursor solution and then gelatinize under certain conditions (Li et al., 2021). Fe_3O_4 is the magnetic particle most frequently blended with a polymer system to formulate a composite hydrogel (Ganguly and Margel, 2021). The blending method has several advantages in the preparation of magnetic hydrogels (Table 8.1). Firstly, magnetic particles with homogeneous size in hydrogels can be obtained by modifying the stirring speed, the concentration of the substances, and the fabrication period. Secondly, the preparation process is easy to perform since the preparation and cross-linking of magnetic particles are conducted separately (Figure 8.2). It has been reported that maghemite was added to magnetic beads made of alginate to create a hydrogel that responds to magnetic fields (Brulé et al., 2011). Another study described the use of magnetite MNPs mixed with methacrylate-functionalized hydrogel to cure hyperthermia (Derakhshankhah et al., 2021).

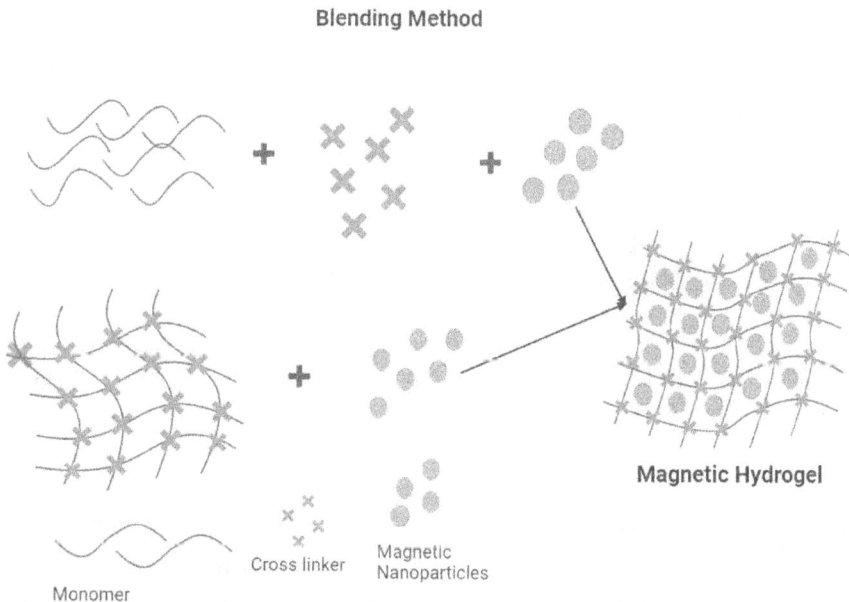

FIGURE 8.2   Representation of blending method to create magnetic hydrogels.

In addition, MNPs can be easily combined with other functional fillers to exert a synergistic effect. For instance, MNPs coated with gold shells can achieve magnetic/photothermal dual functionality (Rittikulsittichai et al., 2016). Moreover, the approach is simple to operate and can be widely applied on many occasions. However, the incorporated MNPs are liable to aggregate and achieving a homogeneous MNP distribution within the hydrogels is challenging.

8.3.2 In Situ Co-precipitation Method

In situ MNP-based composite hydrogel synthesis is more accurate in terms of hydrogel architecture and particle size (Liu et al., 2021). In this approach, MNPs are made inside the gel matrices while the gelation is happening. The hydrogel precursor materials, particularly the monomers, are initially mixed with the precursor metal ions of MNPs (Jahanban-Esfahlan et al., 2021). The co-precipitation method is a convenient and facile way to prepare ferrite nanoparticles from Fe^{3+}/Fe^{2+} aqueous solutions by the addition of a precipitating agent (NaOH or $NH_3·H_2O$) at room or elevated temperature and under an inert atmosphere. Therefore, the magnetic Fe_3O_4 nanoparticles can be easily in situ synthesized by the coprecipitation method taking the preformed hydrogel matrix as a chemical reactor based on the swelling characteristic of hydrogel.

8.3.3 Cross-Linking Method

The cross-linking method is similar to the blending method in preparation process, where the MNP synthesis is separate from the hydrogel formation (Figure 8.3). While the difference between the two methods is that in this method the functionalized MNPs serve as the only or one of the cross-linkers to supply the cross-linking sites via covalently coupling with the polymers. Therefore, the MNPs need to be modified with functional groups before blending with the hydrogel precursor solution (Hu et al., 2019).

Helminger et al. (2014), prepared thermoreversible gelatin–based magnetic hydrogels in water using cross-linking techniques. There was a three-step procedure used. To enable variable mesh sizes in the gelatin gels by concentration-dependent modulation of the cross-linking degree, gelatin hydrogels were initially made at various biopolymer concentrations, ranging from 6 to 18 wt%. To reach the swelling equilibrium, these hydrogels were submerged in a solution of Fe II (0.1 mol L^{-1}) and Fe III (0.2 mol L^{-1}) ions with a molar ratio of ferrous to ferric ions of 1:2. After dipping the gel into a NaOH (0.1 mol L^{-1}) solution, magnetite was produced inside the gelatin network in the third stage.

8.3.4 Incorporation Method

The ways of incorporating MNPs in magnetic micro-/nanogel are nothing more than the three methods presented above. Nevertheless, some auxiliary measures need to be taken to assist the micro-/nanoscale magnetic hydrogel formation. At present, the inverse emulsion polymerization technique is most commonly used for magnetic micro-/nanogel fabrication due to its simplicity and low cost (Rodkate and Rutnakornpituk, 2016).

Cross linking Method

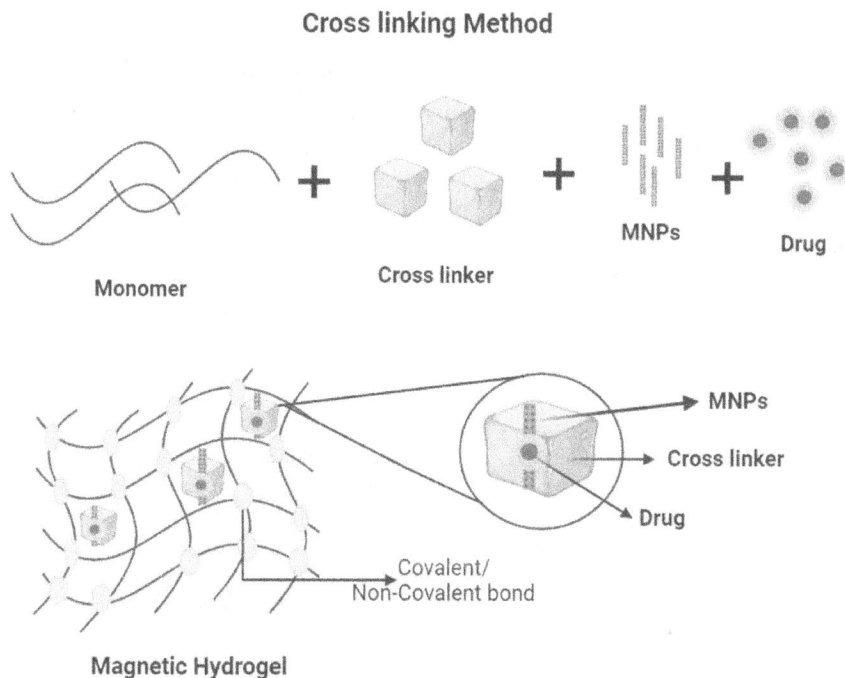

FIGURE 8.3 Representation of preparation of magnetic hydrogel using the cross-linking method.

8.3.5 OTHER METHODS

Lin et al. (2019) reported a simple one-pot method for the synthesis of magnetic β-cyclodextrin (β-CD)/cellulose hydrogel beads, which exhibited rapid swelling–deswelling properties under an external magnetic field (EMF) to remotely control drug release from passive release to stepwise release. The grafted β-CD endows the hydrogel with high drug loading capacity and, simultaneously, the incorporation of Fe_3O_4 nanoparticles provides the force for stepwise drug release through EMF-induced rapid and reversible deformation of the 3D network (Table 8.1). Mohamed et al. (2020), synthesized gum acacia/polyacrylamide (GA/PAAm) magnetic hydrogel using gamma irradiation.

Manish et al. (2022) utilized a novel strategy to create hard magnetic soft hydrogel with a greater water content. Strontium hexaferrite (SFO) particles were incorporated in cross-linked gelatin and water macromolecules that made up the hard magnetic soft hydrogel. First, the auto-combustion process was used to create SFO nanoparticles. The produced particles were then mixed with three different weight percentages of hydrogels. To investigate the mechanical and actuation characteristics, respectively, the hydrogels were loaded mechanically and magnetically.

TABLE 8.1

The Methods of Preparation of Magnetic Hydrogels, Physiological Properties, and Required Magnetic Field

Method	Magnetic Particles	Hydrogels	Physicochemical Properties	Magnetic Field	References
Blending Method	Magnetite	PVA	Biocompatible and biodegradable		Baron et al., 2021
	Fe$_3$O$_4$	Biphosphate-modified HA	Better rheology and rapid heat generation	Alternative magnetic field	Shi et al., 2019
	Magnetite	Hyaluronate hydrogel	Stable and homogenous	-	Tóth et al., 2015
	Magnetic nanoparticles	Collagen	Biocompatible	Static magnetic field	Yuan et al., 2018
In Situ Precipitation Method	Fe$_3$O$_4$	Chitosan	Enhanced mechanical properties	Low-frequency magnetic field	Wang et al., 2018
	Dextran-coated Fe$_3$O$_4$	Bacterial cellulose	-	Neomydium magnet	Arias et al., 2018
	Polydopamine-chelated carbon nanotube-Fe$_3$O$_4$	Acrylamide	Better mechanical properties	Low-static magnetic field	Liu et al., 2019
Cross-Linking Method		PVA	High retention on cell surfaces	External magnetic field	Ishihara et al., 2021
	Ferrous ions	Gelatin	Thermoreversible, biocompatible	External magnetic field	Helminger et al., 2014
	CoFe$_2$O$_4$	Polyacrylamide	Greater stability	-	Messing et al., 2011
Incorporation method	Iron oxide	PEG		External static magnetic field	Filippi et al., 2019
	PEG-magnetic microparticles	Thiolated HA	Uniform distribution of magnetic particles	Applied magnetic field	Tay et al., 2018
Other methods	Magnetic β-cyclodextrin	Cellulose	Simple one-pot method; rapid swelling–deswelling	External magnetic field	Lin et al., 2019
	Ferrous ions	Gum acacia/ Poly-acrylamide	Using gamma irradiation	External magnetic field	Mohamed et al., 2020

8.4 BIOMEDICAL APPLICATION OF MAGNETIC HYDROGELS

The ability to remotely control the mechanical properties of magnetic-responsiveness hydrogels sets them distinctively apart. When ferrogels are compared to nonmagnetic hydrogels, they offer unique features whereby magnetic fields can be delivered remotely to act on ferromagnetic materials, such as ferrogels. Moreover, the typical magnetic fields—at least the stationary or very low frequency fields reported for uses of these materials—are permeable to human tissues and safe for human bodily functions and tissues (Lopez-Lopez et al., 2015; Rodriguez-Arco et al., 2016). Consequently, ferrogels can be directed and modulated within the human body using externally provided noncontact magnetic fields. Because hydrogels are ferromagnetic, magnetic resonance imaging can be used for both visualization and in vivo monitoring and magnetic fields can be used to controllably alter the mechanical properties of ferrogels by up to several orders of magnitude (Ilg, 2013).

Ferrogels are especially intriguing for a wide range of biomedical applications due to their unique features, which include fast magnetic response, temporal and spatial control, and non-invasive remote actuation (Zhang N et al., 2015). Biomedical applications of magnetic gels include the following and are summarized in Figure 8.4:

- Remote-controlled drug/cell delivery, enzyme immobilization (Rodkate and Rutnakornpituk, 2016)
- Artificial muscles and tissue engineering (Huang et al., 2018; Huang et al., 2020)
- Hyperthermia therapy (Zhou et al., 2018)
- Soft actuators or robotics (Kim and Zhao, 2022; Xu and Xu, 2023)

8.4.1 APPLICATIONS IN TISSUE ENGINEERING AND AS SCAFFOLDS FOR TISSUE ENGINEERING

Scaffolds are crucial in advanced tissue engineering because they help to restore damaged tissues and trigger the regeneration process. Scaffolds are frequently

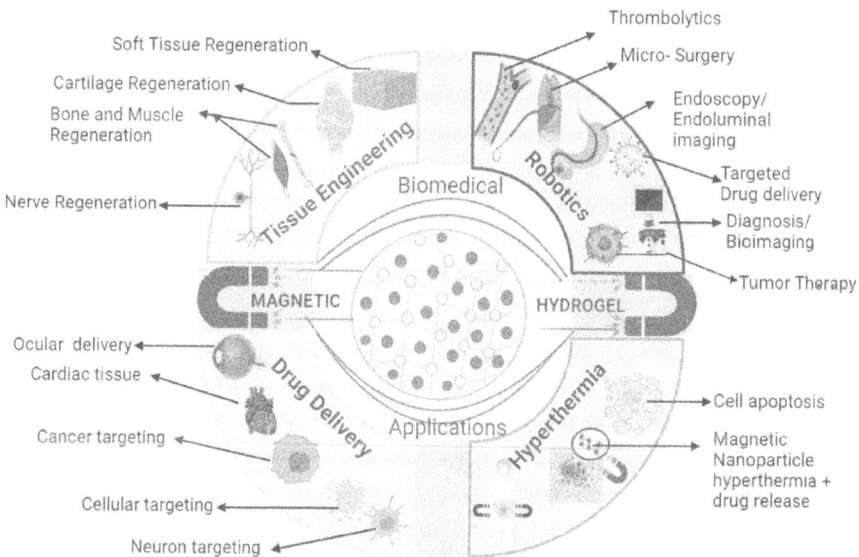

FIGURE 8.4   Biomedical applications of magnetic hydrogels.

loaded with a set quantity of growth factors unique to different tissues to encourage, direct, and speed up the repair and regeneration process (Farzaneh et al., 2021). Magnetic-sensitive hydrogels, by incorporating exogenous paramagnetic or ferromagnetic compounds such as superparamagnetic particles (i.e., $CoFe_2O_4$, Fe_3O_4, and $-Fe_2O_3$) within the polymeric matrix, offer a number of benefits, including fast reaction, appealing mechanical properties, and biocompatibility. They provide hydration to the regenerative microenvironment, promoting the functional matrix for artificial reconstruction, and are now widely explored for cell scaffold in bone tissue, nerve tissue engineering, cartilage tissue engineering, and skin tissue engineering (Bonhome-Espinosa et al., 2020).

The magnetism induced by physical stimulators provides specific spatiotemporal controlled delivery of encapsulated magnetic nanoparticles with exogenous biomolecules and growth factors imparting significant signaling pathways to achieve tissue regeneration, proliferation, healing, and cell adhesion and growth. It has been noted that the presence of an external magnetic field has a favorable impact on cellular osteogenic differentiation and speeds up bone fracture repair, and as a result, it has been recommended as an orthopedic therapy strategy; other advantages include increasing the viability of some cell types, like fibroblasts, in cell culture prompting cell proliferation *in-vitro* as well as bone formation in vivo (Vo et al. 2012; Bonhome-Espinosa et al., 2020, Marin et al., 2014). Various studies have demonstrated that magnetic gel induces stem cell differentiation in muscle cells, adipocytes, osteoblasts, chondrocytes, and nerves cells. Moreover, in varied ways magnetic fields help the alignment of cells in situ, allowing adequate differentiation and growth in the right directions to provide proper tissue 3D structure which is needed in some tissue regeneration such as that of neural tissue (Babaniamansour et al., 2022; Wang et al., 2022). Applications of magnetic hydrogels and scaffolds in tissue engineering are mentioned in Table 8.2.

8.4.2   Applications in Drug Delivery

The use of stimuli-responsive polymeric devices for controlled drug release has drawn a lot of attention, since it has several advantages over existing drug delivery methods. The development of pulsatile release systems that may imitate the bodily release profile of certain peptides or hormones may result in optimal drug administration and enable the ideal zero-order release of drugs over an extended period (Zhao et al., 2015).

Magnetic hydrogels have been tested for their viability in remotely controlling the pulsatile delivery of medications; hence they have the potential to induce on-demand drug release at the ideal concentration and the appropriate region and timing through careful control of the external magnetic field (Liu et al., 2018). Moreover, there are no restrictions on the depth of tissue penetration and little risk to human health when an external magnetic field is applied remotely. Under the control of an external magnetic field, magnetic hydrogels can transport drug molecules to precisely target lesions, dramatically improving drug delivery efficiency and minimizing systemic side effects (Liu et al., 2018). It is generally accepted that an external magnetic field causes the magnetic particles inside the polymer network to get agitated, resulting in the relaxation of the

TABLE 8.2

Summary of the Polymer Constituents, Physicochemical Properties, and Biomedical Applications of Magnetic Hydrogels in Tissue Engineering

Tissue Engineering Application	Functional Molecules	Hydrogels/ Polymer Constituents	Physicochemical Properties	Biomedical Application	References
Bone tissue engineering	Human dental pulp–derived stem cells (hDPSCs)	Cobalt ferrite nanoparticles (CoFe$_2$O$_4$) as magnetic agents in the structure of polyacrylic acid hydrogel	Synthesized magnetic scaffolds possess improved porous structure and surface roughness.	Novel magnetic hydrogel with optimal potential for cell proliferation as well as osteogenic differentiation of hDPSCs in vitro	Farzaneh et al., 2021
Cartilage tissue engineering	Human hyaline–derived chondrocytes (HHC)	3D magnetic cartilage-like fibrin-agarose constructs (3D-MCFAC) containing magnetite (iron oxide) core	Magnetic nanoparticles–embedded polymer resulted in soft viscoelastic solids, which could be implanted by injection.	Encapsulated HHC expressed at least type II collagen, one of the most significant proteins found in a native cartilage's extracellular matrix	Bonhome-Espinosa et al., 2020
Cardiac tissue engineering	H9C2 cardiac cells	Magnetically (iron oxide)–actuated cell alignment in fibrin	High precision and accuracy were advantageously given by magnetic actuation for the control of spatial and temporal patterns.	Enabled cellular alignment, exhibited a favorable environment for nutrient exchange and cell growth, generated centimeter-scale viable tissues	Richard et al., 2020
Soft tissue engineering	Human recombinant tropoelastin (rTE)	Tropoelastin doped with magnetic nanoparticles gelatin hydrogel	The secondary structure of tropoelastin was significantly altered by the presence of magnetic nanoparticles without morphological alterations.	Enhanced cell adhesion and cell viability	Pesqueira et al., 2018

(Continued)

TABLE 8.2 (Continued)

Summary of the Polymer Constituents, Physicochemical Properties, and Biomedical Applications of Magnetic Hydrogels in Tissue Engineering

Tissue Engineering Application	Functional Molecules	Hydrogels/ Polymer Constituents	Physicochemical Properties	Biomedical Application	References
Bone tissue engineering	SAOS-2 human osteoblastic cell culture	Magnetic maghemite (γ-Fe$_2$O$_3$) nanoparticles dispersed in superporous poly(2-hydroxyethyl methacrylate) scaffolds	Enhanced toughness and compressive modulus	Mediocre cell colonization and significantly improved cell adhesion	Zasońska et al., 2021
Neuron tissue engineering	HA-thiol	Magnetic hyaluronic acid hydrogel	Low magnetic microparticle leakage, mimics extracellular matrix of brain and spinal cord, excellent biocompatibility	Expansion of functional neurites and the development of excitatory and inhibitory ion channels, modulated expression of mechanosensitive PIEZO2 channels	Tay et al., 2018
Neuron tissue engineering	Neural stem cell	Magnetic chitosan gel	Matrix stiffness, viscosity, elasticity, and surface topography; showed positive effect on neurogenesis	Facilitated neuronal regeneration, improved locomotor recovery	Zhang et al., 2023

polymer chains and the formation of an enlarged, loose polymer network that improves drug diffusion (Lin et al., 2019). On the other hand, hydrogel mechanical vibration and deformation caused by the dipole orientation of magnetic nanoparticles under an external magnetic field result in local tensile and compression strains, which promote the release of drugs. By modulating magnetic fields, it has been possible to accurately control the release of drugs from drug-loaded magnetic hydrogels (Liu et al., 2015). The applications of magnetic hydrogels in drug delivery are mentioned in Table 8.3.

8.4.3 APPLICATION OF MAGNETIC HYDROGELS AS SOFT ROBOTICS

The development of "stimuli-responsive" soft materials has resulted in the creation of soft actuators or artificial muscles that can be remotely controlled by stimuli

TABLE 8.3

Summary of the Polymer Constituents, Physicochemical Properties, and Biomedical Applications of Magnetic Hydrogel in Drug Delivery

Drug/ Functional Molecules	Polymer Constituents	Physicochemical Properties	Biomedical Application/ Biofunction	References
Caffeine	Magnetic gelatin/alginate (Gel/Alg/MNP) hydrogels/$FeCl_2 \cdot 4H_2O$	Porous macrostructure tunable biodegradability.	Controlled release from fabricated reservoir improved viability of SH-SY5Y cells, improved neuronal cell proliferation	Frachini et al., 2023
Doxorubicin hydrochloride	Alg-Gel/Fe_3O_4 magnetic gel	Higher thermal stabile porous structures, enhanced encapsulation efficiency	Achieved slow drug release profile, pH-dependent drug release behavior, diagnosis through MRI technique, isolation at targeted area	Jahanban-Esfahlan et al., 2020
5-fluorouracil	5-Fu/MGMs/CEC-OAlg (Gel): 5-fluorouracil-loaded magnetic gelatin microspheres dispersed in carboxyethyl chitosan compounded with oxidized alginate	Discrete spherical microspheres with smooth surface, magnetic gel with stable and consistent structure	Composite hydrogel possessed effective self-healing abilities, sustained release for 5-Fu delivery, effective cancer treatment and soft tissue regeneration	Chen et al., 2019
Dexamethasone	Magnetic nano carboxymethylcellulose-alginate-based hydrogel beads	Improved drug loading capacity, better swelling properties, regulated drug release	pH-sensitive drug release profiles, prevented GI tract release, potential delivery system for encapsulated and controlled release delivery	Karzar Jeddi and Mahkam, 2019
Dopamine (DOPA)	Alg-MNP/DOPA magnetic alginate beads (Alg-MNP)	Rough microstructures, increased drug loading efficiency	% release of dopamine from Alg-MNP increased under magnetic field, potential controlled release delivery system	Kondaveeti et al., 2016
Doxorubicin	Alginate ferrogel	Macroporous structure	Controlled release of drug from porous structure under magnetic stimulation, reduced number of cancer cells depicted through in vivo study	Kim et al., 2019

(Continued)

TABLE 8.3 (*Continued*)

Summary of the Polymer Constituents, Physicochemical Properties, and Biomedical Applications of Magnetic Hydrogel in Drug Delivery

Drug/ Functional Molecules	Polymer Constituents	Physicochemical Properties	Biomedical Application/ Biofunction	References
Doxorubicin hydrochloride	Composite hydrogel of $Fe_3O_4@SiO_2$ nanoparticles in Salecan-g-poly(PA-co-HEAA) copolymer	Enhanced amorphous phase of hydrogel, porous hydrogel with irregular pores, significant increase in thermal stability, increase in water uptake or swelling abilities under influence of pH	Enhanced drug release rate in pH-controlled system, promising platform for medication delivery with magnetic field targeting	Hu et al., 2017

like heat, light, solvent, or an electric or magnetic field (Shen et al., 2020). Stimuli-responsive soft materials are not novel in the field of materials science; most are made of soft polymers (such as elastomers or hydrogels) with incorporated micro- or nanoparticles as functional components (Yang et al., 2017). While many stimulus-responsive hydrogels have been successfully used in soft robotics, most of them are unable to handle the challenging requirements of the in vivo physiological environment due to the need for contact stimuli or a shallow tissue penetration depth. Due to the benefits of remote noncontact trigger and inconceivable tissue penetration depth, magnetic hydrogel–based soft robots are receiving more and more attention (Nguyen et al., 2012). These magnetic soft robots have enormous promise for biomedical applications, as tissues and organs are permeable to static and low-frequency magnetic fields, and magnetic fields can be remotely administered throughout a broad range of actuating field strengths and frequencies without having any negative impacts on biological systems (Hines et al., 2017). Moreover, magnetic fields may be accurately and quickly modified in terms of their magnitude, phase, and frequency, making them comparatively simple to regulate. These factors make magnetic fields an attractive option for manipulating free-floating soft robots in confined areas inside the human body (Wu et al., 2020). The biomedical applications include invasive surgeries (cardiac, neuro, vascular interventional procedures) (Wang et al., 2021), targeted drug delivery (Zhou et al., 2021), and diagnosis and cell imaging (biopsy, endoscopy, etc.) (Yim and Sitti, 2012). The intriguing possibilities of magnetic soft robots for biomedical applications are summarized in Table 8.4.

TABLE 8.4
Summary of Biomedical Applications of Magnetic Soft Robots

Proposed In Vivo Application	Constituents/Materials	Target Site	Disease Treatment/ Therapeutic Application	References
Ocular Drug Delivery	Bilayer hydrogel microrobot layer of PEGDA (poly(ethylene glycol) diacrylate) and iron oxide nanoparticles and gelatin/PVA and PLGA–DOX (poly (lactic-co-glycolic acid)-doxorubicin)	Intraocular lesions or the posterior segment	Acute targeting of drug to the lesion site, death of Y79 cells, induced retinoblastoma	Kim et al., 2020
Tumor Therapy	3D-printed microrobot iron oxide nanoparticles embedded in poly(ethylene glycol) diacrylate (PEGDA)	Multiple drugs at multiple sites	Significant synergistic effects of ASA and DOX, accelerated HeLa cell apoptosis, inhibited HeLa cell metastasis	Li et al., 2023
Therapeutic Drug Delivery	Multifunctional nanorobot (MNC@ Au@PDA/PEG/FA)	Hepatocellular carcinoma	X-ray and magnetic resonance imaging for transcatheter liver chemoembolization in vivo, targeted delivery of doxorubicin	Go et al., 2022
Endoluminal Imaging and Delivery	Endoscopy-assisted magnetic actuation with dual imaging system (EMADIS).	Bile duct	Precise magnetic stem cell spheroid microrobot delivery, achieved therapeutic and clinical imaging	Wang et al., 2021
Drug Delivery	Capsule microrobots– magnetic actuated	Blood vessels	Targeted drug delivery in blood vessels, simulated locomotion in the vasculature	Qiao et al., 2023
Cancer Cell Therapy	Natural alginate, N-Isopropylacrylamide (NIPAM) with magnetic nanoparticles spring type alginate/NIPAM hydrogel-based soft microrobot	-	Targeted controlled drug delivery of doxorubicin via swelling and deswelling of soft microrobot	Lee et al., 2018
Drug Delivery	Doxorubicin-loaded (Pd@Au)/Fe_3O_4@ Sp. microrobots	-	Significant synergistic chemo-photothermal therapeutic efficacy, targeted delivery and chemo-photothermal therapeutic efficacy	Wang et al., 2019

8.4.4 APPLICATION FOR CANCER HYPERTHERMIA THERAPY

The term "hyperthermia" refers to a type of cancer treatment that involves applying heat to the area surrounding the tumor which damages the cancerous cells. According to some research, tumor cells are more likely to die between the temperatures of 42°C and 46°C due to a cell apoptosis mechanism (denaturation of proteins) or because they are more vulnerable to radiation and antitumor medications (Dutta et al., 2021). Recently, magnetic hydrogels have been used to remotely heat up target tumors using an external magnetic field and release a regulated amount of anticancer medication. Magnetic nanoparticles act as heat-energy dissipative centers made from hyperthermia (Farzin et al., 2020; Singh and Datta, 2023). Small MNPs can pass through cell walls with ease, and they can be heated in the presence of an external oscillating magnetic field; as a result, cell heating and necrosis may be managed more precisely and intelligently (Suleman and Riaz, 2020).

Target-specific molecules or surface engineering can functionalize MNPs, which tend to adhere to cell walls, facilitating preferable tissue-specific hyperthermia. MNPs can easily cross the blood-brain barrier and can be delivered along with medicinal molecules (Varmazyar et al., 2020). This characteristic may enable MNPs to function as dual model nanoparticles for more effective therapeutic assays (Das et al., 2020). Also, magnetic hydrogels have a great deal of potential as an injectable hydrogel system, particularly for cancer therapy. For instance, several polymers that are thermosensitive, such as poloxamer and chitosan, and a few ionic response polymers have been studied as injectable hydrogels for cancer therapy (Tu et al., 2019). After injection, the MNP-containing polymer solutions go through a sol-gel transition at a higher temperature or calcium ion concentration to create magnetic hydrogels, and the implanted MNPs subsequently permit local heating under the influence of an amplified magnetic field (Shi et al., 2020). Thus, hyperthermia can be used in several areas of medical therapy, including surgery, radiation therapy, gene therapy, chemotherapy, and cancer immunotherapy, as summarized in Table 8.5.

TABLE 8.5
Summary of the Potentials of Magnetic Gel in Hyperthermia Therapy

Drug/ Functional Molecules/ Therapy	Polymer Constituents	Physicochemical Properties	Biomedical Applications/ Therapeutic Attributes	References
Doxorubicin hydrochloride	(HPMC)/Fe_3O_4 hydrogel	High swelling index, significant biocompatibility	Synergistic chemo-magnetic hyperthermia therapy, high efficacy, reduced toxicity	Zhou et al., 2018
Cancer therapy	CS–SF hydrogel/ PVA/Fe_3O_4 nanobio-composite scaffold	Homogenous dispersion of magnetic nanoparticles in gel matrix, significant chemocompatibility	Provided localized heat to desired site, low toxicity and better biological properties	Eivazzadeh-Keihan et al., 2023

(Continued)

TABLE 8.5 (*Continued*)

Summary of the Potentials of Magnetic Gel in Hyperthermia Therapy

Drug/ Functional Molecules/ Therapy	Polymer Constituents	Physicochemical Properties	Biomedical Applications/ Therapeutic Attributes	References
Doxorubicin	Ferumoxytol and medical chitosan hydrogel	Change *in-vivo* into a physical hydrogel	Temperature-dependent gradual release, improved synergistic effectiveness with 32.4% for the in vitro treatment of colon cancer cells, localized magnetic hyperthermia and chemo therapy	Chen et al., 2020
Glucose oxidase (GOx)	(Fe$_3$O$_4$/GOx/ MgCO$_3$@PLGA) magnetic bone repair hydrogels	Ability of liquid to solid phase transition to produce magneto-thermal effects	Synergistic therapy for osteosarcoma treatment, enhanced the effectiveness of starvation therapy, accelerated bone defect repair	Yu et al., 2023
Cancer therapy, bone regeneration	Injectable magnetic hydrogel (ferrimagnetic silk fibroin hydrogel)	Shear thinning behavior, produce heat efficiently, tumor thermal ablation *in-vivo*	Effective destruction of deeper located tumors, potential embolization agents	Qian et al., 2020
Hepatocellular carcinoma (HCC)	Magnetic nanocomposite hydrogel (CG-IM) with nanosheets (iron oxide nanoparticles@ Mica, IM)	Magnetocaloric property, integrated high injectability, mechanical robustness	Postoperative tumor margin treatment, percutaneous locoregional ablation, inhibited unresectable HCC	Gong et al., 2023

8.5 THE CURRENT CHALLENGES FOR THE DEVELOPMENT OF MAGNETIC HYDROGELS

For the development of bioaligned actuators, magnetically responsive smart hydrogels offer enormous potential. Uncommon advancements have been made in the biomedical disciplines by taking advantage of the intriguing qualities of magnetic-responsive hydrogels, which include but are not limited to biocompatibility, biodegradability, porosity, extracellular matrix mimicking, and rapid response to magnetic fields. For magnetic hydrogels to be widely used in healthcare, certain challenges require further attention to be resolved. The in vivo biosafety and long-term degradation mechanisms of magnetic hydrogels are still unknown. After the implantation of magnetic hydrogels, there is frequently concern about the excessive release of free iron ions, which might cause the production of free radicals, trigger oxidative stress, and activate an immune response.

Although most studies have demonstrated the availability of magnetic hydrogels in vitro and in vivo, the therapeutic effects of magnetic hydrogels in the current reported studies have been limited to animal tests and short-term results. The fact that the pharmacokinetics, metabolism, and long-term fate of MNPs embedded in hydrogel matrix have generally been neglected may be one of the reasons the application of magnetic hydrogels in clinic has been restricted. It has been reported that MNPs, after entering into the body, can be absorbed by cells and distributed into various tissues and organs. Small MNPs (<10 nm) can be removed rapidly by the extravasation and renal clearance, whereas large MNPs (>200 nm) can be phagocytized and metabolized through the mononuclear phagocyte system. After cellular uptake, the MNPs in lysosomes may slowly release free iron ions to interfere with the physiological iron metabolism (Liu et al., 2013).

Therefore, long-term metabolism and toxicology research studies on MNPs are vital for evaluating the biosafety of magnetic hydrogels and are of great significance for their clinical translation. On one hand, the magnetic-responsive properties of MNPs are expected to be optimized through component, size, morphology and functionalized modification to reduce the content and increase the stability of MNPs in magnetic hydrogels for improved biosafety. On the other hand, biodegradable synthetic and natural polymers with excellent biocompatibility and nontoxic degradation products can be selected preferentially to prepare the magnetic hydrogels.

8.6 CONCLUSION AND FUTURE PROSPECTS

Magnetic hydrogels with organized structure, in contrast to conventional hydrogels, have demonstrated enormous potential due to their improved functional capabilities. Even though magnetic hydrogels with ordered structures have significantly advanced biomedical research, there are still some issues that need to be resolved. To further expand the functionality and usability of hydrogels, it will be crucial to examine the magnetic properties of organized structure and designs of bionic architecture.

Future perspectives should concentrate on overcoming the current challenges. Additionally, given the importance of these factors in biomedical applications, more emphasis should be given to assessing the pharmacokinetics/toxicokinetics, metabolism, biodegradation in vivo, and other aspects of magnetic hydrogels.

REFERENCES

Abdeen AA, Lee J, Bharadwaj NA, Ewoldt RH, Kilian KA. Temporal modulation of stem cell activity using magnetoactive hydrogels. Adv Healthc Mater. 2016;5:2536–44.

Abrougui Mariem M, Lopez-Lopez Modesto T, Duran Juan DG. Mechanical properties of magnetic gels containing rod-like composite particles. Phil Trans R Soc A. 2019;377:20180218.

Ali A, Saroj S, Saha S, Gupta SK, Rakshit T, Pal S. Glucose-responsive chitosan nanoparticle/poly(vinyl alcohol) hydrogels for sustained insulin release *in vivo*. ACS Appl Mater Interfaces. 2023;15(27):32240–50.

Amani S, Mohamadnia Z, Mahdavi A. pH-responsive hybrid magnetic polyelectrolyte complex based on alginate/BSA as efficient nanocarrier for curcumin encapsulation and delivery. Int J Biol Macromol. 2019;141:1258–70.

Arias SL, Shetty A, Devorkin J, Allain JP. Magnetic targeting of smooth muscle cells *in vitro* using a magnetic bacterial cellulose to improve cell retention in tissue-engineering vascular grafts. Acta Biomater. 2018;77:172–81.

Babaniamansour P, Salimi M, Dorkoosh F, Mohammadi M. Magnetic hydrogel for cartilage tissue regeneration as well as a review on advantages and disadvantages of different cartilage repair strategies. Biomed Res Int. 2022;2022:7230354.

Baron RI, Biliuta G, Socoliuc V, Coseri S. Affordable magnetic hydrogels prepared from biocompatible and biodegradable sources. Polymers (Basel). 2021;13(11):1693.

Bonhome-Espinosa AB, Campos F, Durand-Herrera D, Sánchez-López JD, Schaub S, Durán JDG, et al. In-vitro characterization of a novel magnetic fibrin-agarose hydrogel for cartilage tissue engineering. J Mech Behav Biomed Mater. 2020; 104:103619

Bonhome-Espinosa AB, Campos F, Rodriguez IA, Carriel V, Marins JA, Zubarev A, et al. Effect of particle concentration on the microstructural and macromechanical properties of biocompatible magnetic hydrogels. Soft Matter. 2017;13:2928–41.

Brulé S, Levy M, Wilhelm C, Letourneur D, Gazeau F, Ménager C, Le Visage C. Doxorubicin release triggered by alginate embedded magnetic nanoheaters: A combined therapy. Adv Mater. 2011;23:787–90.

Chen B, Xing J, Li M, Liu Y, Ji M. DOX@Ferumoxytol-medical chitosan as magnetic hydrogel therapeutic system for effective magnetic hyperthermia and chemotherapy in vitro. Colloids Surf B Biointerfaces. 2020;190:110896.

Chen H, Qin Z, Zhao J, He Y, Ren E, Zhu Y, Liu G, Mao C, Zheng L. Cartilage-targeting and dual MMP-13/pH responsive theranostic nanoprobes for osteoarthritis imaging and precision therapy. Biomaterials. 2019;225:119520.

Chen X, Fan M, Tan H, Ren B, Yuan G, Jia Y, et al. Magnetic and self-healing chitosan-alginate hydrogel encapsulated gelatin microspheres via covalent cross-linking for drug delivery. Mater. Sci. Eng. C. 2019;101:619–29.

Cui L, Zhang J, Zou J, Yang X, Guo H, Tian H, et al. Electroactive composite scaffold with locally expressed osteoinductive factor for synergistic bone repair upon electrical stimulation. Biomaterials. 2019;230:119617.

da Silva Fernandes R, Tanaka FN, Angelotti AM, Ferreira Júnior CR, Yonezawa UG, Watanuki Filho A, et al. Properties, synthesis, characterization and application of hydrogel and magnetic hydrogels: A concise review. In Jogaiah S, Singh HB, Fraceto LF, de Lima R. (Eds). Woodhead Publishing Series in Food Science, Technology and Nutrition, Advances in Nano-Fertilizers and Nano-Pesticides in Agriculture, Woodhead Publishing. 2021;437–57.

Das P, Ganguly S, Margel S, Gedanken A. Tailor made magnetic nanolights: Fabrication to cancer theranostics applications. Nanoscale Adv. 2020;3(24):6762–96.

Derakhshankhah H, Jahanban-Esfahlan R, Vandghanooni S, Akbari-Nakhjavani S, Massoumi B, Haghshenas B, et al. A bio-inspired gelatin-based pH-and thermal-sensitive magnetic hydrogel for in-vitro chemo/hyperthermia treatment of breast cancer cells. J Appl Polym Sci. 2021;138:50578.

Dutta J, Kundu B, Yook S. Three-dimensional thermal assessment in cancerous tumors based on local thermal non-equilibrium approach for hyperthermia treatment. Int J Therm Sci. 2021;159:106591.

Ebara M, Kotsuchibashi Y, Narain R, Idota N, Kim Y, Hoffman J, et al. Smart Biomaterials. Springer Tokyo. 2014.

Eivazzadeh-Keihan R, Pajoum Z, Aliabadi HAM, Mohammadi A, Kashtiaray A, Bani MS, et al. Magnetized chitosan hydrogel and silk fibroin, reinforced with PVA: A novel nanobiocomposite for biomedical and hyperthermia applications. RSC Adv. 2023;13(13):8540–50.

Farzaneh S, Hosseinzadeh S, Samanipour R, Hatamie S, Ranjbari J, Khojasteh A. Fabrication and characterization of cobalt ferrite magnetic hydrogel combined with static magnetic field as a potential bio-composite for bone tissue engineering. J Drug Deliv Sci Technol. 2021;64:102525.

Farzin A, Etesami SA, Quint J, Memic A, Tamayol A. Magnetic nanoparticles in cancer therapy and diagnosis. Adv Healthc Mater. 2020;9(9):e1901058.

Filippi M, Dasen B, Guerrero J, Garello F, Isu G, Born G, et al. Magnetic nanocomposite hydrogels and static magnetic field stimulate the osteoblastic and vasculogenic profile of adipose-derived cells. Biomaterials. 2019;223:119468.

Frachini ECG, Selva JSG, Falcoswki PC, Silva JB, Cornejo DR, Bertotti M, et al. Caffeine release from magneto-responsive hydrogels controlled by external magnetic field and calcium ions and its effect on the viability of neuronal cells. Polymers (Basel). 2023;15(7):1757.

Gaharwar AK, Peppas NA, Khademhosseini A. Nanocomposite hydrogels for biomedical applications. Biotechnol Bioeng. 2014;111(3):441–53.

Ganguly S, Margel S. Design of magnetic hydrogels for hyperthermia and drug delivery. Polymers (Basel). 2021;13(23):4259.

Gao F, Xie W, Miao Y, Wang D, Guo Z, Ghosal A, et al. Magnetic hydrogel with optimally adaptive functions for breast cancer recurrence prevention. Adv Healthc Mater. 2019;8:1900203.

Gila-Vilchez C, Bonhome-Espinosa AB, Kuzhir P, Zubarev A, Duran JDG, Lopez-Lopez MT. Rheology of magnetic alginate hydrogels. J Rheol. 2018;62:1083–96.

Go G, Yoo A, Nguyen KT, Nan M, Darmawan BA, Zheng S, et al. Multifunctional microrobot with real-time visualization and magnetic resonance imaging for chemoembolization therapy of liver cancer. Sci Adv. 2022;8(46):eabq8545.

Gong J, Hu J, Yan X, Xiang L, Chen S, Yang H, et al. Injectable hydrogels including magnetic nanosheets for multidisciplinary treatment of hepatocellular carcinoma via magnetic hyperthermia. Small. 2023;20(3):e2300733.

González-Díaz EC, Varghese S. Hydrogels as extracellular matrix analogs. Gels. 2016;2(3):20.

Helminger M, Wu B, Kollmann T, Benke D, Schwahn D, Pipich V, et al. Synthesis and characterization of gelatin-based magnetic hydrogels. Adv Funct Mater. 2014;24(21):3187–96.

Hines L, Petersen K, Lum GZ, Sitti M. Soft actuators for small-scale robotics. Adv Mater. 2017;29(13).

Hu X, Nian G, Liang X, Wu L, Yin T, Lu H, et al. Adhesive tough magnetic hydrogels with high Fe_3O_4 content. ACS Appl Mater Interfaces. 2019;11:10292–300.

Hu X, Wang Y, Zhang L, Xu M, Zhang J, Dong W. Dual-pH/magnetic-field-controlled drug delivery systems based on Fe_3O_4 @SiO_2-incorporated salecan graft copolymer composite hydrogels. Chem Med Chem. 2017;12(19):1600–9.

Huang J, Jia Z, Liang Y, Huang Z, Rong Z, Xiong J, Wang D. Pulse electromagnetic fields enhance the repair of rabbit articular cartilage defects with magnetic nano-hydrogel. RSC Adv. 2020;10(1):541–50.

Huang J, Liang Y, Jia Z, Chen J, Duan L, Liu W, et al. Development of magnetic nanocomposite hydrogel with potential cartilage tissue engineering. ACS Omega. 2018;3(6):6182–9.

Ilg P. Stimuli-responsive hydrogels cross-linked by magnetic nanoparticles. Soft Matter. 2013;9:3465–8.

Ishihara K, Narita Y, Teramura Y, Fukazawa K. Preparation of magnetic hydrogel microparticles with cationic surfaces and their cell-assembling performance. ACS Biomater Sci Eng. 2021;7(11):5107–17.

Jahanban-Esfahlan R, Derakhshankhah H, Haghshenas B, Massoumi B, Abbasian M, Jaymand M. A bio-inspired magnetic natural hydrogel containing gelatin and alginate as a drug delivery system for cancer chemotherapy. Int J Biol Macromol. 2020;156:438–45.

Jahanban-Esfahlan R, Soleimani K, Derakhshankhah H, Haghshenas B, Rezaei A, Massoumi B, Farnudiyan-Habibi A, Samadian H, Jaymand M. Multi-stimuli-responsive magnetic hydrogel based on Tragacanth gum as a de novo nanosystem for targeted chemo/hyperthermia treatment of cancer. J Mater Res. 2021;36:858–69.

Karzar Jeddi M, Mahkam M. Magnetic nano carboxymethyl cellulose-alginate/chitosan hydrogel beads as biodegradable devices for controlled drug delivery. Int J Biol Macromol. 2019;135:829–38.

Kim C, Kim H, Park H, Lee KY. Controlling the porous structure of alginate ferrogel for anti-cancer drug delivery under magnetic stimulation. Carbohydr Polym. 2019;223:115045.

Kim DI, Lee H, Kwon SH, Sung YJ, Song WK, Park S. Bilayer hydrogel sheet-type intra-ocular microrobot for drug delivery and magnetic nanoparticles retrieval. Adv Healthc Mater. 2020;9(13):e2000118.

Kim Y, Zhao X. Magnetic soft materials and robots. Chem Rev. 2022;122(5):5317–64.

Kondaveeti S, Cornejo DR, Petri DF. Alginate/magnetite hybrid beads for magnetically stim-ulated release of dopamine. Colloids Surf B Biointerfaces. 2016;138:94–101.

Lee H, Choi H, Lee M, Park S. Preliminary study on alginate/NIPAM hydrogel-based soft microrobot for controlled drug delivery using electromagnetic actuation and near-infrared stimulus. Biomed Microdevices. 2018;20(4):103.

Lee SC, Kwon IK, Park K. Hydrogels for delivery of bioactive agents: A historical perspec-tive. Adv Drug Deliv Rev. 2013;65:17–20.

Li X, Wang Y, Li A, Ye Y, Peng S, Deng M, et al. A novel pH and salt-responsive n-succinyl-chitosan hydrogel via a one-step hydrothermal process. Molecules. 2019;24:E4211.

Li Y, Dong D, Qu Y, Li J, Chen S, Zhao H, et al. A multidrug delivery microrobot for the synergistic treatment of cancer. Small. 2023;19:e2301889.

Li Z, Chen H, Li B, Xie Y, Gong X, Liu X, et al. Photoresponsive luminescent polymeric hydro-gels for reversible information encryption and decryption. Adv Sci. 2019;6:1901529.

Li Z, Li Y, Chen C, Cheng Y. Magnetic-responsive hydrogels: From strategic design to bio-medical applications. J Control Release. 2021;335:541–56.

Lin F, Zheng J, Guo W, Zhu Z. Smart cellulose-derived magnetic hydrogel with rapid swelling and deswelling properties for remotely controlled drug release. Cellulose. 2019;26:6861–77.

Liu G, Gao J, Ai H, Chen X. Applications and potential toxicity of magnetic iron oxide nanoparticles. Small. 2013;9:1533–45.

Liu Q, Li H, Lam KY. Optimization of deformable magnetic-sensitive hydrogel-based targeting system in suspension fluid for site-specific drug delivery. Mol Pharm. 2018;15(10):4632–42.

Liu TY, Chan TY, Wang KS, Tzou HM. Influence of magnetic nanoparticle arrangement in the ferrogels for tunable biomolecules diffusion. RSC Adv. 2015;5(109):90098–102.

Liu K, Han L, Tang P, Yang K, Gan D, Wang X, et al. An anisotropic hydrogel based on mussel-inspired conductive ferrofluid composed of electromagnetic nanohybrids. Nano Lett. 2019;19:8343–56.

Liu X, Yang Y, Inda ME, Lin S, Wu J, Kim Y, et al. Magnetic living hydrogels for intestinal localization, retention, and diagnosis. Adv Funct Mater. 2021;31:2010918.

Liu Z, Liu J, Cui X, Wang X, Zhang L, Tang P. Recent advances on magnetic sensitive hydro-gels in tissue engineering. Front Chem. 2020;8:124.

Lopez-Lopez MT, Scionti G, Oliveira AC, Duran JD, Campos A, Alaminos M, Rodriguez IA. Generation and characterization of novel magnetic field-responsive biomaterials. PLoS One. 2015;10(7):e0133878.

Maitra J, Shukla V. Cross-linking in hydrogels - a review. Am J Polym Sci. 2014;4:25–31.

Manish V, Siva KV, Arockiarajan A, Tamadapu G. Synthesis and characterization of hard magnetic soft hydrogels. Mater Lett. 2022;320:132323.

Manjua AC, Alves VD, Crespo JG, Portugal CAM. Magnetic responsive PVA hydrogels for remote modulation of protein sorption. ACS Appl Mater Interfaces. 2019;11:21239–49.

Marín T, Ortega D, Montoya P. et al. A new contribution to the study of the electrosyn-thesis of magnetic nanoparticles: The influence of the supporting electrolyte. J Appl Electrochem. 2014;44:1401–10.

Messing R, Frickel N, Belkoura L, Strey R, Rahn H, Odenbach S, et al. (2011). Cobalt ferrite nanoparticles as multifunctional cross-linkers in PAAm ferrohydrogels. Macromolecules. 2011;44:2990–9.

Mohamed AA, Mahmoud GA, Ezz ElDin MR, Saad EA. Synthesis and properties of (Gum acacia/polyacryamide/SiO2) magnetic hydrogel nanocomposite prepared by gamma irradiation. Polymer-Plastics Technol and Mater. 2020;59(4):357–70.

Nguyen VQ, Ahmed AS, Ramanujan RV. Morphing soft magnetic composites. Adv Mater. 2012;24:4041–54.

Pesqueira T, Costa-Almeida R, Mithieux SM, Babo PS, Franco AR, Mendes BB, et al. Engineering magnetically responsive tropoelastin spongy-like hydrogels for soft tissue regeneration. J Mater Chem B. 2018;6(7):1066–75.

Qian KY, Song Y, Yan X, Dong L, Xue J, Xu Y, et al. Injectable ferrimagnetic silk fibroin hydrogel for magnetic hyperthermia ablation of deep tumor. Biomaterials. 2020;259:120299.

Qiao S, Ouyang H, Zheng X, Qi C, Ma L. Magnetically actuated hydrogel-based capsule micro-robots for intravascular targeted drug delivery. J Mater Chem B. 2023;11(26):6095–105.

Richard S, Silva AK, Mary G, Ragot H, Perez JE, Ménager C, et al. 3D magnetic alignment of cardiac cells in hydrogels. ACS Applied Bio Materials. 2020;3(10):6802–10.

Rittikulsittichai S, Kolhatkar AG, Sarangi S, Vorontsova MA, Vekilov PG, Brazdeikis A, Randall Lee T. Multi-responsive hybrid particles: Thermo-, pH-, photo-, and magneto-responsive magnetic hydrogel cores with gold nanorod optical triggers. Nanoscale. 2016;8(23):11851–61.

Rodkate N, Rutnakornpituk M. Multi-responsive magnetic microsphere of poly(N-isopropylacrylamide)/carboxymethylchitosan hydrogel for drug controlled release. Carbohydr Polym. 2016;151:251–9.

Rodriguez-Arco L, Rodriguez IA, Carriel V, Bonhome-Espinosa AB, Campos F, Kuzhir P, et al. Biocompatible magnetic core-shell nanocomposites for engineered magnetic tissues. Nanoscale. 2016;8(15):8138–50.

Shen Z, Chen F, Zhu X, Yong K, Gu G. Stimuli-responsive functional materials for soft robotics. J Mater Chem B. 2020;8(39):8972–91.

Shi L, Zeng Y, Zhao Y, Yang B, Ossipov D, Tai CW, et al. Biocompatible injectable magnetic hydrogel formed by dynamic coordination network. ACS Appl Mater Interfaces. 2019;11(49),46233–40.

Shi W, Huang J, Fang R, Liu M. Imparting functionality to the hydrogel by magnetic-field-induced nano-assembly and macro-response. ACS Appl Mater Interfaces. 2020;12:5177–94.

Singh R, Datta B. Advances in biomedical and environmental applications of magnetic hydrogels. ACS Appl Polym Mater. 2023;5(7):5474–94.

Suleman M, Riaz S. In silico study of enhanced permeation and retention effect and hyperthermia of porous tumor. Med Eng Phys. 2020;86:128–37.

Tay A, Sohrabi A, Poole K, Seidlits S, Di Carlo D. A 3D magnetic hyaluronic acid hydrogel for magnetomechanical neuromodulation of primary dorsal root ganglion neurons. Adv Mater. 2018;10:e1800927.

Tóth IY, Veress G, Szekeres M, Illés E, Tombácz E. Magnetic hyaluronate hydrogels: Preparation and characterization. J Magn Magn Mater. 2015;380:175–80.

Tu Y, Chen N, Li C, Liu H, Zhu R, Chen S, et al. Advanced in injectable self-healing biomedical hydrogels. Acta Biomater. 2019;90:1–20.

Van Bemmelen JM. Das hydrogel und das krystallinische hydrat des kupferoxyds. Zeitschrift Für Anorganische Chemie. 1894;5(1):466–83.

Varmazyar M, Habibi M, Amini M, Pordanjani A, Afrand M, Vahedi SM. Numerical simulation of magnetic nanoparticle-based drug delivery in presence of atherosclerotic plaques and under the effects of magnetic field. Powder Technol. 2020;366:164–74.

Vo TN, Kasper FK, Mikos AG. Strategies for controlled delivery of growth factors and cells for bone regeneration. Adv Drug Deliv Rev. 2012;64(12):1292–309.

Wang B, Chan KF, Yuan K, Wang Q, Xia X, Yang L, et al. Endoscopy-assisted magnetic navigation of biohybrid soft microrobots with rapid endoluminal delivery and imaging. Sci Robot. 2021;6(52):eabd2813.

Wang B, Kostarelos K, Nelson BJ, Zhang L. Trends in micro-/nanorobotics: Materials development, actuation, localization, and system integration for biomedical applications. Adv Mater. 2021;33(4):e2002047.

Wang M, Deng Z, Guo Y, Xu P. Designing functional hyaluronic acid-based hydrogels for cartilage tissue engineering. Mater Today Bio. 2022;17:100495.

Wang X, Cai J, Sun L, Zhang S, Gong D, Li X, Yue S, Feng L, Zhang D. Facile fabrication of magnetic microrobots based on spirulina templates for targeted delivery and synergistic chemo-photothermal therapy. ACS Appl Mater Interfaces. 2019;11(5):4745–56.

Wang Y, Li B, Xu F, Han Z, Wei D, Jia D, Zhou Y. Tough magnetic chitosan hydrogel nanocomposites for remotely stimulated drug release. Biomacromolecules. 2018;19(8):3351–60.

Wang Y, Li B, Zhou Y, Jia D. In situ mineralization of magnetite nanoparticles in chitosan hydrogel. Nanoscale Res Lett. 2009;4:1041–6.

Wichterle O, Lím D. Hydrophilic gels for biological use. Nature. 1960;185:117–8.

Wu S, Hu W, Ze Q, Sitti M, Zhao R. Multifunctional magnetic soft composites: A review. Multifunct Mater. 2020;3(4):042003.

Xu R, Xu Q. Design of a bio-inspired untethered soft octopodal robot driven by magnetic field. Biomimetics (Basel). 2023;8(3):269.

Yang GZ, Fischer P, Nelson B. New materials for next-generation robots. Sci Robot. 2017;2(10):eaap9294.

Yim S, Sitti M. Design and rolling locomotion of a magnetically actuated soft capsule endoscope. IEEE Trans Robot. 2012;28:183–94.

Yu K, Zhou H, Xu Y, Cao Y, Zheng Y, Liang B. Engineering a triple-functional magnetic gel driving mutually-synergistic mild hyperthermia-starvation therapy for osteosarcoma treatment and augmented bone regeneration. J Nanobiotechnol. 2023;21(1):201.

Yuan P, Ding X, Yang YY, Xu QH. Metal nanoparticles for diagnosis and therapy of bacterial infection. Adv Healthc Mater. 2018;7(13):e1701392. doi: 10.1002/adhm.201701392.

Zasońska BA, Brož A, Šlouf M, Hodan J, Petrovský E, Hlídková H, Horák D. Magnetic superporous poly(2-hydroxyethyl methacrylate) hydrogel scaffolds for bone tissue engineering. Polymers (Basel). 2021;13(11):1871.

Zhang J, Wang Y, Shu X, Deng H, Wu F, He J. Magnetic chitosan hydrogel induces neuronal differentiation of neural stem cells by activating RAS-dependent signal cascade. Carbohydr Polym. 2023;314:120918.

Zhang NY, Lock J, Sallee A, Liu HN. Magnetic nanocomposite hydrogel for potential cartilage tissue engineering: Synthesis, characterization, and cytocompatibility with bone marrow derived mesenchymal stem cells. ACS Appl Mater Interfaces. 2015;7(37):20987–98.

Zhang Y, Chen K, Li Y, Lan J, Yan B, Shi LY, et al. High-strength, self-healable, temperature-sensitive, MXene-containing composite hydrogel as a smart compression sensor. ACS Appl Mater Interfaces. 2019;11(50):47350–7.

Zhao W, Odelius K, Edlund U, Zhao C, Albertsson AC. In situ synthesis of magnetic field-responsive hemicellulose hydrogels for drug delivery. Biomacromolecules. 2015;16(8):2522–8.

Zhao C, Qazvini NT, Sadati M, Zeng Z, Huang S, De La Lastra, et al. A pH-triggered, self-assembled, and bioprintable hybrid hydrogel scaffold for mesenchymal stem cell based bone tissue engineering. ACS Appl Mater Interfaces. 2019;11(9):8749–62. https://doi.org/10.1021/acsami.8b19094.

Zhou H, Mayorga-Martinez CC, Pané S, Zhang L, Pumera M. Magnetically driven micro and nanorobots. Chem Rev. 2021;121(8):4999–5041.

Zhou X, Wang L, Xu Y, Du W, Cai X, Wang F, et al. pH and magnetic dual-response hydrogel for synergistic chemo-magnetic hyperthermia tumor therapy. RSC Adv. 2018;8(18):9812–21.

9 Magnetic Polymer Composites for Wastewater Treatment

Bambesiwe M. May, Nonkululeko Miya, and Sivuyisiwe Mapukata

9.1 INTRODUCTION

Population growth and economic development have increased the demand for everyday resources. This has led to rapid growth in industrialization and an increase in cases of complex water pollution, which consists of various pollutants, making wastewater treatment processes more complicated [1–3]. The various pollutants commonly found in wastewater include heavy metals, organic contaminants (e.g., dyes, pharmaceuticals, pesticides), and other harmful substances. Some of these pollutants are not biodegradable; thus, streams and agricultural land are easily contaminated if untreated effluents are discharged into the respective receiving water bodies. They might have devastating effects on the environment and human health even at low concentrations. Hence, organizations such as the World Health Organization (WHO) and the United States Environmental Protection Agency (U.S. EPA) have established maximum concentration levels for some pollutants of concern [2–4]. Thus, there has been increased interest in research towards developing appropriate wastewater treatment methods for the removal of these pollutants to keep them within acceptable levels. Wastewater treatment methods such as chemical precipitation, ion exchange, membrane filtration, electro-dialysis, and adsorption have been explored in the past [5, 6]. However, adsorption is considered the most promising method because the process is low-cost, simple, flexible, and environmentally friendly. Adsorption is the accumulation of molecules on the surface of the adsorbents by taking advantage of physical and chemical interactions such as van der Waals forces, hydrogen bonding, electrostatic attraction, and chelation [7]. The preferred adsorbent materials are those with a high surface area, fast adsorption rate, and short adsorption equilibrium time [8]. Extensive research has been devoted to demonstrating the removal of various pollutants in water using sulphur-bearing compounds, polymers, porous silica, carbon-based nanomaterials (NMs), and metal-organic frameworks (MOFs) [8–13]. Despite these improvements, researchers have faced challenges in separating and recovering adsorbent materials from water. Thus, research interest shifted to magnetic adsorbents due to their easy separation from water after treatment. Once the magnetic adsorbent is saturated with pollutants, an external magnetic field can be applied, causing it to aggregate and be quickly removed from the water, simplifying the filtration process. Additionally, the magnetic adsorbent can be collected, regenerated, and reused for subsequent treatment cycles,

DOI: 10.1201/9781003454236-9

reducing both operational costs and waste generation. Furthermore, its recyclability reduces the amount of waste produced throughout the process [14].

The most studied magnetic materials for the treatment of contaminant are magnetic versions of iron oxide NMs such as Fe_3O_4. Indeed, such magnetic materials have been widely applied for pollutant removal from water owing to their superparamagnetic properties, easy surface modification, electronic conductivity, and biocompatibility [6, 15–18]. Due to the continuous increase in the water complexity of wastewater, the quest for better functional materials has motivated the development of magnetic polymeric composite materials in which the strengths and complementary and synergetic functions of magnetic NMs and polymers are utilized for achieving improved water remediation performance. Magnetic polymer composites have gained considerable attention in recent years for various applications, including wastewater treatment. The magnetic nanoparticles provide the magnetic properties required for easy separation of the composite from water using an external magnetic field. The polymer matrix serves as a supporting structure that stabilizes the nanoparticles and enhances their dispersibility in the water. The polymers commonly used in the synthesis of magnetic polymers include synthetic polymers [e.g., polystyrene (PS) and polyvinyl alcohol (PVA)], naturally occurring polymers (e.g., xanthan gum and cellulose), and others (e.g., chitosan and cyclodextrin) [17, 19–26]. The high surface area and porous nature of magnetic polymer composites contribute to their impressive adsorption capacity. They can effectively capture and remove from water various pollutants, such as heavy metals, organic contaminants (e.g., dyes, pharmaceuticals, pesticides), and other harmful substances [2, 15, 16, 27–30]. The choice of synthesis approaches influences the properties and performance of the composite, so researchers carefully tailor the preparation techniques to meet specific requirements. The various synthesis approaches for fabricating magnetic polymer composites include in situ polymerization, emulsion polymerization, and co-precipitation, which commonly result in co-encapsulation of individually synthesized magnetic NMs in polymers or core/shell formations, where the polymer is deposited around the magnetic NMs to form a shell [31–34]. The application of magnetic polymer composites in wastewater treatment has shown great promise and so has been studied extensively in recent years. This chapter presents some of the key applications of magnetic polymer composites in wastewater treatment, including the removal of pollutants such as heavy metals, oil, dyes, pharmaceuticals, and multipollutants. The challenges, conclusions, and future outlooks associated with the use of magnetic polymer composites for wastewater treatment are also fully elucidated.

9.2 APPLICATION OF MAGNETIC POLYMER COMPOSITES FOR WASTEWATER TREATMENT

9.2.1 HEAVY METALS REMOVAL

Heavy metals are a class of metallic elements with relatively high densities in comparison to water and high atomic weights [35]. Although many metals are necessary for life, they can be hazardous at some level, causing harm to plants, humans, and animals.

Some of the most hazardous heavy metals present in wastewater are As, Cd, Pb, Ni, Cr, Cu, Zn, and Hg, and their solubility in aquatic environments renders them easily adsorbed by living organisms [6, 36]. Heavy metals are released into the environment through direct and indirect human activities: mining, agricultural activities, rapid industrialization, waste disposal, pharmaceuticals, electroplating, and so on [37, 38]. The heavy metal ions in wastewater come from different sources. Since heavy metals are non-biodegradable, their accumulation quickly exceeds permitted levels and therefore usually leads to adverse consequences for people, animals, and the environment as a whole [39, 40]. Therefore, wastewater must be properly treated to mitigate the harmful effects of heavy metal toxicity.

As a result of an increase in interest in composites, researchers have been investigating novel ways to engineer materials that combine the unique properties of nanomaterials and polymers to form nanocomposites [18, 41]. Magnetic polymer composites have become a promising solution for removing heavy metals from wastewater, offering an effective and environmentally friendly method of removing heavy metals. The pollutants can be selectively collected and extracted from the water matrix by harnessing the magnetic responsiveness of the composite materials.

Vunain et al. (2013) demonstrated improved surface area due to the addition of magnetite nanoparticles, which resulted in better adsorption of As(III) from contaminated water [19]. The polymer nanocomposite consisted of ethylene-vinyl acetate (EVA) (70%), polycaprolactone (PCL) (15%), and Fe_3O_4 (15%) and achieved a maximum adsorption capacity of 2.83 mg/g at pH 8.6. EVA and PCL's hydrophobic qualities enable their chemical and mechanical properties to be stable in water systems, which encourages the usage of these materials in aquatic environments [42]. The nanocomposites investigated in this study showed good potential for As(III) removal from contaminated water.

The use of bio-based materials has seen growing interest and progress in research. Chitosan is a natural, biodegradable, and hydrophilic polymer with adsorption properties [28, 43]. However, chitosan has several drawbacks, such as mechanical instability, a challenging separation, and low equilibrium capacity; therefore, incorporating other materials into chitosan-based materials can enhance its properties [22]. Magnetic chitosan composites (MCCs) have gained increased attention for their potential to remove metal ions due to their strong binding ability. Peralta and co-workers developed a magnetic chitosan composite by embedding a chitosan matrix with magnetite/maghemite for the removal of toxic Cu(II), Pb(II), and Ni(II) from water [17]. The composite exhibited extremely high removal capacity toward Cu(II) and Pb(II), with the adsorption of the heavy metals to MCC reaching equilibrium within 120 min with maximum uptakes of 108.9 mg/g, 216.8 mg/g, and 220.9 mg/g for Ni(II), Cu(II), and Pb(II), respectively. These results demonstrated that the amino and hydroxyl functional groups of chitosan were involved in the adsorption process. The MCC could be easily separated from the aqueous media by the application of an external magnetic field and reused for up to six cycles.

Modifying the functionalization of Fe_3O_4 with various materials such as silica and polymer can help mitigate drawbacks such as particle aggregation due to its high surface energy and susceptibility to air oxidation due to high chemical activity on its surface, which can cause a decline in its magnetic properties [44]. Peng and co-workers fabricated a magnetic Fe_3O_4@silica–xanthan gum hybrid adsorbent for the removal and recovery of Pb(II) [20]. Xanthan gum (XG), a naturally occurring polysaccharide, was fixed on the magnetic Fe_3O_4 microspheres' surface via a sol-gel process. Active sites for the selective removal of Pb(II) ions from the aqueous solution were made accessible by the condensation of the XG molecule. The maximum adsorption capacity was found to be 21.32 mg/g, with a sorption-desorption cycle over 21 days without a significant capacity loss. The magnetically switchable aspect of the composite made solid-liquid separation possible and regenerated using HCl. Based on the results, the composite could potentially be utilized in the treatment of lead-containing industrial wastewater. Recently, Wang et al. utilized a combination of Fe_3O_4 and SiO_2 in the synthesis of a magnetic xanthate-modified polyvinyl alcohol and chitosan composite (XMPC) from polyvinyl alcohol, chitosan, and magnetic Fe_3O_4@SiO_2 nanoparticles [21]. The interaction between chitosan and heavy metals in solution can be improved by introducing xanthate, which increases the adsorption capacity [45]. Magnetic hydrogel beads of the XMPC were prepared for their investigation in the adsorption of Pd(II), Cu(II), and Cd(II) ions. Adsorption equilibrium was reached after adsorption for about 120 min at 303 K, and the removal rates of Pd(II), Cu(II), and Cd(II) ions were 67 mg/g, 100 mg/g, and 307 mg/g, respectively. The removal rate of metal ions after five cycles of use remained above 50%, and the study demonstrated the adsorption capacity of chitosan as well as the ease and speed of solid–liquid separation after adsorption.

Zhang and co-workers successfully prepared a chitosan-stabilized iron sulfide (FeS) magnetic composite (MC-FeS) to enhance the removal of Cr(VI) from water [46]. FeS has been reported to effectively reduce Cr(VI) because of its ability to generate strong reducing species of S(–II) and Fe(II) [47]. The results showed that the MC-FeS composite enhanced the Cr(VI) removal capacity compared to its level with FeS on its own. The composite performed well in a long-term reaction system and a typical natural water environment and presented high stability against aging. Chitosan addressed the drawbacks of easy oxidation and aggregation by acting as a stabilizer. MC-FeS showed great potential in remediation of wastewater contaminated by heavy metal ions.

The non-toxicity, environmental stability, and affordability of conducting polymers such as those based on polypyrrole (PPy) have caused them to gain increased attention as adsorbents [48].

The polymer most commonly encountered in nature, cellulose nanofibers are a renewable and environmentally friendly material [49]. Carboxylated cellulose nanofibrils (CCNFs) are used in the preparation of composites, as in the preparation of novel magnetic hydrogel beads with amine-functionalized magnetite nanoparticles, polyvinyl alcohol, and blended chitosan, forming m-CS/PVA/CCNFs [23]. The composite was employed in the removal of Pb(II)ions from aqueous solutions. The carboxylate groups on the CCNFs surface were found to play a key role in the adsorption of Pb(II), with a capacity of 171.0 mg/g. Due to the high adsorption capacity and the ability to be separated from aqueous solutions in a rapid and biodegradable manner, m-CS/PVA/CCNFs could be considered for potential use in the removal of Pb(II) ions. Another biodegradable, water-soluble, and environmentally friendly polymer is carboxymethylcellulose (CMC), which has been used in the synthesis of superabsorbent hydrogels for various applications [50]. Wu et al. (2021) constructed a novel magnetic polysaccharide composite hydrogel using sodium alginate (SA) and carboxymethyl cellulose (CMC) as the backbone, which was then filled with Fe_3O_4 nanoparticles in situ [51]. The composite was employed in the removal of heavy metal ions in aqueous solution. The maximal adsorption capacities of the composite as calculated from the Langmuir model were 71.83 mg/g, 89.49 mg/g, and 105.93 mg/g for Mn(II), Pb(II), and Cu(II), respectively. The adsorption process was attributed to ion exchange and chemical adsorption. This composite offers potential for practical application in the removal of heavy metal ions, as it exhibited high efficiency after four cycles.

The development and optimization of magnetic polymer composites for the removal of heavy metals is a current topic of research, and researchers are always working to improve their effectiveness. Examples of other studies conducted on the efficient removal of heavy metals using a variety of magnetic polymer composites are listed in Table 9.1.

TABLE 9.1
Adsorption Capacity of Magnetic Polymer Composites for Heavy Metal Ion Removal

Magnetic Polymer Composite	Heavy Metal Pollutant	Adsorption Capacity (mg g^{-1})	References
Graphene/ Fe_3O_4@polypyrrole	Cr(VI)	348.4	[5]
Magnetic Chitosan Composite	Cu(II)	229.6	[17]
(MCC)	Ni(II)	108.9	
	Pb(II)	220.9	
EVA-PCL	As(III)	2.83	[19]
Fe_3O_4@silica-XG	Pb(II)	21.32	[11]
Fe_3O_4@SiO$_2$/PVA/CS	Cd(II)	307	[21]
	Cu(II)	100	
	Pb(II)	67	
m-chitosan/PVA/CCNFs	Pb(II)	171.0	[14]
MC-FeS	Cr(VI)		[46]
SA/CMC	Cu(II)	105.93	[51]
	Mn(II)	71.83	
	Pb(II)	89.49	
MCGO	Pb(II)	76.94	[52]
MCh-Fe	Cr(VI)	144.9	[53]
Fe_3O_4@glyPPy NC	Cr(VI)	238	[54]
Ppy-Fe_3O_4/rGO	Cr(VI)	293.3	[55]
Fe_3O_4-PVBC	Cu(II)	61.2	[56]
	Ni(II)	57.18	
	Zn(II)	67.29	
MCGO	Pb(II)	112.35	[57]
MGOCS	Ni(II)	80.48	[14]
β-CD@ Fe_3O_4/MWCNT	Ni(II)	103	[58]
MCTP	Hg(II)	247.02	[59]

9.2.2   OIL REMOVAL

One of the current environmental challenges is wastewater that contains oils in the form of fats, lubricants, petroleum products, and other oils [60]. Oil refineries, petrochemical facilities, transportation, lubricant and cooling agents, car washing, restaurants, and manufacturing processes can all be sources of oil pollution in wastewater [61]. Oil pollution in wastewater can also be caused by oil spills, incorrect oil disposal, and leakage from storage tanks. Oil pollution has persistent adverse effects on the environment, making its eradication one of the major environmental issues affecting the petrochemical and oil industries, where polluted effluents must be treated for contaminants. Due to potential hazards to the environment, oil spills and the treatment of oily water have recently received increased attention [62].

Several investigations have been conducted on a new approach to treating oily water using magnetic polymer composites, enabling the magnetic recovery of the oily phase. Magnetic materials such as Fe_3O_4 have, however, no affinity for oils. As a result, they need to be functionalized by species that can facilitate the adsorption of oil [63]. Collagen is an abundant naturally occurring matrix polymer that is found in many applications [64]. Thanikaivelan et al. (2012) utilized collagen-containing waste fibers in the synthesis of a cheap, stable, and magnetic nanocomposite with superparamagnetic iron oxide nanoparticles (SPIONs) [65]. The nanocomposite was used to selectively remove oil and was found to adsorb up to 2 times its own weight in oil, with a maximum oil sorption capacity of about 2 g/g. The nanoparticles were also used to stabilize the collagen waste fibers suspended in aqueous medium.

As mentioned earlier, it is considered easy for cyclodextrins to form clathrates with organic pollutants. In a small-scale laboratory experiment, Kumar and co-workers reported the synthesis of SPIONs and their nanocomposites with β-cyclodextrin (SPION/β-CD) [24]. Cyclodextrins are also reported to be hydrophobic inside, allowing them to form complexes with hydrophobic compounds, and their hydrophilic exterior make them compatible in water phase. The nanocomposite was used to absorb and selectively remove oil from a lubricating oil–water mixture. The nanocomposite had an oil retention capacity of 7.2 g/g and was capable of magnetic separation.

Chen and co-workers prepared a Fe_3O_4@polystyrene (PS) nanocomposite for the removal of oils and organic contaminants from water under an external magnetic field [25]. The oleophilic and hydrophobic properties of polystyrene increase the oil adsorption capacity of magnetic polymer composites. The nanocomposite selectively adsorbed up to 3 times its weight in lubricating oil and had excellent recyclability in the oil-absorbent capacity. Similarly, Ju et al. (2015) synthesized a Fe_3O_4/PS composite with a rough surface and different coating rates via emulsion polymerization [66]. The composite could adsorb up to 2.492 times its own weight in diesel oil. The oil could easily be removed from the surface of the nanoparticles by a simple ultrasonic treatment while the composite's hydrophobic and superoleophilic properties remained. Fe_3O_4/PS showed potential for application in the recovery of spilled oil.

Jiang et al. (2015) reported the incorporation of Fe_3O_4 nanoparticles in a magnetic nanofibrous composite mat composed of PS/polyvinylidene fluoride (PVDF) prepared via an electrospinning process for oil-in-water separation [67]. The PVDF exhibited oleophilic and hydrophobic properties similar to those of PS. The PVDF/Fe_3O_4@PS composite mat showed an adsorption capacity of 35–46 g/g and an improved mechanical property due to the presence of PVDF. This composite could be potentially useful for the efficient removal of oil in water, with the magnetic nanoparticles helping in the recovery of the sorbent material. Recently, Damavandi and Soares proposed polystyrene magnetic nanocomposite blends to remediate oil spills on water [68]. The composite was composed of polystyrene chains grafted on the surface of silica coated on iron oxide nanoparticles (IONP) and polystyrene (PS). The PS-SiO$_2$-IONP nanocomposite was also blended with polystyrene to form PS/PS-SiO$_2$-IONP, and both blends were employed in the removal of high- and low-viscosity (diesel and diluted bitumen) oils in water. The silica coating was employed to help protect and stabilize the IONP. These adsorbents were able to adsorb up to 5 times their own weight in oil in only 5 minutes and could be easily spread over the contaminated water in their powder form and removed using an external magnetic field.

Several properties can be exploited to fabricate an effective oil-adsorbing magnetic polymer composite. Sharma and co-workers successfully synthesized a magnetic nanosorbent by exploiting distinctive physiochemical properties of zincoxide tetrapod (ZnO-T), polydimethylsiloxane (PDMS), and iron oxide nanorod (Fe$_2$O$_3$-NR) for the removal of oil in wastewater [29]. The PDMS@ZnO-T@Fe$_2$O$_3$-NR nanohybrid (PZF nanohybrid) possessed a high surface area, low surface energy, and supermagnetism. It exhibited an adsorption capacity of 1135 mg/g within 50 minutes and could separate diesel oil from water with an oil removal percentage of 96. From these results it can be inferred that the PZF nanohybrid has the potential to be employed for effective removal of oil from oily wastewater.

Magnetic polymeric nanocomposites have been considered promising materials due to their high oil adsorption capacity, simple regeneration for reuse, and ability to quickly remove oil from the environment by means of an external magnetic field. Examples of other studies conducted on the efficient removal of oil using a variety of magnetic polymer composites are listed in Table 9.2.

TABLE 9.2
Adsorption Capacity of Magnetic Polymer Composites for Oil Removal

Magnetic Polymer Composite	Oil Pollutant	Adsorption Capacity (g g⁻¹)	References
SPION/β-CD	Lubricating	7.2	[24]
Fe$_3$O$_4$@PS	Lubricating	-	[25]
PZF nanosorbent	Diesel	1.135	[29]
Collagen-SPION	Premium motor oil	2.0	[65]
	Used motor oil		
Fe$_3$O$_4$/PS	Diesel	-	[66]
PVDF/ Fe$_3$O$_4$@PS	Sunflower	35–46	[67]
	Soybean		
	Motor		
	Diesel		
PS-SiO$_2$-IONP &	Diesel	-	[68]
PS/PS-SiO$_2$-IONP	Diluted bitumen		

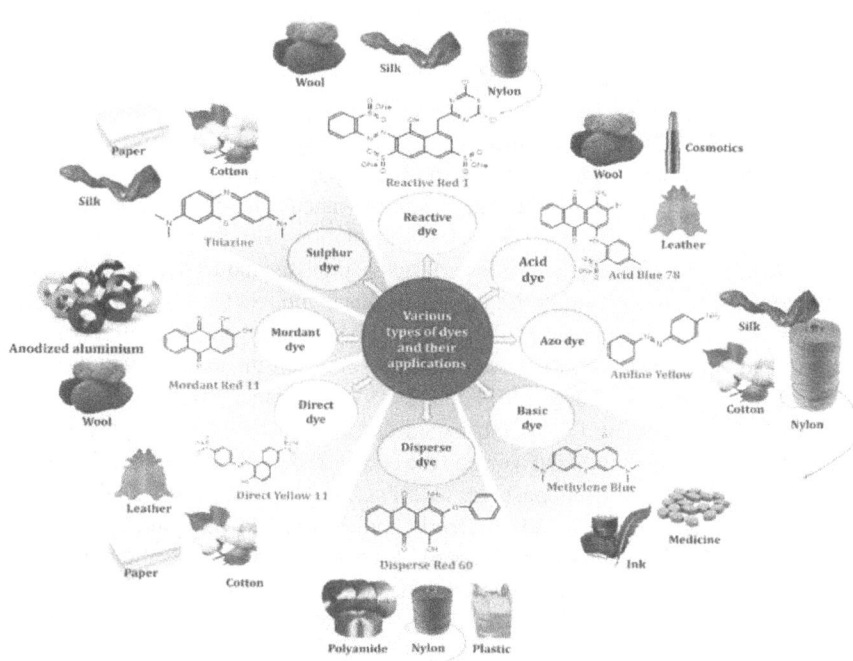

FIGURE 9.1   Various categories of dyes and their possible industrial applications. (Reprinted with permission from [73].)

9.2.3   DYE REMOVAL

Dyes are colored organic compounds, commonly used to impart color onto various substrates [69]. As shown in Figure 9.1, exposure to them is inevitable, as they are widely utilized for a range of applications including electronics, textiles, tanneries, food, paper and pulp, cosmetics; and many other fields of domestic and industrial interest [70, 71]. Consequently, dye effluents are released into the environment and water reservoirs, posing a threat to the ecosystem, plants, and humans [71, 72].

Upon ingestion, many of these dyes are toxic, mutagenic, and carcinogenic [74]. In addition to making aesthetic alterations to water bodies, dyes also hinder the penetration of sunlight in the water, leading to a drop in photosynthesis and water oxygen levels and thus affecting aquatic fauna and flora [73]. Moreover, most dyes are soluble and non-biodegradable, making them last for a long time in the environment and difficult to remove by conventional methods [75]. This has thus prompted research in devising efficient technologies for their treatment and eradication to ensure sustainability of the environment for future generations.

Although adsorption is the most common method, various other methods have been proposed and employed for the treatment of dyes in wastewater, including photodegradation, chemical precipitation, filtration, and electrochemical treatment [76–80]. A range of materials have been applied to carry out these treatment processes, such as semiconductors, carbon materials, and nanoparticles [81–83]. Due to their attractive properties, however, research interest in the use of magnetic polymer composites for the treatment of dyes has soared in recent years and is elaborated on below.

For instance, by impregnating polyaniline onto manganese ferrite, Das et al. (2022) fabricated a novel magnetic manganese ferrite/polyaniline nanocomposite ($MnFe_2O_4$-PANI-NC) for eradication of methyl orange (MO) dye from wastewater. The composite was characterized with a variety of techniques, and the MO dye removal efficiency was studied through sono-assisted adsorption. The researchers reported a removal efficiency of 90.03% at solution pH 6.0 and sonication time 15 min. The adsorption kinetics followed the pseudo–second order kinetics model and the linear progression fit the Langmuir isotherm model. They ascribed the MO dye adsorption mechanism onto $MnFe_2O_4$-PANI-NC to the synergistic effect of electrostatic interaction and $\pi-\pi$ dispersive bonding between the anionic MO dye molecules and the positively charged $MnFe_2O_4$-PANI-NC surface [84].

Modisha and Nyokong (2014) fabricated electrospun fibers using polyamide-6 (PA-6) which was functionalized using a zinc octacarboxy phthalocyanine (ZnOCPc) as well as Fe_3O_4 magnetic nanoparticles (MNPs). They reported good singlet oxygen quantum yields for the fibers and efficiently applied them for the photodegradation of orange-G (OG). They found that the photodegradation was in agreement with both first-order and Langmuir–Hinshewood kinetics. They attributed the good activity of the fibers to also be a result of MNPs inducing adsorption of the dye onto the catalyst surface, as was shown by comparing adsorption coefficients [85].

Similarly, Mapukata et al. (2017) electrospun polystyrene (PS) fibers modified with different phthalocyanines (ZnTCPPc and ZnTAPPc) as well as $CoFe_2O_4$ MNPs. The resulting composites (ZnTCPPc-$CoFe_2O_4$/PS and ZnTAPPc-$CoFe_2O_4$/PS) were compared based on their photophysical properties and photocatalytic efficiencies in degrading OG, using a laser as the photoexcitation source. They observed better singlet oxygen and reactive oxygen species (ROS) production for the composites than for the individual constituents. Kinetic analyses of the photodecomposition of OG showed that the degradation was in agreement with pseudo–first order kinetics and followed the Langmuir–Hinshelwood model [86].

Amiralian et al. (2020) grafted Fe_3O_4 MNPs onto the surface of cellulose nanofibers via in situ hydrolysis of metal precursors at room temperature. Flexible magnetic membranes containing a high concentration of MNPs (83–60 wt%) were generated, and they showed superparamagnetic behavior. Fenton-like catalytic oxidation of rhodamine B (RhB) was carried out. In the sulphate radical–based advanced oxidation process, peroxymonosulphate (PMS) was used to generate the sulphate radicals through the catalytic reaction driven by the MNPs. The membrane therefore successfully activated PMS to remove RhB, and a decrease in the concentration of Fe_3O_4 in the membrane resulted in a decrease in the catalytic efficiency of the membrane. A degradation efficiency of 94.9% was ultimately achieved in 300 min at room temperature [87].

Bober et al. (2020) reported a one-step procedure for the preparation of polyaniline/poly(vinyl alcohol) composite aerogels containing hexaferrite particles. They reported that the conducting and magnetic composites were macroporous, with good mechanical properties. They reported that the composite aerogel could adsorb 99% of the organic dye reactive black 5 after 4 h. They also found that the hexaferrite particles enhanced the activity of the materials and, with a high value of coercive force, allowed for the easy separation of adsorbent from the aqueous medium [88].

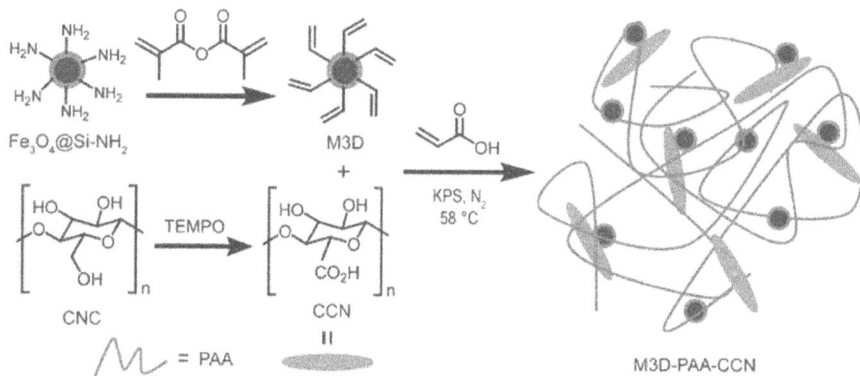

SCHEME 9.1   Schematic illustration of the synthesis of the M3D–PAA–CCN nanocomposite. (Reprinted with permission from [27].)

Samadder et al. (2020) developed a nanocomposite (M3D–PAA–CCN) based on polyacrylic acid (PAA) cross-linked with magnetic 3D cross-linkers (M3D) and carboxylated cellulose nanocrystals (CCN), for the removal of cationic methylene blue (MB). As depicted in Scheme 9.1, acrylic-functionalized Fe_3O_4 nanoparticles were covalently linked to the polymer chains to introduce magnetic properties into the as-synthesized nanocomposite. The addition of highly dispersive CCN reduced the gel-like properties of the nanocomposite and instead incorporated a diffusive nature, which was more desirable for adsorbents. The adsorption capacity of the M3D–PAA–CCN was found to increase with the increase in pH, owing to the greater negative charge as indicated by the higher zeta potential. The adsorption kinetics of MB on the composite was found to follow the pseudo–second order model [27].

Moreover, Mahto et al. (2014) synthesised polyaniline (PANI)–modified magnetic nanocomposites which were used as adsorbents for the removal of the cationic malachite green (MG) dye. They studied the effect of the pH, contact time, and initial dye concentration on the removal of the dye through batch experiment, wherein almost 85% of the dye was removed within 30 min. They also reported that the rate of adsorption followed pseudo–second order kinetics and fit the Langmuir isotherm model well [89].

Huang et al. (2017) encapsuled $Fe_3O_4@SiO_2$ nanoparticles in disordering porous organic polymers through a coupling reaction, generating magnetic porous organic polymer composites (MOPs) with abundant free phenolic hydroxyl groups. The prepared MOPs possessed excellent stability, high special surface areas, controllable magnetism, and high adsorption capacity toward MB. Moreover, the MOPs could easily be recovered and reused at least 5 times without decreasing their adsorption capacity. They attributed the efficiency of the MOPs to the phenolic-OH groups, which were effective for capturing MB through hydrogen bonding as well as electrostatic interaction [33].

Numerous other studies were conducted on the efficient removal of dyes using a variety of magnetic polymer composites, as listed in Table 9.3. Generally, these research findings demonstrate the versatility, durability, and flexibility of magnetic polymer

TABLE 9.3
Magnetic Polymer Composites for the Treatment of Dyes

Composite	Fabrication Method	Dye	Treatment Method	Surface Area (m^2/g)	Efficiency (%)	Isotherm	Kinetics	Ref.
MnFe$_2$O$_4$-PANI	Impregnation	Methyl orange	Sono-assisted adsorption	22.01	90.03	Langmuir	Pseudo–2nd order	[84]
PA-6/ZnOCPc-MNPs	Electrospinning	Orange G	Photodegradation	–	–	Langmuir–Hinshelwood	1st order	[85]
ZnTCPPc-CoFe$_2$O$_4$/PS and ZnTAPPc-CoFe$_2$O$_4$/PS	Electrospinning	Orange G	Photodegradation	108.62 and 27.70	–	Langmuir–Hinshelwood	Pseudo–1st order	[86]
Magnetic nanocellulose	In situ hydrolysis	Rhodamine B	Catalysis	–	94.90	–	–	[87]
PANI/PVAL/F	Oxidation	Reactive black 5	Adsorption	–	99.00	–	–	[88]
M3D–PAA–CCN	In situ polymerization	Methylene blue	Adsorption	–	–	Langmuir	Pseudo–2nd order	[27]
Fe$_3$O$_4$@PANI	In situ polymerization	Malachite green	Adsorption	–	85.00	Langmuir	Pseudo–2nd order	[89]
MOPs	Azo-coupling reaction	Methylene blue	Adsorption	321	100.0	–	Pseudo–2nd order	[33]
Fe$_3$O$_4$/MRCS	Solvothermal, cross-linking	Acid red 18	Adsorption	226.6	99. 65	Redlich–Peterson	Pseudo–2nd order	[90]
MPMWCNT	Multi-step process	Orange II	Adsorption	–	–	Langmuir	Pseudo–2nd order	[91]
MMGO	Multi-step process	Methyl violet	Adsorption	226 .0	–	Langmuir	Pseudo–2nd order	[92]
Fe$_3$O$_4$@polydopamine	Multi-step process	Toluidine blue	Adsorption	–	85.00	Freundlich	Pseudo–2nd order	[93]
MCPEI	Cross-linking	Reactive black 5	Adsorption	–	89.1	Langmuir	Pseudo–2nd order	[94]
GOs/Fe$_3$O$_4$/PANI	Hummers method, ultrasonication	Methyl orange	Adsorption	60.38	–	Langmuir	Pseudo–2nd order	[95]
NZF@PANi	Facile polymerization	Orange ll	Photocatalysis	–	100.0	–	–	[96]

Note: MRCS = magnetic porous melamine-formaldehyde resin-chitosan polymer networks composite, MPMWCNT = magnetic polymer multi-wall carbon nanotube, MMGO = modified magnetic graphene oxide by metformin, MCPEI = magnetic-cellulose polyethyleneimine, and NZF@PANi = Ni$_{0.5}$Zn$_{0.5}$Fe$_2$O$_4$@polyaniline composites in the treatment of dyes. This is a promising breakthrough because purification of water from dyes is of tremendous importance from the perspectives of water purification and materials reusability, both leading to enhanced sustainability.

9.2.4 REMOVAL OF PHARMACEUTICAL COMPOUNDS

Although pharmaceuticals are essential to the livelihood and well-being of humans and animals alike, their release into the environment is a rising concern. Pharmaceutical residues are often found in water due to discharge during drug manufacturing, excretion of human and animal waste (urine and feces), and hospital discharges, as well as inappropriate disposal of unused drugs [97].

Environmental emissions of antibiotics are particularly worrisome from a human health perspective because once released into the environment, they can cause antimicrobial resistance [98]. Antibiotic residues can be absorbed by plants, thereby interfering with physiological processes and potentially leading to ecotoxicological effects [97]. Additional repercussions associated with ingesting of some of these pharmaceutical compounds include an assortment of reproductive disturbances and behavioral changes as well as disruption of organ development [99]. A plethora of research has thus been conducted to combat the effects of pollution resulting from pharmaceutical compounds, as it is a major global threat to public health and economic development.

Numerous researchers have explored a range of treatment materials and methodologies for the eradication of pharmaceutical compounds in water. These include biological, photochemical, and sonochemical treatments, as well as ozonation among others [100–103]. Due to the complex nature of pharmaceutical compounds, the use of magnetic polymer composites for their treatment is still emerging, although promising results have been acquired from the little work that has been reported, as shown in Table 9.4.

Kumar et al. (2020) demonstrated the versatility of magnetic polymer composites by applying them to the detection, removal, and degradation of the antibiotic ciprofloxacin. They fabricated bismuth phosphate@graphene oxide–based magnetic nano-sized molecularly imprinted polymers ($BiPO_4$@GO-MMIPs) for electrochemical detection as well as adsorption and photocatalytic treatment of ciprofloxacin. They efficiently detected trace levels of ciprofloxacin, in addition to using the magnetic feature of the prepared polymer, which enhanced the fast removal by and high adsorption capacity of the polymer composite [104].

Mohammadi et al. (2020) used a reversible addition fragmentation transfer (RAFT) method to synthesize poly(styrene-block-acrylic acid) diblock copolymer/Fe_3O_4 magnetic nanocomposite P(St-b-AAc)/Fe_3O_4 for the removal of ciprofloxacin. After characterization, the effects on the process of various parameters, such as initial drug concentration, solution pH, adsorbent dosage, and contact time, were extensively studied. Upon optimization, a maximum removal efficiency of 97.5% was obtained. The adsorption studies were better fitted to the Langmuir isotherm and the kinetics were better fitted to the second-order kinetic equation [105].

Tamaddon et al. (2020) synthesized a novel $CuFe_2O_4$@methyl cellulose magnetic photocatalyst using a facile, rapid, and green new microwave-assisted method, as

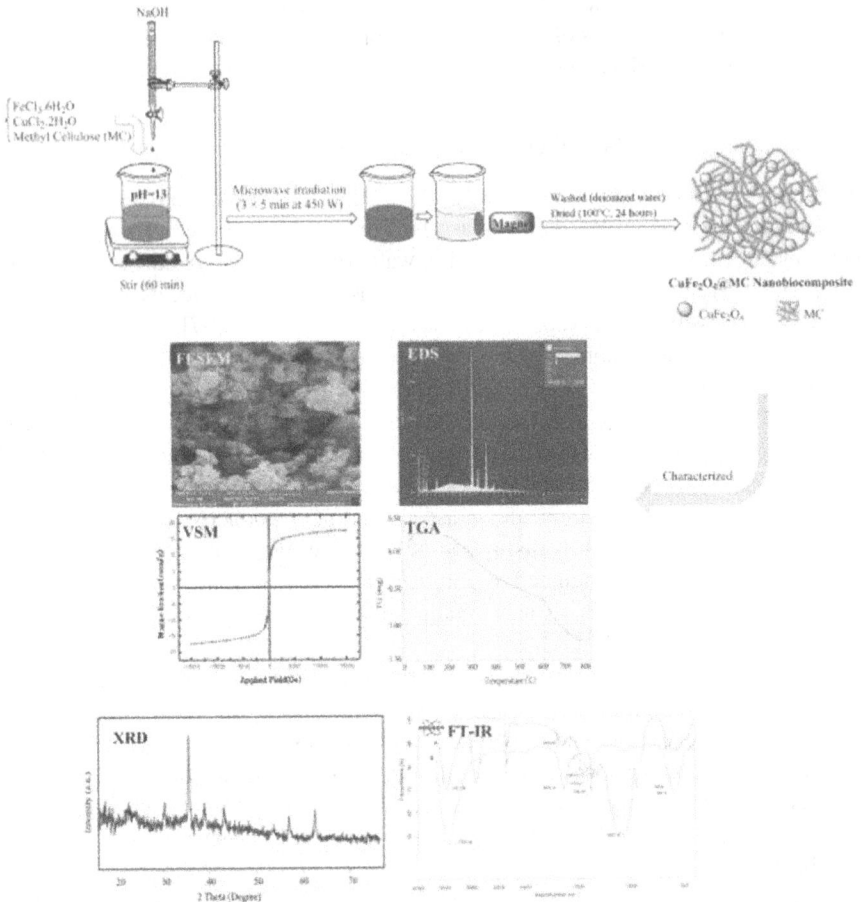

SCHEME 9.2   Synthesis and characterization of CuFe$_2$O$_4$@MC as a new magnetic nanobio-composite. (Reprinted with permission from [106].)

depicted in Scheme 9.2. The magnetic composite was characterized with a range of instrumentation, and its photocatalytic activity was evaluated based on its efficiency in the removal of ciprofloxacin. They reported maximum ciprofloxacin removal efficiencies of 72.87% and 80.74% from real and synthetic samples, respectively, at optimal conditions. Their kinetics studies were in agreement with the pseudo–first order and the Langmuir-Hinshelwood models [106].

Mohammadi and Pourmoslemi (2018) fabricated a polymer-ZnO composite by incorporating Fe$_3$O$_4$ and ZnO nanoparticles in the structure of an adsorbent polymer. The polymerization was used for synthesizing the adsorbent polymer, and its efficiency in extracting doxycycline from an aqueous solution was optimized according to several parameters, including time, pH, and amount of the polymer. Results showed 76.5% degradation of doxycycline in 6 h, which was significantly higher than the degradation observed for an equivalent amount of ZnO nanoparticles. The photocatalytic degradation of doxycycline fit the pseudo–first order kinetic model [107].

Moreover, Malakootian et al. (2019) synthesized a zinc ferrite-carboxymethyl cellulose carbohydrate ($ZnFe_2O_4$@CMC) composite using the hydrothermal method and studied its photocatalytic efficiency. They successfully characterized the photocatalyst using a range of instrumentation and optimized the photocatalysis conditions by evaluating the effect of reaction time, initial ciprofloxacin concentration, pH (3–11), and photocatalyst dosage. Optimal conditions for the maximum removal efficiency in the synthetic (87%) and real (79%) samples were achieved, and the kinetic studies showed that the degradation followed the pseudo–first order kinetic and Langmuir-Hinshelwood equation. Additionally, the new magnetic $ZnFe_2O_4$@CMC nanobiocomposite demonstrated good chemical stability and reusability after five runs [108].

Although the results depict the efficiency of magnetic polymer composites in the degradation of pharmaceuticals, extensive research on selectivity and co-existing pharmaceuticals of a different nature remains to be done.

9.2.5 MULTI-POLLUTANT REMOVAL

The feasibility of magnetic polymer composites for real-life applications is also highly influenced by their efficiency in treating complex water systems with a range of pollutants. Ideally, multi-pollutant (i.e., organic, inorganic, emerging pollutants. or pollutants of the same molecular species but with different charges) adsorbent

TABLE 9.4
Magnetic Polymer Composites for the Treatment of Pharmaceutical Compounds

Composite	Fabrication Method	Pharmaceutical Compound	Treatment Method	Surface Area (m²/g)	Efficiency (%)	Isotherm	Kinetics	References
BiPO₄@ GO-MMIPs	Atom transfer radical polymerization	Ciprofloxacin	Adsorption, photocatalysis	-	-	-	Pseudo–2nd order	[104]
P(St-*b*-AAc)/ Fe₃O₄	RAFT, adsorption	Ciprofloxacin	Adsorption	-	97.50	Langmuir	2nd order	[105]
CuFe₂O₄@ methyl cellulose	Microwave-assisted	Ciprofloxacin	Photocatalysis	-	80.74	Langmuir–Hinshelwood	Pseudo–1st order	[106]
Polymer-ZnO	Precipitation polymerization	Doxycycline	Photocatalysis	-	76.50	-	Pseudo–1st order	[107]
ZnFe₂O₄@ CMC	Hydrothermal	Ciprofloxacin	Photocatalysis	147.75	87.00	Langmuir–Hinshelwood	Pseudo–1st order	[108]

materials will be used to combat the degree of complexity in the treatment of wastewater [109–112]. For instance, a polypyrrole-modifed $Fe_3O_4@SiO_2$ (PPR@ $Fe_3O_4@$ SiO_2) magnetic polymer composite was evaluated for the removal of Congo red dye (CRD) and Cr^{6+} ions in distilled water, tap water, ground water, and sewage water [34]. The adsorption of CRD and Cr^{6+} ions onto a polypyrrole-modifed $Fe_3O_4@$ SiO_2 composite followed the Langmuir isotherm, suggesting a monolayer adsorption process. Moreover, the composite demonstrated a maximum adsorption capacity of 361 mg/g and 298 mg/g for the CRD and Cr^{6+} ions, respectively, with successful reusability for six cycles. The study displayed a slight decline in the removal efficiency (from ≈98% to 80%, Figure 9.2a) with an increase in the complexity of the type of water (i.e., distilled water > tap water > ground water > sewage water). This was attributed to the competitive adsorption between the targeted pollutants and the other contaminants existing in each type of water tested [34]. Although the competing contaminants were not mentioned in the report, some authors have reported the interference of sulphate, chlorides, and carbonates in the adsorption with the Cr^{6+} ions. For example, Alsaiari et al. (2021) reported simultaneous adsorption of Cr^{6+} and methyl orange (MO) onto polypyrrole-coated $Fe_3O_4@chitosan$ (PPR@$Fe_3O_4@$ chitosan) composites and the effect of the presence of carbonates, sulphates, and chloride ions on the adsorption efficiency of MO and Cr^{6+} [16]. The adsorption of MO and Cr^{6+} fitted well with the Langmuir isotherms, suggesting monolayer adsorption processes. FTIR and XPS studies suggested that the adsorption mechanism of MO was based on the adsorption of MO via binding sites on the composite. In acidic conditions, some Cr^{6+} ions were reduced to Cr^{3+}, which bonded both to nitrogen species and via electrostatic interactions with other functionalities on the surface of the composite. A maximum adsorption capacity of 95 mg/g and 105 mg/g was obtained for MO and Cr^{6+}, respectively, with up to five successful cycles of reusability. The sulfate and chloride ions greatly affected the adsorption of MO and Cr^{6+}, while bicarbonate had the least effect on both pollutants (Figure 9.2b). The decline in

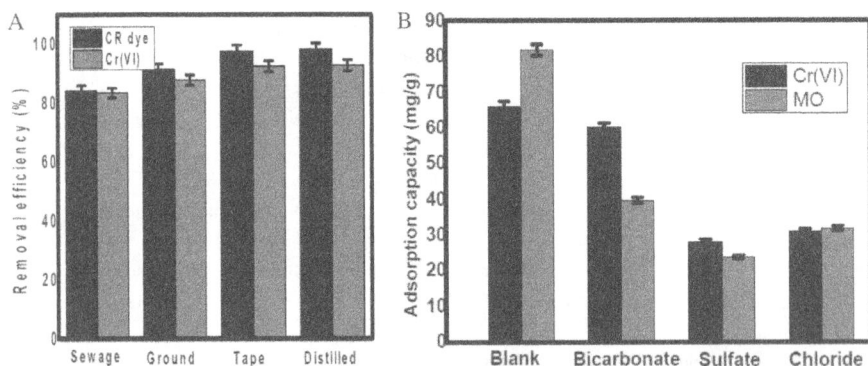

FIGURE 9.2   (a) Percentage removal efficiency of Congo red dye and Cr(VI) ions in distilled water, tap water, ground water, and sewage water, using the polypyrrole-modifed $Fe_3O_4@$ SiO_2 magnetic polymer composite. (Reprinted with permission from [34].) (b) The effect of carbonate, sulfate, and chloride in the adsorption performance of a polypyrrole-coated $Fe_3O_4@chitosan$ composite toward MO and Cr^{6+}. (Reprinted with permission from [16].)

the adsorption capacity in the presence of sulfate and chloride ions was attributed to the high affinity of sulfate groups for amino groups on the composite surface and the greater negative density charge of chloride ions compared to other negative ions [16].

Furthermore, by combining novel materials previously reported to perform well in the removal of different types of pollutants, researchers have designed multi-functional magnetic composites capable of removing multiple pollutants. For example, Yadav et al. (2019) demonstrated the fabrication of magnetic polymer gels based on Fe_3O_4 NPs–activated charcoal particles (AC), β-cyclodextrin (CD), and sodium alginate (SA) (Fe_3O_4/CD/AC/SA) for the simultaneous removal of two cationic dyes [i.e., methyl violet (MV) and brilliant green (BG)], one metal (i.e., Cu^{2+}), and two drugs [i.e., norfloxacin (NOX) and ciprofloxacin (CPF)] [26]. A previous report by Zhao et al. (2015) had showed that EDTA-cross-linked β-cyclodextrin performed well in the simultaneous removal of metals and cationic dyes [113]. Activated charcoal is suitable for adsorption of several pollutants, including pharmaceutical drugs and dyes, at large scale owing to its low cost, high surface area, and porous surfaces [114, 115]. Enhanced adsorption capacity has been reported when β-cyclodextrin was conjugated to sodium alginate to form a new SA-CD compound [116]. Sodium alginate is a non-toxic natural polymer, highly hydrophilic due to the numerous carboxylic acid and hydroxyl groups [117]. Sodium alginate is commonly used to fabricate hydrogels/beads for wastewater treatment because it is a good entrapping agent for other adsorbent material in the fabrication of multi-functional composites and has the ability to bind metal ions [117, 118]. The magnetic polymer gel Fe_3O_4/CD/AC/SA displayed a maximum adsorption capacity of 5.882 mg/g, 2.283 mg/g, 2.551 mg/g, 3.125 mg/g, and 10.10 mg/g for MV, BG, NOX, CPX, and Cu^{2+} ions, respectively. The adsorption mechanism was attributed to a combination of hydrogen bond, π–π interaction, electrostatic interaction, electron-donor acceptor interaction, and complex formation, with evidence from FTIR, EDX, and SEM analysis (Figure 9.3). Furthermore, the magnetic polymer gel was easily generated without weight loss for up to four cycles of adsorption [26].

In another development, Phiri et al. (2019) demonstrated the synthesis of magnetic alginate beads based on zeolites, activated carbon (AC), layered double hydroxides (LDHs), and Fe_3O_4 NPs bound in xanthan gum, for simultaneous removal of Cu^{2+}, PO_4^{2-}, and toluene [119]. In a previous report, Choi et al. (2009) had developed beads based on zeolite and activated carbon to produce highly porous adsorbent materials in order to increase the kinetics of the adsorption process for the removal of multiple pollutants [120]. Zeolites are natural mineral materials based on hydrated aluminosilicate minerals that contain alkali and alkaline–earth metals. Zeolites have been widely used as adsorbents for pollutants [121–123]. Further modification of some mineral-based materials by entrapping them in sodium alginate beads has led to enhanced adsorption capacity and easier collection of composite after treatment [118, 124–126]. Activated carbon is another adsorbent that has been used for the removal of inorganic and organic pollutants in water, both on its own and as a component in composites [127–130]. Xanthum gum is often used as a binder and was applied in the bead to bind Fe_3O_4 NPs, AC, LDHs, and zeolites to avoid their leaching outside the beads. It is also a promising adsorbent for metal ions [131]. LDHs occur naturally and can be easily fabricated at large -scale. The LDH structure exhibits positively charged sheets or layers of hydroxides. The positive charges are balanced with anions such as nitrates, which provide unique anionic exchange properties, competing with anion exchange resins—hence their wide applications in the removal of anions in water. They display high surface area and good thermal stability [132–134]. Additionally, their functionalized derivatives have been reported for simultaneous removal of dyes and heavy metals [112]. The zeolite@AC@LDH@ xanthan gum magnetic alginate beads exhibited maximum adsorption capacities of 60.24 mg g^{-1}, 120.77 mg g^{-1}, and 25.52 mg g^{-1} for Cu^{2+}, PO_4^{3-}, and toluene, respectively, demonstrating a greater affinity for PO_4^{2-} than for Cu^{2+} and toluene (PO_4^{3-} > Cu^{2+} > toluene) [119]. The greater affinity of the composite for PO_4^{3-} (Figure 9.4) is possibly due to the multiple beneficial binding sites of PO_4^{3-} such as (1) electrostatic attraction of the PO_4^{3--} ions and the positively charged LDH surface, (2) ligand complexation, and (3) anionic exchange between NO_3^- ions and the PO_4^{3-} in the basal space [135–138]. Isotherm and kinetic studies suggested Langmuir isotherm model and second-order kinetics governed the adsorption processes. The recyclability studies showed that the magnetic alginate beads can be reused for up to 10 cycles with minimal loss in adsorption performance [119].

Although only a few reports have been disseminated (Table 9.5), the multifunctional magnetic polymer composites show great promise for the simultaneous removal of various types of pollutants, which is a close simulation of the scenario in real wastewater.

TABLE 9.5

Adsorption Capacity of Magnetic Polymer Composites for Multi-pollutant Removal

Magnetic Polymer Composite	Pollutants	Adsorption Capacity (mg/g)	References
PPR-modified Fe$_3$O$_4$@chitosan	MO	95	[16]
	Cr^{6+}	105	
Fe$_3$O$_4$/CD/AC/SA	MV	5.882	[26]
	BG	2.283	
	NOX	2.551	
	CPX	3.125	
	Cu^{2+}	10.10	
PPR-modified Fe$_3$O$_4$@SiO$_2$	CRD	361	[34]
	Cr^{6+}	298	
Zeolite@AC@ LDH@xanthan gum	Cu^{2+}	60.24	[119][139]
	PO$_4^{3-}$	120.77	
Magnetic alginate beadPolydopamin@ Fe$_3$O$_4$	Toluene	25.52	
	MB	204.1	
	Tartrazine	100	
	Cu^{2+}	112.9	
	Ag$^+$	259.1	
	Hg^{2+}	467.3	

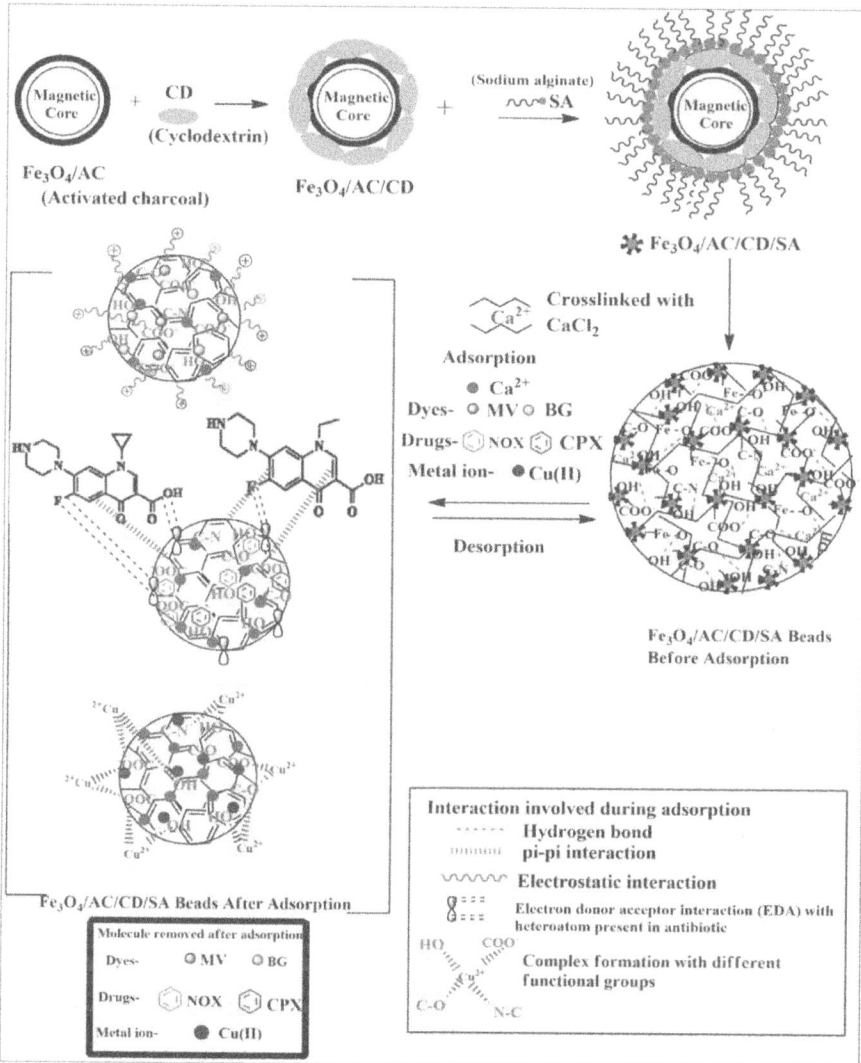

FIGURE 9.3 Schematic diagram of the proposed mechanism for the adsorption of MV, BG, NOX, CPX, and Cu^{2+} ions onto Fe_3O_4/CD/AC/SA magnetic composites. (Reprinted with permission from [26].)

(a)

Layers of double hydroxides

Basal space

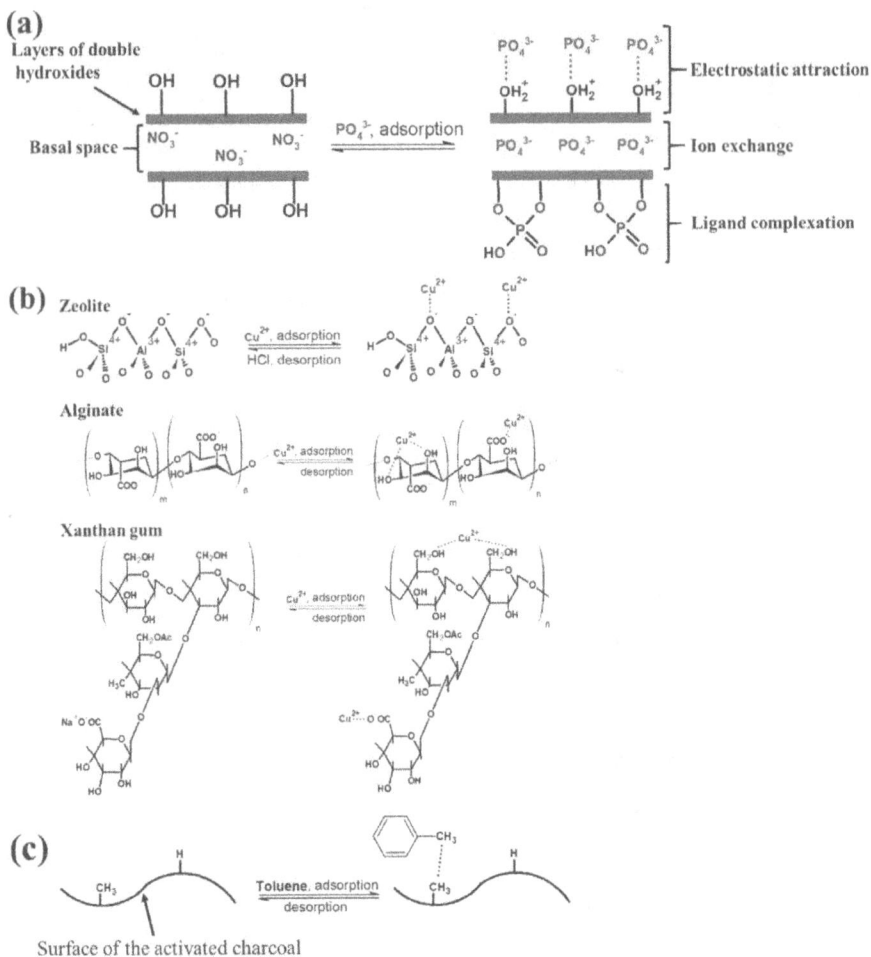

PO$_4^{3-}$, adsorption

Electrostatic attraction

Ion exchange

Ligand complexation

(b) Zeolite

Cu^{2+}, adsorption
HCl, desorption

Alginate

Cu^{2+}, adsorption
desorption

Xanthan gum

Cu^{2+}, adsorption
desorption

(c)

Toluene, adsorption
desorption

Surface of the activated charcoal

FIGURE 9.4 (a–c) Schematic diagram of the proposed adsorption mechanism for the removal of Cu^{2+}, PO$_4^{3-}$, and toluene using zeolite@AC@LDH@xanthan gum magnetic alginate bead. (Reprinted with permission from [119].)

9.3 CONCLUSIONS, CHALLENGES, AND FUTURE OUTLOOK

In summary, magnetic polymer composites offer a promising approach to removing pollutants from wastewater efficiently. Their unique properties, such as magnetic responsiveness and high adsorption capacity, make them attractive for practical applications in the field of environmental remediation. However, further research and development are necessary to optimize their performance and ensure their safe and sustainable use in wastewater treatment processes.

Some of the key challenges for magnetic polymer composites are synthesis, scale-up, and cost-effectiveness. The synthesis of magnetic polymer composites can be complex and requires careful control of parameters to achieve the desired properties. For instance, the adsorption capacity and selectivity of magnetic polymer composites for specific pollutants might not be as high as desired for some contaminants. Tailoring the composite's properties to target a wide range of pollutants with high efficiency remains a challenge. In some cases, magnetic polymer composites suffer from problems associated with the compatibility of the polymer with the magnetic NM, as well as obtaining proper homogeneity of magnetic NM in the polymer host. Even though a lot of research on synthesis methods has been published in the previous decade, scaling up the production of magnetic polymer composites for real-world applications while maintaining consistency and reproducibility is still a challenge. The cost of synthesizing magnetic polymer composites can be higher than that of traditional water treatment materials, which could impact their widespread adoption in wastewater treatment because cost-effectiveness will be a critical consideration for large-scale applications.

Although research has shown the efficiency of magnetic polymer composites in the removal of various types of pollutants, selectivity and co-existing pollutants of different natures remain understudied. Most reports demonstrate the removal of single pollutants or multiple pollutants of the same type (i.e., multi-metal or multi-dye removal), which does not solve the problem of remediation of complex wastewater systems. A few reports have been dedicated to the fabrication of multi-functional magnetic composites for the removal of multiple pollutants of different types, but oil is rarely included as one of the target pollutants. Some of these reports have not studied the effect of co-existing anionic pollutants, such as sulfate, phosphate and carbonate, which could compete for adsorption sites with target pollutants, and the composites were rarely tested in real wastewater samples.

In reality, the chemical composition of wastewater can differ significantly depending on the location, industrial activities, and the wastewater treatment process. Thus, it would be crucial to initially profile the chemical composition of the water before developing an adsorbent for remediation. Furthermore, real-world testing and demonstration should be conducted; carrying out extensive pilot-scale and field trials of the magnetic polymers in actual wastewater treatment plants will provide valuable insights into their performance, challenges, and practical implementation.

As previously discussed in Section 9.2.5, innovative adsorbents can be fabricated by combining previously reported materials that claim to remove pollutants. So multi-functional magnetic polymer composites can be developed to remediate wastewater of a particular profile. Moreover, the application of magnetic polymer composites is not limited to adsorption; they can be utilized in other treatment processes, such as membrane filtration or advanced oxidation, to create hybrid systems that offer improved overall wastewater treatment efficiency.

Other issues include nanoparticle leaching. Although one of the benefits of polymerizing magnetic NPs is to avoid NP leaching, the possibility of NP leaching cannot be ruled out, especially considering the complexity of the components of wastewater, and this may pose environmental risks. The release of magnetic NPs into the treated water needs to be carefully assessed to ensure that it does not lead to secondary contamination. This can be achieved by covalent conjugation between the polymer and magnetic NPs. Furthermore, the potential long-term environmental impact of using magnetic polymer composites in wastewater treatment needs to be thoroughly assessed. This includes understanding their fate in the environment, potential toxicity, and overall sustainability.

REFERENCES

1. M. Elsayed Abdel-Raouf, N.E. Maysour, R. Kamal Farag, A.-R. Mahmoud Abdul-Raheim, C. Wastewater treatment methodologies, review article, Int. J. Environ. Agric. Sci. 3 (2019) 18.

2. A. Sharma, D. Mangla, S.A. Chaudhry, Recent advances in magnetic composites as adsorbents for wastewater remediation, J. Environ. Manage. 306 (2022). https://doi.org/10.1016/j.jenvman.2022.114483.

3. M. Sharma, P. Kalita, K.K. Senapati, A. Garg, Study on magnetic materials for removal of water pollutants, Intech, 2018: pp. 1–19. https://doi.org/10.5772/intechopen.75700

4. M. Khodakarami, M. Bagheri, Recent advances in synthesis and application of polymer nanocomposites for water and wastewater treatment, J. Clean. Prod. 296 (2021) 126404. https://doi.org/10.1016/j.jclepro.2021.126404.

5. W.J. Weber Jr., P.M. Mcginley, L.E. Katz, Sorption phenomena in subsurface systems: Concepts, models and effects on contaminant fate and transport, Water Res. 25 (1991) 499–528.

6. M.A. Barakat, New trends in removing heavy metals from industrial wastewater, Arab. J. Chem. 4 (2011) 361–377. https://doi.org/10.1016/j.arabjc.2010.07.019.

7. H. Patel, Fixed-bed column adsorption study: A comprehensive review, Appl. Water Sci. 9 (2019) 1–17. https://doi.org/10.1007/s13201-019-0927-7.

8. E. Vunain, A.K. Mishra, B.B. Mamba, Dendrimers, mesoporous silicas and chitosan-based nanosorbents for the removal of heavy-metal ions: A review, Int. J. Biol. Macromol. 86 (2016) 570–586. https://doi.org/10.1016/j.ijbiomac.2016.02.005.

9. R.E. Morsi, A.M. Al-Sabagh, Y.M. Moustafa, S.G. ElKholy, M.S. Sayed, Polythiophene modified chitosan/magnetite nanocomposites for heavy metals and selective mercury removal, Egypt. J. Pet. 27 (2018) 1077–1085. https://doi.org/10.1016/j.ejpe.2018.03.004.

10. M. Najafi, Y. Yousefi, A.A. Rafati, Synthesis, characterization and adsorption studies of several heavy metal ions on amino-functionalized silica nano hollow sphere and silica gel, Purif. Technol. 85 (2012) 193–205. https://doi.org/10.1016/j.seppur.2011.10.011.

11. S.S. Kotsyuda, V.V. Tomina, Y.L. Zub, I.M. Furtat, A.P. Lebed, M. Vaclavikova, I.V. Melnyk, Bifunctional silica nanospheres with 3-aminopropyl and phenyl groups. Synthesis approach and prospects of their applications, Appl. Surf. Sci. 420 (2017) 782–791. https://doi.org/10.1016/j.apsusc.2017.05.150.

12. C. Yu, X. Han, Adsorbent material used in water treatment-a review, Proceedings of the 2015 2nd International Workshop on Materials Engineering and Computer Sciences (2015) 286–289. https://doi.org/10.2991/iwmecs-15.2015.55.

13. A.O. Ibrahim, K.A. Adegoke, R.O. Adegoke, Y.A. AbdulWahab, V.B. Oyelami, M.O. Adesina, Adsorptive removal of different pollutants using metal-organic framework adsorbents, J. Mol. Liq. 333 (2021) 115593. https://doi.org/10.1016/j.molliq.2021.115593.

14. T.T.N. Le, V.T. Le, M.U. Dao, Q.V. Nguyen, T.T. Vu, M.H. Nguyen, D.L. Tran, H.S. Le, Preparation of magnetic graphene oxide/chitosan composite beads for effective removal of heavy metals and dyes from aqueous solutions, Chem. Eng. Commun. 206 (2019) 1337–1352. https://doi.org/10.1080/00986445.2018.1558215.

15. W. Yao, T. Ni, S. Chen, H. Li, Y. Lu, Graphene/Fe_3O_4 at polypyrrole nanocomposites as a synergistic adsorbent for Cr(VI) ion removal, Compos. Sci. Technol. 99 (2014) 15–22. https://doi.org/10.1016/j.compscitech.2014.05.007.

16. N.S. Alsaiari, A. Amari, K.M. Katubi, F.M. Alzahrani, F. Ben Rebah, M.A. Tahoon, Innovative magnetite based polymeric nanocomposite for simultaneous removal of methyl orange and hexavalent chromium from water, Processes. 9 (2021). https://doi.org/10.3390/pr9040576.

17. M.E. Peralta, R. Nisticò, F. Franzoso, G. Magnacca, L. Fernandez, M.E. Parolo, E.G. León, L. Carlos, Highly efficient removal of heavy metals from waters by magnetic chitosan-based composite, Adsorption. 25 (2019) 1337–1347. https://doi.org/10.1007/s10450-019-00096-4.

18. G. Lofrano, M. Carotenuto, G. Libralato, R.F. Domingos, A. Markus, L. Dini, R.K. Gautam, D. Baldantoni, M. Rossi, S.K. Sharma, M.C. Chattopadhyaya, M. Giugni, S. Meric, Polymer functionalized nanocomposites for metals removal from water and wastewater: An overview, Water Res. 92 (2016) 22–37. https://doi.org/10.1016/j.watres.2016.01.033.

19. E. Vunain, A.K. Mishra, R.W. Krause, Fabrication, characterization and application of polymer nanocomposites for arsenic(III) removal from water, J. Inorg. Organomet. Polym. Mater. 23 (2013) 293–305. https://doi.org/10.1007/s10904-012-9775-8.

20. X. Peng, F. Xu, W. Zhang, J. Wang, C. Zeng, M. Niu, E. Chmielewská, Magnetic Fe_3O_4 @ silica-xanthan gum composites for aqueous removal and recovery of Pb2+, Colloids Surfaces A Physicochem. Eng. Asp. 443 (2014) 27–36. https://doi.org/10.1016/j.colsurfa.2013.10.062.

21. S. Wang, Y. Liu, A. Yang, Q. Zhu, H. Sun, P. Sun, B. Yao, Y. Zang, X. Du, L. Dong, Xanthate-modified magnetic Fe_3O_4@SiO_2-based polyvinyl alcohol/chitosan composite material for efficient removal of heavy metal ions from water, Polymers (Basel). 14 (2022). https://doi.org/10.3390/polym14061107.

22. M. Ahmadi, M. Foladivanda, N. Jaafarzadeh, Z. Ramezani, B. Ramavandi, S. Jorfi, B. Kakavandi, Synthesis of chitosan zero-valent iron nanoparticles-supported for cadmium removal: Characterization, optimization and modeling approach, J. Water Supply Res. Technol. - AQUA. 66 (2017) 116–130. https://doi.org/10.2166/aqua.2017.027.

23. Y. Zhou, S. Fu, L. Zhang, H. Zhan, M.V. Levit, Use of carboxylated cellulose nano-fibrils-filled magnetic chitosan hydrogel beads as adsorbents for Pb(II), Carbohydr. Polym. 101 (2014) 75–82. https://doi.org/10.1016/j.carbpol.2013.08.055.

24. A. Kumar, G. Sharma, M. Naushad, S. Thakur, SPION/β-cyclodextrin core-shell nano-structures for oil spill remediation and organic pollutant removal from waste water, Chem. Eng. J. 280 (2015) 175–187. https://doi.org/10.1016/j.cej.2015.05.126.

25. M. Chen, W. Jiang, F. Wang, P. Shen, P. Ma, J. Gu, J. Mao, F. Li, Synthesis of highly hydrophobic floating magnetic polymer nanocomposites for the removal of oils from water surface, Appl. Surf. Sci. 286 (2013) 249–256. https://doi.org/10.1016/j.apsusc.2013.09.059.

26. S. Yadav, A. Asthana, A.K. Singh, R. Chakraborty, S.S. Vidya, M.A.B.H. Susan, S.A.C. Carabineiro, Adsorption of cationic dyes, drugs and metal from aqueous solutions using a polymer composite of magnetic/β-cyclodextrin/activated charcoal/Na alginate: Isotherm, kinetics and regeneration studies, J. Hazard. Mater. 409 (2021). https://doi.org/10.1016/j.jhazmat.2020.124840.

27. R. Samadder, N. Akter, A.C. Roy, M.M. Uddin, M.J. Hossen, M.S. Azam, Magnetic nanocomposite based on polyacrylic acid and carboxylated cellulose nanocrystal for the removal of cationic dye, RSC Adv. 10 (2020) 11945–11956. https://doi.org/10.1039/d0ra00604a.

28. R.P. Dhavale, R.P. Dhavale, S.C. Sahoo, P. Kollu, S.U. Jadhav, P.S. Patil, T.D. Dongale, A.D. Chougale, P.B. Patil, Chitosan coated magnetic nanoparticles as carriers of anti-cancer drug telmisartan: pH-responsive controlled drug release and cytotoxicity studies, J. Phys. Chem. Solids. 148 (2021) 109749. https://doi.org/10.1016/j.jpcs.2020.109749.

29. M. Sharma, M. Joshi, S. Nigam, D.K. Avasthi, R. Adelung, S.K. Srivastava, Y.K. Mishra, Efficient oil removal from wastewater based on polymer coated superhydro-phobic tetrapodal magnetic nanocomposite adsorbent, Appl. Mater. Today. 17 (2019) 130–141. https://doi.org/10.1016/j.apmt.2019.07.007.

30. S. Zhang, Y. Dong, Z. Yang, W. Yang, J. Wu, C. Dong, Adsorption of pharmaceuticals on chitosan-based magnetic composite particles with core-brush topology, Chem. Eng. J. 304 (2016) 325–334. https://doi.org/10.1016/j.cej.2016.06.087.

31. Q. Lu, K. Choi, J.D. Nam, H.J. Choi, Magnetic polymer composite particles: Design and magnetorheology, Polymers (Basel). 13 (2021) 1–21. https://doi.org/10.3390/polym13040512.

32. A. Jazzar, H. Alamri, Y. Malajati, R. Mahfouz, M. Bouhrara, A. Fihri, Recent advances in the synthesis and applications of magnetic polymer nanocomposites, J. Ind. Eng. Chem. 99 (2021) 1–18. https://doi.org/10.1016/j.jiec.2021.04.011.

33. L. Huang, M. He, B. Chen, Q. Cheng, B. Hu, Facile green synthesis of magnetic porous organic polymers for rapid removal and separation of methylene blue, ACS Sustainable Chem. Eng. (2017). http://pubs.acs.org.

34. F.M. Alzahrani, N.S. Alsaiari, K.M. Katubi, A. Amari, F. Ben Rebah, M.A. Tahoon, Synthesis of polymer-based magnetic nanocomposite for multi-pollutants removal from water, Polymers (Basel). 13 (2021) 1–16. https://doi.org/10.3390/polym13111742.

35. P.B. Tchounwou, C.G. Yedjou, A.K. Patlolla, D.J. Sutton, Molecular, clinical and environmental toxicicology: Volume 3: Environmental toxicology, Springer, 2012. https://doi.org/10.1007/978-3-7643-8340-4.

36. R.M.A. Shmeis, Nanotechnology in wastewater treatment, Elsevier, 2022. https://doi.org/10.1016/bs.coac.2021.11.002.

37. N. Abdu, A.A. Abdullahi, A. Abdulkadir, Heavy metals and soil microbes, Environ. Chem. Lett. 15 (2017) 65–84. https://doi.org/10.1007/s10311-016-0587-x.

38. E.M.S. Azzam, G. Eshaq, A.M. Rabie, A.A. Bakr, A.A. Abd-Elaal, A.E. El Metwally, S.M. Tawfik, Preparation and characterization of chitosan-clay nanocomposites for the removal of Cu(II) from aqueous solution, Int. J. Biol. Macromol. 89 (2016) 507–517. https://doi.org/10.1016/j.ijbiomac.2016.05.004.

39. M. Yadav, R. Gupta, R.K. Sharma, Green and sustainable pathways for wastewater purification, Elsevier, 2019. https://doi.org/10.1016/B978-0-12-814790-0.00014-4.

40. G.K. Kinuthia, V. Ngure, D. Beti, R. Lugalia, A. Wangila, L. Kamau, Levels of heavy metals in wastewater and soil samples from open drainage channels in Nairobi, Kenya: Community health implication, Sci. Rep. 10 (2020) 1–13. https://doi.org/10.1038/s41598-020-65359-5.

41. A.C. Balazs, T. Emrick, T.P. Russell, Nanoparticle polymer composites: Where two small worlds meet, Science. 314 (2006) 1107–1110. https://doi.org/10.1126/science.1130557.

42. D.S. Dlamini, A.K. Mishra, B.B. Mamba, Adsorption behaviour of ethylene vinyl acetate and polycaprolactone-bentonite composites for pb 2+ uptake, J. Inorg. Organomet. Polym. Mater. 22 (2012) 342–351. https://doi.org/10.1007/s10904-011-9640-1.

43. F.-C. Wu, R.-L. Tseng, R.-S. Juang, Kinetic modeling of liquid-phase adsorption of reactive dyes and metal ions on chitosan, Water Res. 35 (2001) 613–618. https://doi.org/10.1016/S0043-1354(00)00307-9.

44. A. Nikmah, A. Taufiq, A. Hidayat, Synthesis and characterization of Fe_3O_4/SiO_2 nanocomposites, Earth Environ. Sci. 276 (2019). https://doi.org/10.1088/1755-1315/276/1/012046.

45. C. Guo, Y. Wang, F. Wang, Y. Wang, Adsorption performance of amino functionalized magnetic molecular sieve adsorbent for effective removal of lead ion from aqueous solution, Nanomaterials. 11 (2021). https://doi.org/10.3390/nano11092353.

46. H. Zhang, L. Peng, A. Chen, C. Shang, M. Lei, K. He, S. Luo, J. Shao, Q. Zeng, Chitosan-stabilized FeS magnetic composites for chromium removal: Characterization, performance, mechanism, and stability, Carbohydr. Polym. 214 (2019) 276–285. https://doi.org/10.1016/j.carbpol.2019.03.056.

47. Y. Liu, Z. Zhang, N. Bhandari, Z. Dai, F. Yan, G. Ruan, A.Y. Lu, G. Deng, F. Zhang, H. Al-Saiari, A.T. Kan, M.B. Tomson, A new approach to study iron sulfide precipitation kinetics, solubility, and phase transformation, Ind. Eng. Chem. Res. 56 (2017) 9016–9027. https://doi.org/10.1021/acs.iecr.7b01615.

48. M. Karthikeyan, K.K. Satheeshkumar, K.P. Elango, Removal of fluoride ions from aqueous solution by conducting polypyrrole, J. Hazard. Mater. 167 (2009) 300–305. https://doi.org/10.1016/j.jhazmat.2008.12.141.

49. X. He, F. Guo, T. Ge, K. Tang, S. Shi, M. Geng, Preparation and properties of carboxylated cellulose nanofibers/monomer casting nylon composites, Polym. Eng. Sci. 62 (2022) 3462. https://doi.org/10.1002/pen.26118.

50. G.F. de Lima, A.G. de Souza, D.S. Rosa, Nanocellulose as reinforcement in carboxy-methylcellulose superabsorbent nanocomposite hydrogels, Macromol. Symp. 394 (2020) 1–9. https://doi.org/10.1002/masy.202000126.

51. S. Wu, J. Guo, Y. Wang, C. Huang, Y. Hu, Facile preparation of magnetic sodium alginate/carboxymethyl cellulose composite hydrogel for removal of heavy metal ions from aqueous solution, J. Mater. Sci. 56 (2021) 13096–13107. https://doi.org/10.1007/s10853-021-06044-4.

52. L. Fan, C. Luo, M. Sun, X. Li, H. Qiu, Highly selective adsorption of lead ions by water-dispersible magnetic chitosan/graphene oxide composites, Colloids Surfaces B Biointerfaces. 103 (2013) 523–529. https://doi.org/10.1016/j.colsurfb.2012.11.006.

53. Z. Yu, X. Zhang, Y. Huang, Magnetic chitosan-iron(III) hydrogel as a fast and reusable adsorbent for chromium(VI) removal, Ind. Eng. Chem. Res. 52 (2013) 11956–11966. https://doi.org/10.1021/ie400781n.

54. N. Ballav, H.J. Choi, S.B. Mishra, A. Maity, Synthesis, characterization of Fe3O4@glycine doped polypyrrole magnetic nanocomposites and their potential performance to remove toxic Cr(VI), J. Ind. Eng. Chem. 20 (2014) 4085–4093. https://doi.org/10.1016/j.jiec.2014.01.007.

55. H. Wang, X. Yuan, Y. Wu, X. Chen, L. Leng, H. Wang, H. Li, G. Zeng, Facile synthesis of polypyrrole decorated reduced graphene oxide-Fe$_3$O$_4$ magnetic composites and its application for the Cr(VI) removal, Chem. Eng. J. 262 (2015) 597–606. https://doi.org/10.1016/j.cej.2014.10.020.

56. S. Lapwanit, T. Trakulsujaritchok, P.N. Nongkhai, Chelating magnetic copolymer composite modified by click reaction for removal of heavy metal ions from aqueous solution, Chem. Eng. J. 289 (2016) 286–295. https://doi.org/10.1016/j.cej.2015.12.073.

57. M.S. Samuel, S.S. Shah, J. Bhattacharya, K. Subramaniam, N.D. Pradeep Singh, Adsorption of pb(II) from aqueous solution using a magnetic chitosan/graphene oxide composite and its toxicity studies, Int. J. Biol. Macromol. 115 (2018) 1142–1150. https://doi.org/10.1016/j.ijbiomac.2018.04.185.

58. S. Lin, C. Zou, H. Liang, H. Peng, Y. Liao, The effective removal of nickel ions from aqueous solution onto magnetic multi-walled carbon nanotubes modified by β-cyclodextrin, Colloids Surfaces A Physicochem. Eng. Asp. 619 (2021) 126544. https://doi.org/10.1016/j.colsurfa.2021.126544.

59. Y. Fu, Y. Sun, Y. Zheng, J. Jiang, C. Yang, J. Wang, J. Hu, New network polymer functionalized magnetic-mesoporous nanoparticle for rapid adsorption of hg(II) and sequential efficient reutilization as a catalyst, Purif. Technol. 259 (2021) 118112. https://doi.org/10.1016/j.seppur.2020.118112.

60. A. Abass O, A.T. Jameel, S.A. Muyubi, M.I. Abdul Karim, A.M.Z. Alam, Removal of oil and grease as emerging pollutants of concern (EPC) in wastewater stream, IIUM Eng. J. 12 (2011) 161–169. https://doi.org/10.31436/iiumej.v12i4.218.

61. D. Tonelli, E. Scavetta, I. Gualandi, Electrochemical deposition of nanomaterials for electrochemical sensing, Sensors (Switzerland). 19 (2019). https://doi.org/10.3390/s19051186.

62. G. Hu, J. Li, G. Zeng, Recent development in the treatment of oily sludge from petroleum industry: A review, Elsevier, 2013. https://doi.org/10.1016/j.jhazmat.2013.07.069.

63. S. Kalia, S. Kango, A. Kumar, Y. Haldorai, B. Kumari, R. Kumar, Magnetic polymer nanocomposites for environmental and biomedical applications, Colloid Polym. Sci. 292 (2014) 2025–2052. https://doi.org/10.1007/s00396-014-3357-y.

64. U. Cheema, M. Ananta, V. Muder, Collagen: Applications of a natural polymer in regenerative medicine, Regen. Med. Tissue Eng. - Cells Biomater. (2011). https://doi.org/10.5772/24165.

65. P. Thanikaivelan, N.T. Narayanan, B.K. Pradhan, P.M. Ajayan, Collagen based magnetic nanocomposites for oil removal applications, Sci. Rep. 2 (2012). https://doi.org/10.1038/srep00230.

66. L. Yu, G. Hao, J. Gu, S. Zhou, N. Zhang, W. Jiang, Fe_3O_4/PS magnetic nanoparticles: Synthesis, characterization and their application as sorbents of oil from waste water, J. Magn. Magn. Mater. 394 (2015) 14–21. https://doi.org/10.1016/j.jmmm.2015.06.045.

67. Z. Jiang, L.D. Tijing, A. Amarjargal, C.H. Park, K.J. An, H.K. Shon, C.S. Kim, Removal of oil from water using magnetic bicomponent composite nanofibers fabricated by electrospinning, Compos. Part B Eng. 77 (2015) 311–318. https://doi.org/10.1016/j.compositesb.2015.03.067.

68. F. Damavandi, J.B.P. Soares, Polystyrene magnetic nanocomposite blend: An effective, facile, and economical alternative in oil spill removal applications, Chemosphere. 286 (2022). https://doi.org/10.1016/j.chemosphere.2021.131611.

69. A. Kumar, U. Dixit, K. Singh,, S.P. Gupta, M.S. Jamal Beg. Structure and properties of dyes and pigments, in: Dyes and pigments - novel applications and waste treatment, IntechOpen, 2016: p. 13. https://www.intechopen.com/books/advanced-biometric-technologies/liveness-detection-in-biometrics.

70. S. Hussain, N. Khan, S. Gul, S. Khan, H. Khan, Contamination of water resources by food dyes and its removal technologies, in: Water Chemistry, IntechOpen, 2020. https://doi.org/10.5772/intechopen.90331.

71. E. Gayathiri, P. Prakash, K. Selvam, M.K. Awasthi, R. Gobinath, R.R. Karri, M.G. Ragunathan, J. Jayanthi, V. Mani, M.A. Poudineh, S.W. Chang, B. Ravindran, Plant microbe based remediation approaches in dye removal: A review, Bioengineered. 13 (2022) 7798–7828. https://doi.org/10.1080/21655979.2022.2049100.

72. Q. Liu, Pollution and treatment of dye waste-water, in: IOP Conf. Ser. Earth Environ. Sci., IOP Publishing Ltd, 2020. https://doi.org/10.1088/1755-1315/514/5/052001.

73. R. Al-Tohamy, S.S. Ali, F. Li, K.M. Okasha, Y.A.G. Mahmoud, T. Elsamahy, H. Jiao, Y. Fu, J. Sun, A critical review on the treatment of dye-containing wastewater: Ecotoxicological and health concerns of textile dyes and possible remediation approaches for environmental safety, Ecotoxicol. Environ. Saf. 231 (2022). https://doi.org/10.1016/j.ecoenv.2021.113160.

74. R.O. Alves de Lima, A.P. Bazo, M.F. Salvadori, M. Rech, D. de Palma Oliveira, G. de Aragão Umbuzeiro, Mutagenic and carcinogenic potential of a textile azo dye processing plant effluent that impacts a drinking water source, Mutat. Res. - Genet. Toxicol. Environ. Mutagen. 626 (2007) 53–60. https://doi.org/10.1016/j.mrgentox.2006.08.002.

75. B. Lellis, C.Z. Fávaro-Polonio, J.A. Pamphile, J.C. Polonio, Effects of textile dyes on health and the environment and bioremediation potential of living organisms, Biotechnol. Res. Innov. 3 (2019) 275–290. https://doi.org/10.1016/j.biori.2019.09.001.

76. Z.Z. Vasiljevic, M.P. Dojcinovic, J.D. Vujancevic, I. Jankovic-Castvan, M. Ognjanovic, N.B. Tadic, S. Stojadinovic, G.O. Brankovic, M.V. Nikolic, Photocatalytic degradation of methylene blue under natural sunlight using iron titanate nanoparticles prepared by a modified sol-gel method: Methylene blue degradation with Fe2TiO5, R. Soc. Open Sci. 7 (2020). https://doi.org/10.1098/rsos.200708.

77. M. Akter, F.B.A. Rahman, M.Z. Abedin, S.M.F. Kabir, Adsorption characteristics of banana peel in the removal of dyes from textile effluent, Textiles. 1 (2021) 361–375. https://doi.org/10.3390/textiles1020018.

78. B.H. Tan, T.T. Teng, A.K.M. Omar, Removal of dyes and industrial dye wastes by magnesium chloride, Water Res. 34 (2000) 597–601.

79. P. Sheeba David, A. Karunanithi, N.N. Fathima, Improved filtration for dye removal using keratin–polyamide blend nanofibrous membranes, Environ. Sci. Pollut. Res. (2020). https://doi.org/10.1007/s11356-020-10491-y.

80. D. Dogan, H. Turkdemir, Electrochemical treatment of actual textile indigo dye effluent, Pol. J. Environ. Stud. 21 (2012) 1185–1190.

81. Q. Sun, K. Li, S. Wu, B. Han, Remarkable improvement of TiO_2 for dye photocatalytic degradation by a facile post-treatment, New J. Chem. 22 (2020) 1942–1952. https://doi.org/10.1039/c9nj05120a.

82. A. Kheddo, L. Rhyman, M.I. Elzagheid, P. Jeetah, P. Ramasami, Adsorption of synthetic dyed wastewater using activated carbon from rice husk, SN Appl. Sci. 2 (2020). https://doi.org/10.1007/s42452-020-03922-5.

83. M. Ismail, Z. Wu, L. Zhang, J. Ma, Y. Jia, Y. Hu, Y. Wang, High-efficient synergy of piezocatalysis and photocatalysis in bismuth oxychloride nanomaterial for dye decomposition, Chemosphere. 228 (2019) 212–218. https://doi.org/10.1016/j.chemosphere.2019.04.121.

84. P. Das, S. Nisa, A. Debnath, B. Saha, Enhanced adsorptive removal of toxic anionic dye by novel magnetic polymeric nanocomposite: Optimization of process parameters, J. Dispers. Sci. Technol. 43 (2022) 880–895. https://doi.org/10.1080/01932691.2020.1845958.

85. P. Modisha, T. Nyokong, Fabrication of phthalocyanine-magnetic nanoparticles hybrid nanofibers for degradation of Orange-G, J. Mol. Catal. A Chem. 381 (2014) 132–137. https://doi.org/10.1016/j.molcata.2013.10.012.

86. S. Mapukata, F. Chindeka, K.E. Sekhosana, T. Nyokong, Laser induced photodegradation of Orange G using phthalocyanine – cobalt ferrite magnetic nanoparticle conjugates electrospun in polystyrene nanofibers, Mol. Catal. 439 (2017) 211–223. https://doi.org/10.1016/j.mcat.2017.06.028.

87. N. Amiralian, M. Mustapic, M.S.A. Hossain, C. Wang, M. Konarova, J. Tang, J. Na, A. Khan, A. Rowan, Magnetic nanocellulose: A potential material for removal of dye from water, J. Hazard. Mater. 394 (2020). https://doi.org/10.1016/j.jhazmat.2020.122571.

88. P. Bober, I.M. Minisy, U. Acharya, J. Pfleger, V. Babayan, N. Kazantseva, J. Hodan, J. Stejskal, Conducting polymer composite aerogel with magnetic properties for organic dye removal, Synth. Met. 260 (2020). https://doi.org/10.1016/j.synthmet.2019.116266.

89. T.K. Mahto, A.R. Chowdhuri, S.K. Sahu, Polyaniline-functionalized magnetic nanoparticles for the removal of toxic dye from wastewater, J. Appl. Polym. Sci. 131 (2014). https://doi.org/10.1002/app.40840.

90. F. Lu, A. Dong, G. Ding, K. Xu, J. Li, L. You, Magnetic porous polymer composite for high performance adsorption of acid red 18 based on melamine resin and chitosan, J. Mol. Liq. 294 (2019). https://doi.org/10.1016/j.molliq.2019.111515.

91. H. Gao, S. Zhao, X. Cheng, X. Wang, L. Zheng, Removal of anionic azo dyes from aqueous solution using magnetic polymer multi-wall carbon nanotube nanocomposite as adsorbent, Chem. Eng. J. 223 (2013) 84–90. https://doi.org/10.1016/j.cej.2013.03.004.

92. G. Abdi, A. Alizadeh, J. Amirian, S. Rezaei, G. Sharma, Polyamine-modified magnetic graphene oxide surface: Feasible adsorbent for removal of dyes, J. Mol. Liq. 289 (2019). https://doi.org/10.1016/j.molliq.2019.111118.

93. J. Zolgharnein, S. Feshki, M. Rastgordani, S. Ravansalar, Simultaneous removal of basic blue and toluidine blue O dyes by magnetic Fe3O4@polydopamine nanoparticle as an efficient adsorbent using derivative spectrophotometric determination and central composite design optimization, Inorg. Chem. Commun. 146 (2022). https://doi.org/10.1016/j.inoche.2022.110203.

94. A.H. Nordin, S. Wong, N. Ngadi, M. Mohammad Zainol, N.A.F. Abd Latif, W. Nahgan, Surface functionalization of cellulose with polyethyleneimine and magnetic nanoparticles for efficient removal of anionic dye in wastewater, J. Environ. Chem. Eng. 9 (2021). https://doi.org/10.1016/j.jece.2020.104639.

95. J. Li, Z. Shao, C. Chen, X. Wang, Hierarchical GOs/Fe3O4/PANI magnetic composites as adsorbent for ionic dye pollution treatment, RSC Adv. 4 (2014) 38192–38198. https://doi.org/10.1039/c4ra05800c.

96. A. Pant, R. Tanwar, B. Kaur, U.K. Mandal, A magnetically recyclable photocatalyst with commendable dye degradation activity at ambient conditions, Sci. Rep. 8 (2018). https://doi.org/10.1038/s41598-018-32911-3.

97. J. Fick, H. Söderström, R.H. Lindberg, C. Phan, M. Tysklind, G.J. Larsson, Contamination of surface, ground, and drinking water from pharmaceutical production, Environ. Toxicol. Chem. 28 (2009) 2522–2527. https://doi.org/10.1897/09-073.1.

98. S.A. Kraemer, A. Ramachandran, G.G. Perron, Antibiotic pollution in the environment: From microbial ecology to public policy, Microorganisms. 7 (2019). https://doi.org/10.3390/microorganisms7060180.

99. M. Ortúzar, M. Esterhuizen, D.R. Olicón-Hernández, J. González-López, E. Aranda, Pharmaceutical pollution in aquatic environments: A concise review of environmental impacts and bioremediation systems, Front. Microbiol. 13 (2022). https://doi.org/10.3389/fmicb.2022.869332.

100. O. Lefebvre, X. Shi, C.H. Wu, H.Y. Ng, Biological treatment of pharmaceutical wastewater from the antibiotics industry, Water Sci. Technol. 69 (2014) 855–861. https://doi.org/10.2166/wst.2013.729.

101. H.R. Andersen, K.M.S. Hansen, T. Kosjek, E. Heath, P. Kaas, A. Ledin, Photochemical treatment of pharmaceuticals, Abstract from 6th IWA World Water Congress and Exhibition, Vienna, Austria. (2008).

102. N. Tran, P. Drogui, S.K. Brar, Sonochemical techniques to degrade pharmaceutical organic pollutants, Environ. Chem. Lett. 13 (2015) 251–268. https://doi.org/10.1007/s10311-015-0512-8.

103. F. Javier Benitez, J.L. Acero, F.J. Real, G. Roldán, Ozonation of pharmaceutical compounds: Rate constants and elimination in various water matrices, Chemosphere. 77 (2009) 53–59. https://doi.org/10.1016/j.chemosphere.2009.05.035.

104. S. Kumar, P. Karfa, K.C. Majhi, R. Madhuri, Photocatalytic, fluorescent BiPO4@ Graphene oxide based magnetic molecularly imprinted polymer for detection, removal and degradation of ciprofloxacin, Mater. Sci. Eng. C. 111 (2020). https://doi.org/10.1016/j.msec.2020.110777.

105. L. Mohammadi, A. Rahdar, R. Khaksefidi, A. Ghamkhari, G. Fytianos, G.Z. Kyzas, Polystyrene magnetic nanocomposites as antibiotic adsorbents, Polymers (Basel). 12 (2020). https://doi.org/10.3390/polym12061313.

106. F. Tamaddon, A. Nasiri, G. Yazdanpanah, Photocatalytic degradation of ciprofloxacin using CuFe2O4@methyl cellulose based magnetic nanobiocomposite, MethodsX. 7 (2020) 74–81. https://doi.org/10.1016/j.mex.2019.12.005.

107. A. Mohammadi, S. Pourmoslemi, Enhanced photocatalytic degradation of doxycycline using a magnetic polymer-ZnO composite, Water Sci. Technol. 2017 (2018) 791–801. https://doi.org/10.2166/wst.2018.237.

108. M. Malakootian, A. Nasiri, A. Asadipour, E. Kargar, Facile and green synthesis of ZnFe2O4@CMC as a new magnetic nanophotocatalyst for ciprofloxacin degradation from aqueous media, process, Saf. Environ. Prot. 129 (2019) 138–151. https://doi.org/10.1016/j.psep.2019.06.022.

109. L.N. Pincus, F. Melnikov, J.S. Yamani, J.B. Zimmerman, Multifunctional photoactive and selective adsorbent for arsenite and arsenate: Evaluation of nano titanium dioxide-enabled chitosan cross-linked with copper, J. Hazard. Mater. 358 (2018) 145–154. https://doi.org/10.1016/j.jhazmat.2018.06.033.

110. G.N. Hlongwane, P.T. Sekoai, M. Meyyappan, K. Moothi, Simultaneous removal of pollutants from water using nanoparticles: A shift from single pollutant control to multiple pollutant control, Sci. Total Environ. 656 (2019) 808–833. https://doi.org/10.1016/j.scitotenv.2018.11.257.

111. J.J. Nataša, MN Aleksandra, Z Marija, M Zorica, B Predrag, B Dojčinović, Organobentonites as multifunctional adsorbents of organic and inorganic water pollutants, J. Serbian Chem. Soc. 79 (2014) 253–263. https://doi.org/10.2298/JSC130125065J.

112. L. Li, G. Qi, B. Wang, D. Yue, Y. Wang, T. Sato, Fulvic acid anchored layered double hydroxides: A multifunctional composite adsorbent for the removal of anionic dye and toxic metal, J. Hazard. Mater. 343 (2018) 19–28. https://doi.org/10.1016/j.jhazmat.2017.09.006.

113. F. Zhao, E. Repo, D. Yin, Y. Meng, S. Jafari, H. Normal, EDTA-cross-linked β-cyclodextrin: An environmentally friendly bifunctional adsorbent for simultaneous adsorption of metals and cationic dyes. Environ. Sci. Technol. 49 (2015) 10570–10580.

114. X. Bao, Z. Qiang, J. Chang, W. Ben, J. Qu, Synthesis of carbon-coated magnetic nanocomposite (Fe$_3$O$_4$ @ C) and its application for sulfonamide antibiotics removal from water, J. Environ. Sci. 26 (2014) 962–969. https://doi.org/10.1016/S1001-0742(13)60485-4.

115. C.S. Mustafa Özdemir, Ö. Durmuş, Removal of methylene blue, methyl violet, rhodamine B, alizarin red, and bromocresol green dyes from aqueous solutions on activated cotton stalks, Desalin. Water Treat. (2015) 1–13. https://doi.org/10.1080/19443994.2015.1085916.

116. L. Fan, C. Luo, M. Sun, H. Qiu, X. Li, Synthesis of magnetic β-cyclodextrin–chitosan/graphene oxide as nanoadsorbent and its application in dye adsorption and removal, Colloids Surfaces B Biointerfaces. 103 (2013) 601–607. https://doi.org/10.1016/j.colsurfb.2012.11.023.

117. S. Thakur, B. Sharma, A. Verma, J. Chaudhary, S. Tamulevicius, V.K. Thakur, Recent progress in sodium alginate based sustainable hydrogels for environmental applications, J. Clean. Prod. 198 (2018) 143–159. https://doi.org/10.1016/j.jclepro.2018.06.259.

118. S. Tlili, E. Gómez, S. Gligorovski, H. Wortham, Adsorption behavior of two model airborne organic contaminants on wafer surfaces, Chem. Eng. J. 187 (2012) 239–247. https://doi.org/10.1016/j.cej.2012.01.067.

119. I. Phiri, J.M. Ko, P. Mushonga, J. Kugara, M. Opiyo Onani, S. Msamadya, S.J. Kim, C. Yeajoon Bon, S. Mugobera, K. Siyaduba-Choto, A. Madzvamuse, Simultaneous removal of cationic, anionic and organic pollutants in highly acidic water using magnetic nanocomposite alginate beads, J. Water Process Eng. 31 (2019). https://doi.org/10.1016/j.jwpe.2019.100884.

120. J. Choi, K. Yang, D. Kim, C. Eui, Adsorption of zinc and toluene by alginate complex impregnated with zeolite and activated carbon, Curr. Appl. Phys. 9 (2009) 694–697. https://doi.org/10.1016/j.cap.2008.06.008.

121. S.F. Chua, A. Nouri, W.L. Ang, E. Mahmoudi, A.W. Mohammad, A. Benamor, M. Ba-Abbad, The emergence of multifunctional adsorbents and their role in environmental remediation, J. Environ. Chem. Eng. 9 (2021) 104793. https://doi.org/10.1016/j.jece.2020.104793.

122. S. Wang, Y. Peng, Natural zeolites as effective adsorbents in water and wastewater treatment, Chem. Eng. J. 156 (2010) 11–24. https://doi.org/10.1016/j.cej.2009.10.029.

123. M. Rakanovi, A. Vukojevi, Zeolites as adsorbents and photocatalysts for removal of dyes from the aqueous environment, Molecules 27 (2022) 6582.

124. S. Zhang, F. Xu, Y. Wang, W. Zhang, X. Peng, F. Pepe, Silica modified calcium alginate – xanthan gum hybrid bead composites for the removal and recovery of pb (II) from aqueous solution, Chem. Eng. J. 234 (2013) 33–42. https://doi.org/10.1016/j.cej.2013.08.102.

125. S.K. Papageorgiou, E.P. Kouvelos, E.P. Favvas, A.A. Sapalidis, G.E. Romanos, F.K. Katsaros, Metal – carboxylate interactions in metal – alginate complexes studied with FTIR spectroscopy, Carbohydr. Res. 345 (2010) 469–473. https://doi.org/10.1016/j.carres.2009.12.010.

126. S.K. Papageorgiou, F.K. Katsaros, E.P. Kouvelos, J.W. Nolan, H. Le, N.K. Kanellopoulos, Heavy metal sorption by calcium alginate beads from *Laminaria digitata*, J. Hazard Mater. 137 (2006) 1765–1772. https://doi.org/10.1016/j.jhazmat.2006.05.017.

127. M. Ghaedi, A.G. Nasab, S. Khodadoust, M. Rajabi, S. Azizian, Application of activated carbon as adsorbents for efficient removal of methylene blue: Kinetics and equilibrium study, J. Ind. Eng. Chem. 20 (2014) 2317–2324. https://doi.org/10.1016/j.jiec.2013.10.007.

128. W.A. Khanday, F. Marrakchi, M. Asif, B.H. Hameed, Mesoporous zeolite – Activated carbon composite from oil palm ash as an effective adsorbent for methylene blue, J. Taiwan Inst. Chem. Eng. 70 (2017) 32–41. https://doi.org/10.1016/j.jtice.2016.10.029.

129. S.M. Anisuzzaman, C.G. Joseph, D. Krishnaiah, V.V. Tay, Modification of commercial activated carbon for the removal of 2, 4-dichlorophenol from simulated wastewater, J. King Saud Univ. - Sci. 27 (2015) 318–330. https://doi.org/10.1016/j.jksus.2015.01.002.

130. M.A. Islam, A. Benhouria, M. Asif, B.H. Hameed, Methylene blue adsorption on factory-rejected tea activated carbon prepared by conjunction of hydrothermal carbonization and sodium hydroxide activation processes, J. Taiwan Inst. Chem. Eng. 52 (2015) 57–64. https://doi.org/10.1016/j.jtice.2015.02.010.

131. S. Ghorai, A. Sinhamahpatra, A. Sarkar, A. Baran, S. Pal, Bioresource technology novel biodegradable nanocomposite based on XG-g-PAM/SiO$_2$: Application of an efficient adsorbent for Pb^{2+} ions from aqueous solution, Bioresour. Technol. 119 (2012) 181–190. https://doi.org/10.1016/j.biortech.2012.05.063.

132. Z. Ping, Y. Jin, S. Liu, Z. Ping, G. Qing, M. Lu, Surface charging of layered double hydroxides during dynamic interactions of anions at the interfaces, J. Colloid Interface Sci. 326 (2008) 522–529. https://doi.org/10.1016/j.jcis.2008.06.062.

133. K. Goh, T. Lim, Z. Dong, Application of layered double hydroxides for removal of oxyanions: A review, Water Res. 42 (2008) 1343–1368. https://doi.org/10.1016/j.watres.2007.10.043.

134. S. Wang, C. Hua, M. Kuang, Y. Hui, P. Neng, Applied clay science arsenate adsorption by Mg/Al – NO 3 layered double hydroxides with varying the Mg/Al ratio, Appl. Clay Sci. 43 (2009) 79–85. https://doi.org/10.1016/j.clay.2008.07.005.

135. H. Yan, Q. Chen, J. Liu, Y. Feng, K. Shih, Phosphorus recovery through adsorption by layered double hydroxide nano-composites and transfer into a struvite-like fertilizer, Water Res. 145 (2018) 721–730. https://doi.org/10.1016/j.watres.2018.09.005.

136. H. Hatami, A. Fotovat, A. Halajnia, Comparison of adsorption and desorption of phosphate on synthesized Zn-Al LDH by two methods in a simulated soil solution, Appl. Clay Sci. 152 (2018) 333–341. https://doi.org/10.1016/j.clay.2017.11.032.

137. C. Novillo, D. Guaya, A.A. Avendaño, C. Armijos, J.L. Cortina, I. Cota, Evaluation of phosphate removal capacity of Mg/Al layered double hydroxides from aqueous solutions, Fuel. 138 (2014) 72–79. https://doi.org/10.1016/j.fuel.2014.07.010.

138. S. Iftekhar, M. Emin, V. Srivastava, E. Repo, Application of zinc-aluminium layered double hydroxides for adsorptive removal of phosphate and sulfate: Equilibrium, kinetic and thermodynamic, Chemosphere. 209 (2018) 470–479. https://doi.org/10.1016/j.chemosphere.2018.06.115.

139. S. Zhang, Y. Zhang, G. Bi, J. Liu, Z. Wang, Q. Xu, H. Xu, X. Li, Mussel-inspired polydopamine biopolymer decorated with magnetic nanoparticles for multiple pollutants removal, J. Hazard. Mater. 270 (2014) 27–34. https://doi.org/10.1016/j.jhazmat.2014.01.039.

10 Magnetic Polymer Composites for Hyperthermia Therapy

Sreeja Nath Chowdhury

10.1 INTRODUCTION

The use of hyperthermia in the treatment of malignant tumors has received a lot of attention. Through the action of so-called heat shock proteins, hyperthermia modifies cell walls and boosts blood flow to the affected area [1, 2], which can improve drug delivery [3]. Additionally, hyperthermia increases the transport of oxygen, a potent radiosensitizer, to the region, over-acidifying tumor cells and depriving them of nutrition. Cell death (apoptosis) can then begin as a result of the disruption to the metabolism of the cells. By producing DNA-damaging free radicals, higher oxygen concentrations can further increase the likelihood that radiation will harm and kill cells, as well as inhibit cells from repairing the damage brought on by the radiation session. It is important to note that normal cells are not necessarily more sensitive to the effects of heat than malignant cells are [3]. Cancer cells' temperature increases between 41°C and 46°C for narcosis of cancer cells, while the surrounding healthy cells are unaffected [4]. A solid tumor's chaotic circulatory system, however, generates an adverse microenvironment inside the tumor (Figure 10.1). The tumor cells are therefore much less able to withstand the additional stress of heat than a healthy

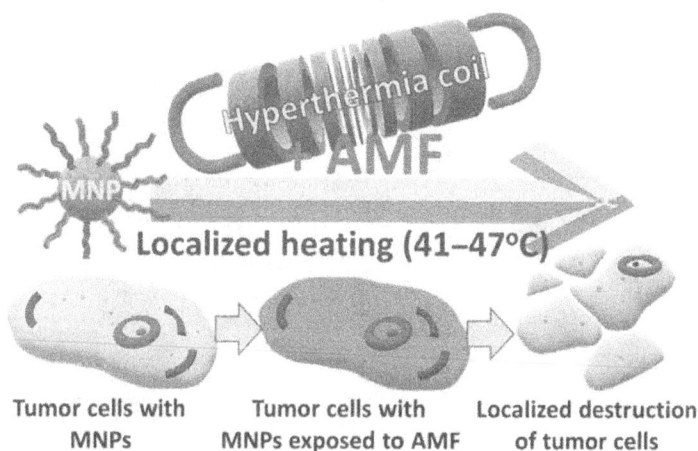

FIGURE 10.1 Graphic diagram of MNP-mediated magnetic hyperthermia for selective tumor cell damage [5].

DOI: 10.1201/9781003454236-10

cell in normal tissue, since they are already under stress from low oxygen levels, higher than normal acid concentrations, and a lack of nutrients.

Since it is impossible to prevent damaging the surrounding healthy tissue when using the hyperthermia technique, heating the local tumor zone to the desired temperature is a significant difficulty (Figure 10.2). As a novel approach to cancer therapy and drug delivery, nanotechnology-based targeting systems are being developed [6–8]. Compared to earlier approaches, they provide a host of impressive advantages, overcoming difficulties from localized hyperthermia, such as temperature heterogeneity, tumor mass distribution, and inability to prevent the overheating of the region of a deep-seated tumor. The role of magnetic nanoparticles is crucial in causing heat to treat tumors. This method, in which the heat is focused on the affected cells, is called magnetic hyperthermia or magnetic nanofluid hyperthermia therapy. Magnetic hyperthermia involves the local or intravascular injection of ferromagnetic, ferrimagnetic, or superparamagnetic particles into target tissues. The particles produce heat in the vicinity of alternating electromagnetic fields. Due to improved nanoparticle suspension properties, U.S. Food and Drug Administration (FDA) approval for magnetic resonance imaging (MRI) applications, and repeatable use in biosystems, superparamagnetic nanoparticles have garnered increased attention [9].

Superparamagnetism in magnetic nanoparticles refers to the ability with which magnetic nanoparticles gain magnetism in the presence of a magnetic field and lose magnetism when the magnetic field is eliminated. Superparamagnetic particles have no residual magnetism; they use two different power loss methods, hysteresis and relaxation loss, to transform the magnetic field's energy into heat. The particles employ several mechanisms to convert the magnetic field's energy into heat, including eddy current loss (a minor effect in ferromagnetic particles like Fe_3O_4 due to their low electrical conductivity), hysteresis loss during reversal magnetization (usually a significant contribution), and relaxation loss, including

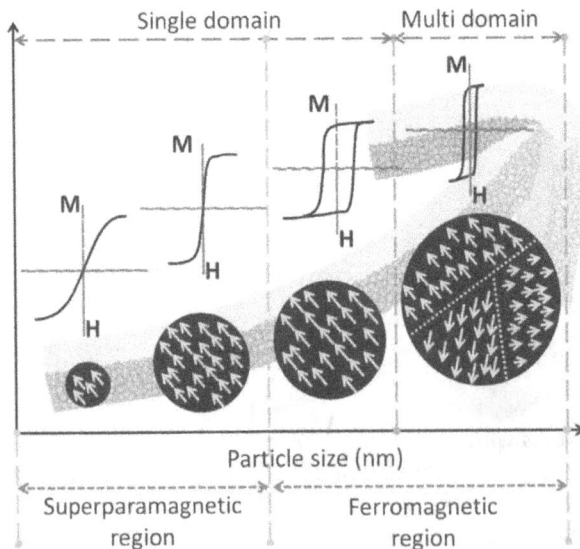

FIGURE 10.2   Size-dependent alteration of coercivity of MNPs [5].

Brownian and Neel relaxation (important contributions in superparamagnetic materials due to zero remanence).

Ferrites in the form of MFe_2O_4 (M is a divalent metal like Mn, Ni, Co, Zn, or Fe) are frequent sources of magnetic materials and have been extensively studied in vitro and in vivo [10–13]. While having high saturation magnetization (Ms), pure metal nanoparticles (Fe, Co, and Ni) suffer oxidation and are exceptionally hazardous, making them challenging for biomedical applications without surface modifications. Iron oxides, on the other hand, are less susceptible to oxidation and can therefore provide steady magnetic responses. Two crystalline forms of iron oxide, maghemite (Fe_2O_3) and magnetite (Fe_3O_4), have remarkable biological characteristics. Thus, they are frequently utilized in magnetic hyperthermia as well. Nowadays many researchers are working on substituted iron oxide (magnetite) nanoparticles, where other divalent metal cations like Co^{2+}, Mn^{2+}, Ni^{2+}, or Zn^{2+} are substituted for Fe^{2+}. This substitution is responsible for tuning the magnetocrystalline anisotropy, thereby increasing the magnetic saturation and decreasing remanent magnetization, which will further aid the superparamagnetic behavior of the final nanoparticles. Along with the total substitution of the divalent cation, researchers have further explored the effect of percentages of divalent metal cation substitution ranging from 0.1% to 0.9% ($M_xFe_{1-x}O_4$; $x = \%$ of substitution) for efficient tuning of the magnetic property in the final nanoparticles [14]. Two distinct classes of superparamagnetic IONP-based materials are currently used for medical applications: superparamagnetic iron oxide nanoparticles (SPIONs) with a mean particle diameter of 50–100 nm and ultra-small superparamagnetic iron oxide nanoparticles (USPIONs) with a size below 50 nm. The superparamagnetic property of magnetic nanoparticles highly depends on the size of the nanoparticles. The major drawback of using naked magnetic nanoparticles is the agglomeration of the nanoparticles, reducing the heating efficiency and high toxicity in a biological medium.

To address these issues, surface-modified magnetic nanoparticles and magnetic nanoparticle polymer composites came into the picture. Due to the highly tailorable properties of polymers, these types of composites are suitable for a wide range of biomedical applications. Properties of polymers such as resemblance to natural living tissue, stimuli-responsive behavior, and inherent biocompatibility make them suitable for a range of biomedical applications. One of the smart polymeric materials is hydrogels, which can undergo abrupt changes in volume without dissolving in the immersed medium; so called smart gels can swell or shrink up to 1000 times in response to small changes in temperature, pH, electric field, or solvent and ionic composition. Magnetic hyperthermia is mainly used to treat solid cancer limited to small areas, like breast, cervical, prostate, head and neck, melanoma, soft-tissue sarcoma, and rectal cancer.

10.2 SYNTHESIS OF MAGNETIC NANOPARTICLES

The efficiency of magnetic nanoparticles (MNPs) is highly dependent on the chemical and physical characteristics of the magnetic nanoparticles, along with their size, shape, surface area, and surface morphology. Therefore, the method of synthesis will allow the tailoring of the size and surface chemistry of MNPs to meet specific demands for biological application. The most commonly practiced methods for fabricating MNPs are co-precipitation, hydrothermal, sol-gel, and polyol.

10.3 CO-PRECIPITATION

In the co-precipitation method, each component is mixed in stoichiometric ratio to get precise control of the doping content. This type of reaction takes place in highly alkaline solution, pH = 9–14. The aqueous mixture contains M^{2+} and Fe^{3+} with a molar ratio of 1:2.

For precipitation of Fe_3O_4 nanoparticles, we need to ensure the formation and stability of Fe^{+2}. To prevent the oxidation of Fe^{2+}, the reaction is carried out under an N_2-blanket. A possible mechanism for the preparation is as follows [15–17]:

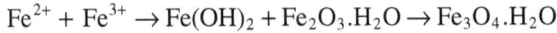

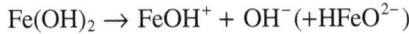

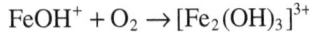

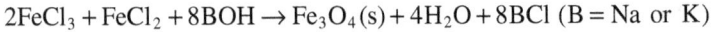

$$Fe^{2+} + Fe^{3+} \rightarrow Fe(OH)_2 + Fe_2O_3.H_2O \rightarrow Fe_3O_4.H_2O$$

$$Fe(OH)_2 \rightarrow FeOH^+ + OH^- (+HFeO_2^-)$$

$$FeOH^+ + O_2 \rightarrow [Fe_2(OH)_3]^{3+}$$

$$[Fe_2(OH)_3]^{3+} + FeOH^+ \rightarrow Fe_3O_4$$

$$2FeCl_3 + FeCl_2 + 8BOH \rightarrow Fe_3O_4(s) + 4H_2O + 8BCl \ (B = Na \ or \ K)$$

pH plays a crucial role in the particle size of the nanoparticles prepared through the co-precipitation method. $ZnFe_2O_4$ nanoparticles were produced by Yang et al. [18, 19] using the alcohol-water co-precipitation technique. The experiment looked at a number of variables, including the temperature at which the alcohol crystallized and the ratio of alcohol to water. The outcome demonstrated how the crystallite size changed as the ratio of alcohol to water varied, and it was found that a ratio of 2:6 (alcohol:water) produced the smallest sample diameter. The author also investigated the connection between grain size and crystallizing temperature.

By using the co-precipitation approach, Lee et al. [20] created $ZnFe_2O_4$ nanoparticles at pH levels ranging from 3 to 12. The pH of the solution was altered using the sodium hydroxide solution. The XRD pattern showed that a pH range of 8 to 12 was used to produce the pure $ZnFe_2O_4$ nanoparticles. Lee et al. [21] completed similar work while manipulating the pH with greater accuracy. The aqueous buffer solutions with a pH range of 6 to 12 were used in place of the sodium hydroxide solution. The author found that a pH of 7–12 was necessary to produce pure $ZnFe_2O_{4n}$ nanoparticles using the buffer solutions.

The major drawback of a wet chemical process (e.g., co-precipitation) lies in the lack of control over the size of the magnetic nanoparticles. Thus, synthesis in a confined environment, like a microemulsion, has been suggested. For instance, the hydrolysis of metal surfactant complexes in water-in-oil emulsions and the traditional co-precipitation process have been carried out using reverse micelles. The microstructure and content of the emulsion, the temperature, and the type of counter ions are the factors that have an impact on MNPs' size. Reverse micelles have been employed to successfully mediate the synthesis of iron oxide magnetic nanoparticles as well as substituted ferrite with an enhanced size distribution in the 4 –12 nm range.

10.4 HYDROTHERMAL

Co-precipitation can be used in hydrothermal conditions and enhances the magnetic characteristics of the generated MNPs. For hydrolytic processes beginning with iron complexes, hydrothermal methods have also been used. For example, iron polyolates have been aged in an aqueous acidic/basic solution before being digested for several days at 80–150°C in an autoclave. The reaction conditions in this synthetic method, such as the solvent temperature and time, typically have a significant impact on the synthetic result. The processes of nucleation and particle growth, which compete with each other, are primarily responsible for controlling the MNP size during crystallization (Figure 10.3). Generally, the hydrothermal method uses a pressure higher than 1 atm and a low synthesis temperature (100–500°C), without subsequent heat treatments. Characteristics of this process include not requiring the use of an organic reagent, being relatively cost-effective, having good particle crystallinity, and having the ability to adjust the efficiency of the size and desired morphology, as well as having high yield.

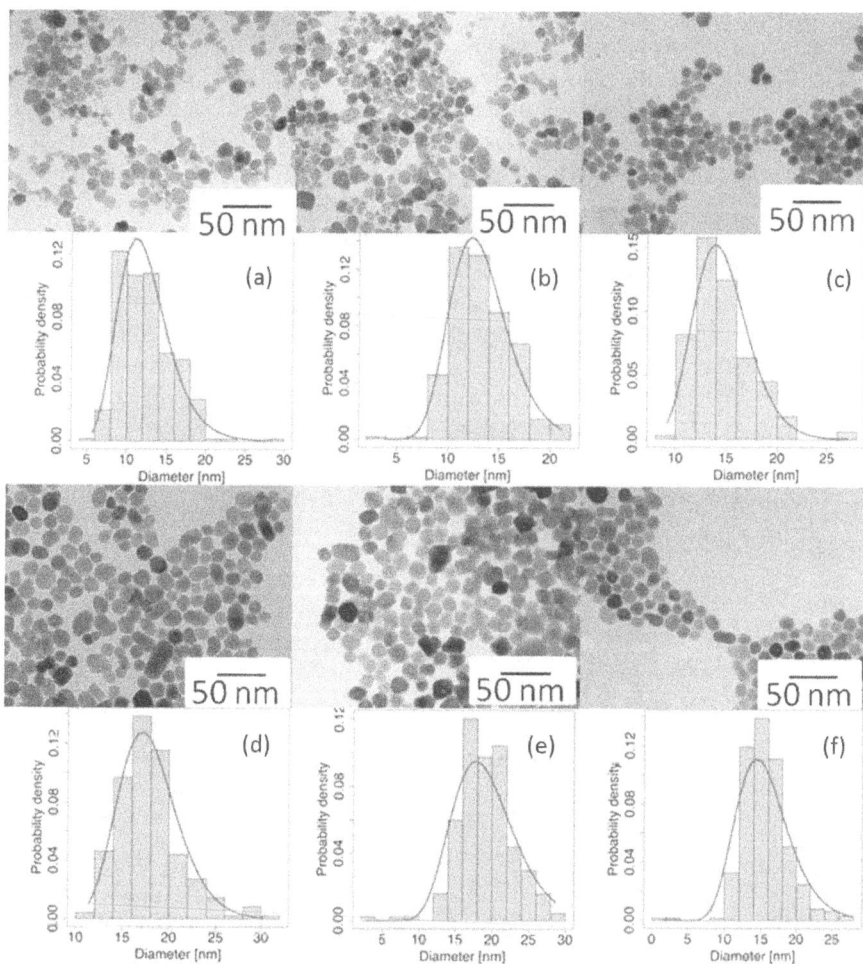

FIGURE 10.3   TEM pictures and corresponding histograms. (a) Sample A: classical heating at 100°C for 1 h. (b) Sample B: hydrothermal treatment at 100°C for 1 h. (c) Sample C: hydrothermal treatment at 200°C for 1 h. (d) Sample D: hydrothermal treatment at 200°C for 24 h. (e) Sample E: hydrothermal treatment at 200°C for 12 h. (f) Sample F: hydrothermal treatment at 200°C for 24 h in the presence of citrate ions (5%) [22].

Valeric Cabuil et al. studied the time/temperature effect on the size of magnetic nanoparticles via hydrothermal reaction. With an increase in either time or temperature, nanoparticle size increases [22].

10.5   MAGNETIC POLYMER COMPOSITE

On the colloidal scale, organic and inorganic matter combine to produce new and sometimes unexpected features. In order to change the physical properties of the polymer, solid fillers are often solid additions. Three types of fillers can be distinguished. First is those that strengthen the polymer and enhance its mechanical properties. The second category includes those that are employed to occupy space and hence save material costs. The third and less frequent category is filler particles that enhance the material's responsiveness (Figure 10.4).

For any polymer, the glass transition temperature (T_g) has to play an important role in the end-use application of the polymer. G. K. Thirunavukkarasu et al. worked on a specific property of p polymers for hyperthermia and drug delivery applications. They worked on a poly(lactic-co-glycolic acid) (PLGA)-Fe_3O_4 nanocomposite where Fe_3O_4 was embedded in the PLGA matrix. In this study, they used PLGA with $M_w = 30,000k–60,000k$ g/mol, 50:50 lactic glycolide content, which helps to prevent aggregation of nanoparticles as well as increase their biocompatibility. On the other hand, PLGA has a T_g above 37°C (i.e., above physiological temperature). When the PLGA-Fe_3O_4 nanocomposite loaded with drug is placed in an alternating magnetic field (AMF), the temperature will increase to 46°C due to the superparamagnetic nature of Fe_3O_4 (Figure 10.5). At this temperature, segmental motion of PLGA starts, which is responsible for the slow release of the drug, promoting nonspecific toxicity and therapeutic effectiveness. Also, the lower critical solution temperature (LCST) of PLGA is 37°C, which means that below this temperature PLGA is in a hydrated

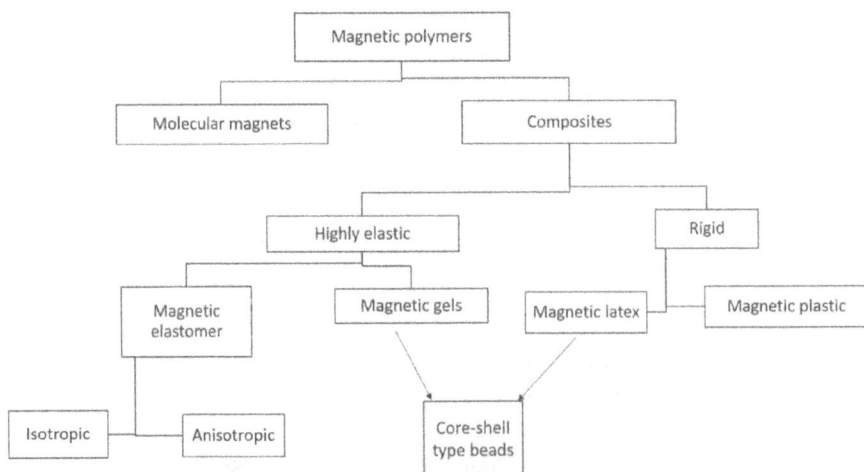

FIGURE 10.4   Schematic representation of a variety of magnetic polymers.

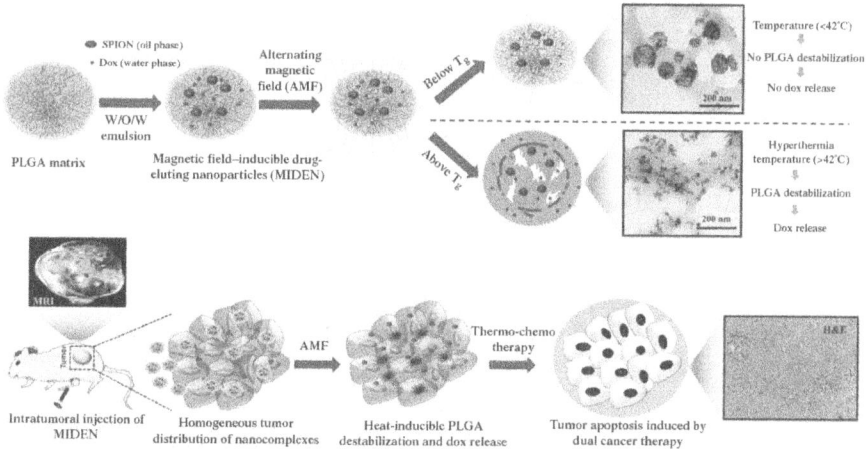

FIGURE 10.5 Schematic illustration showing the formation of magnetic field–inducible drug-eluting nanoparticles (MIDENs) and their application in image-guided thermochemo-therapy [23].

state whereas above this temperature it becomes dehydrated and shrinks (sol-gel transition), further assisting the drug release mechanism. The nanocomposite shows a change in temperature upon application of AMF (4.4 kW, H = 2 KA/m-f = 205 kHz) of 5.1–5.2°C; that is, the temperature goes from 37°C to 42.2°C when the nanoparticle concentration is 1 mg/mL. This result suggests that the nanocomposite can increase the medium temperature nearly to the range of hyperthermia with a specific absorption rate (SAR) of 35.73 W/g [23].

Polyurethanes (PUs) are a heterogeneous class of polymers consisting of organic units joined by urethane links. PUs used in biomedical applications typically have a segmented structure composed of alternating polydispersed "soft" and "hard" segments. These two segments are thermodynamically incompatible and phase-segregate, resulting in discrete, crystalline domains of the associated hard segments surrounded by a continuous, amorphous phase of soft segments. The segregated domains are stabilized by interchain hydrogen bonds and are responsible for the materials' mechanical properties. Mechanical properties of PUs can be easily tailored by playing with the hard and soft segments. Due to biocompatibility and tailorable mechanical properties, PUs are widely used in the preparation of many kinds of medical devices, including wound dressings, artificial organs, and vascular stents.

M. D. Rezoanur Rahman and group are trying to use these advantageous properties of PUs in hyperthermia applications. For their work they have prepared a composite material with PUs and Fe_3O_4. They have adopted an easily scalable solvent casting process for composite preparation. In this process, batches of Fe_3O_4 nanopowder were mixed with 10 mL THF in glass bottles. Into separate glass bottles, with 20 mL THF for every gram of PU, PU pellets were added. For 6 to 8 hours, the solutions were sonicated. The two solutions were mixed with calculated weight ratios of Fe_3O_4:PU (providing theoretical Fe_3O_4 weight percentages of 9, 13.1, and 16.6 wt%, respectively). In order to make sure that the Fe_3O_4 nanopowder was evenly distributed throughout the composite solution, the resulting combination was subjected to 4 further hours of sonication. In the end, Fe_3O_4-PU was produced by casting the sonicated composite solution into a Petri dish and letting the THF solvent evaporate there as air flowed over the dish. Scanning electron microscopy was performed on the sample surfaces to assess the uniformity and distribution of the Fe_3O_4 nanoparticles incorporated in the Fe_3O_4-PU composite. This was combined with use of the EDS technique to perform the elemental mapping. Targeted magnetic inductive heating requires polymers that are stable at high temperatures. Thermogravimetric analysis (TGA) was performed to assess the thermal stability of the Fe_3O_4-PU composites. TGA results showed no mass loss of the polyurethane until ~200C, whereupon there was a two-step mass loss as the material underwent pyrolysis. The remaining mass after this point is a measure of the Fe_3O_4, which is stable to high temperatures. The maximum (saturation) magnetization of pure Fe_3O_4 nanopowder was ~70 emu/g, whereas for the 7 wt% Fe_3O_4-PU composite it was 5 emu/g. This result showed that the PU itself is non-magnetic, and magnetism therefore scales with the percentage of Fe_3O_4 in the composite. To detect whether heating could be achieved in the Fe_3O_4-PU composite, a known mass of the Fe_3O_4-PU composite was inserted in the sample vial with 15 mL of distilled water and exposed to a weak H_{max} = 5 Oe (398 A/m) amplitude, 85 kHz AC magnetic field. The product of field amplitude and frequency $H_{max}f$ must remain less than 5×10^8 Hz A/m in hyperthermia treatments on patients [24], and the product for the composite satisfies this requirement with $H_{max}f$ = 3.4 × 107 Hz A/m (Figure 10.6). The temperature vs. time characteristic shows peak temperatures of 45–50°C were achieved for the pure Fe_3O_4 nanopowder and for the 13 wt% Fe_3O_4:PU composite after heating for approximately 18 minutes (Figure 10.7) [25].

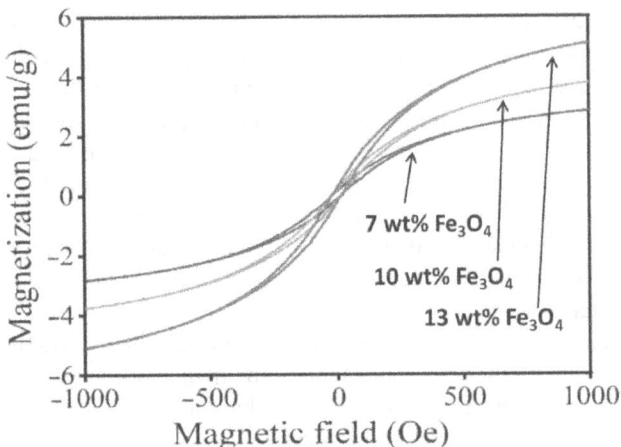

FIGURE 10.6 Room temperature magnetic hysteresis loops of nanocomposites with 7, 10, and 13 wt% Fe_3O_4 loading [25].

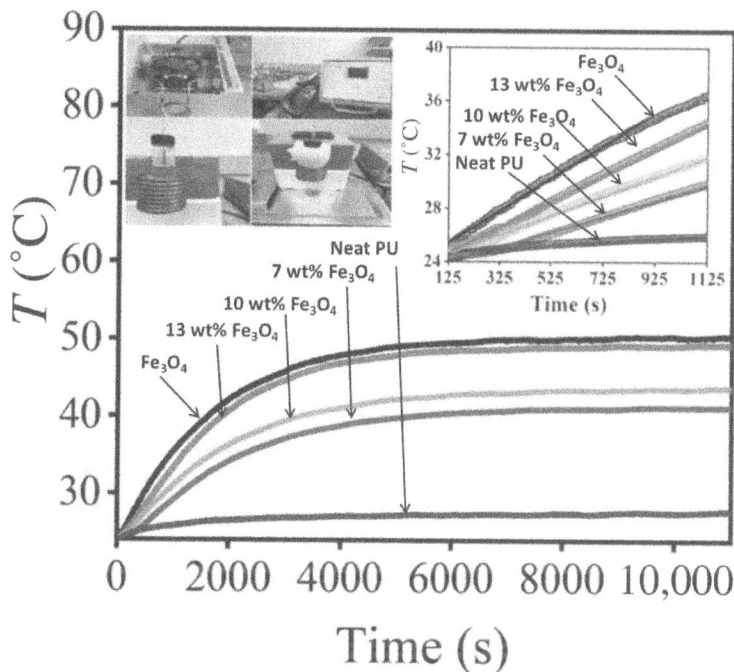

FIGURE 10.7 Temperature increase with respect to time of AC magnetic field applied pure Fe_3O_4 nanopowder and pure PU and PU composites with 7–13 wt% Fe_3O_4. The inset figure shows the rate of temperature increase for pure Fe_3O_4 nanopowder, neat PU, and PU composites with 7–13 wt% Fe_3O_4. The inset photograph shows the apparatus setup [25].

10.6 MAGNETIC GELL COMPOSITES

A novel class of composites, the new generation of magnetic elastomers and gels consists of microscopic (mostly nano- and micro-sized) magnetic particles dispersed in a highly elastic polymeric matrix. These materials are quite recent and display a wide range of intriguing attributes that are the focus of extensive theoretical and experimental study. A variety of motions can be produced and shape change and movement can be controlled using the distinctive magnetoelastic characteristics. Engineers' ability to create new types of switches, sensors, micromachines, biomimetic energy-transducing devices, and controlled delivery systems will be accelerated by their knowledge of magneto-elastic coupling in polymers. The hydrophilic three-dimensional cross-linked hydrogels have exceptional properties such as easy synthesis, super-absorption, and storage of water and biological fluids, with a toxicity avoidance nature, and create a flexible network through physical or chemical cross-linkage. Magnetic gels are frequently referred to as magnetostrictive polymers, magnetorheological polymers, or magnetoelasts, among other synonyms (Figure 10.8).

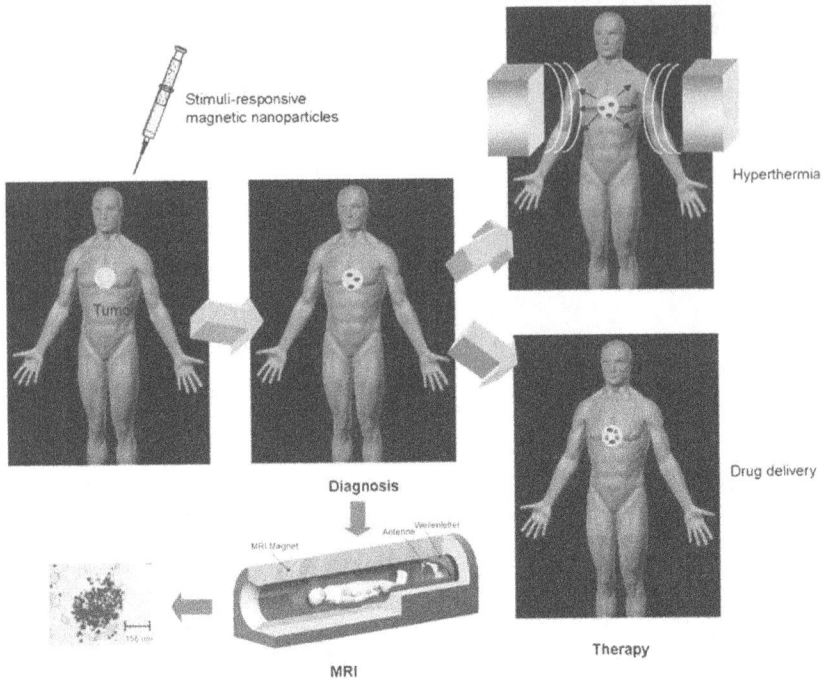

FIGURE 10.8 Schematic representation of the drug-loaded magnetic nanoparticles' localization by MRI, which is followed by treatment of the tumor either by hyperthermia or by the drug release [17].

In the flexible cross-linked polymers that make up magnetic polymer gels, which are a subtype of magnetic elastomers, there are both magnetizable particles and a sizable amount of swelling liquid.

The usefulness of synthetic or natural-based hydrogels in numerous domains, including the medical, pharmaceutical, chemical, and other industries, has increased as a result of the foregoing specific qualities. Proteins and polysaccharides, two naturally occurring polymeric structures, have been exploited to create scaffolds that are more compatible with human tissues [26, 27].

10.7   METHOD OF PREPARATION OF FERROGELS

i. By Physical Association

Physical association refers to the secondary forces such as hydrogen bonding, hydrophobic interaction, and ionic bonding, along with physical entanglements. In this context, the classic example is chitosan hydrogel. An aqueous solution of chitosan is acidic in nature. It can turn into gel if the positive charge on the chitosan can be neutralized by a negative one. Thus, addition of a base (pH = 9–12) to chitosan turns it into a gel due to charge neutralization. Therefore, a ferrogel can be easily formed by an in situ oxidation of iron salt in a chitosan matrix. In a typical process, iron cations are solubilized in an aqueous solution of chitosan, followed by the oxidation of iron salt by addition of a base. This process leads to the formation of magnetic nanoparticles and is also responsible for the gelation of the chitosan matrix [28].

ii. **By Chemical Cross-Linking**

To enhance the hydrogel's poor mechanical stability, cross-linking agents are used to create a flexible hydrogel network with advanced mechanical stability. It is necessary to wash the hydrogel to get rid of unreacted chemicals, since the cross-linking agents, which are typically tiny molecules, emerge with relative cytotoxicity [29, 30].

Reza Eivazzadeh-Kehian et al. worked with alginate-tannic acid base hydrogel, which shows excellent hemocompatability along with desirable mechanical properties. They used silk fibroin (SF) to increase the mechanical property of the scaffold. They prepared the hydrogel by chemical cross-linking of sodium alginate with Ca^{2+} ions. Fe_3O_4 nanoparticles were used as a hypothermic agent. Bulk Fe_3O_4 shows a magnetization value of 90 emu/g, whereas a sodium alginate–tannic acid hydrogel/SF/Fe_3O_4 magnetic nanocomposite shows a magnetization value of 0.90 emu/g (Figure 10.9). This value suggests the increase in the nonmagnetic shell in the nanocomposite [31].

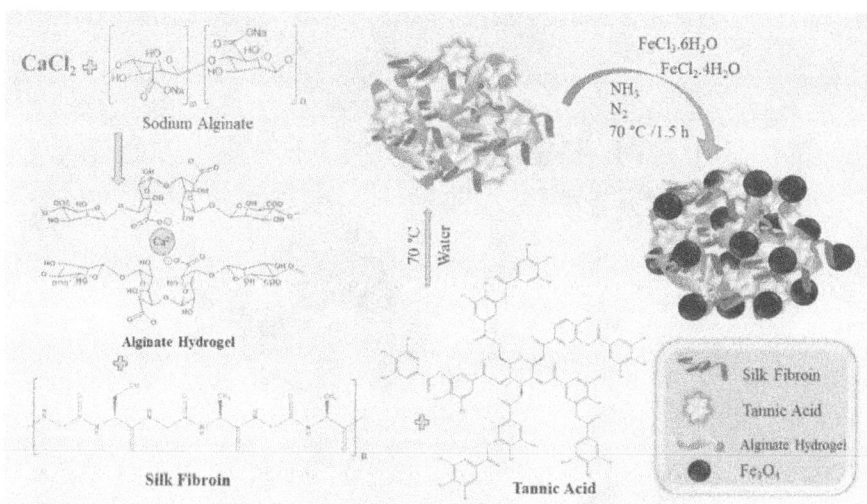

FIGURE 10.9   Schematic illustration of preparation of SA-TA hydrogel/SF/Fe_3O_4 magnetic nanocomposite [31].

The same team did further work on another type of natural polymer– based nanocomposite made up of chitosan-based hydrogel. As in their previously discussed work, they used silk fibroin as an agent to improve the mechanical property of the hydrogel. A hybrid nanoparticle of graphene oxide (GO) and Fe_3O_4 was used as a hyperthermic agent. with the actual role of graphene oxide being to reinforce the hydrogel composite. The chitosan (CS)–based hydrogel was cross-linked by terephthaloyl thiourea, with polyethylene glocol-400 (PEG-400) used as a phase transfer catalyst. The cross-linked CS/SF/GO/Fe_3O_4 scaffold was prepared by a mixture of 1:1 CS hydrogel and SF with 0.02 g GO. Fe_3O_4 nanoparticles were synthesized in situ in the hydrogel matrix by a co-precipitation reaction of the mixture of $FeCl_2$ and $FeCl_3$ salts with ammonium hydroxide.

From FE-SEM it is suggested that the Fe_3O_4 nanoparticles have been covered by the hydrogel, forming a core-shell structure on the GO palates. The mean diameter of the magnetic nanoparticle was measured to be 52.86 nm, with nanoparticle sizes ranging from 39.9 nm to 73.3 nm.

From the VSM analysis the saturation magnetization of the synthesized nanocomposite was found to be 17.09 emu/g, whereas the saturation magnetization of Fe_3O_4 ferro fluid was 44.2 emu/g. The noticeable decrease was observed due to the core-shell structure of the nanocomposite (Figure 10.10). Thermogravimetric studies showed that the hydrogel nanocomposite was stable up to 200°C. Magnetic hyperthermia measurement showed that the maximum temperature rise $\Delta T = 5.2$°C occurs under a field frequency of 200 kHz for 10 min (Figure 10.11). The maximum SAR value obtained was 54.7 W/g (Figure 10.12, 10.13) [32].

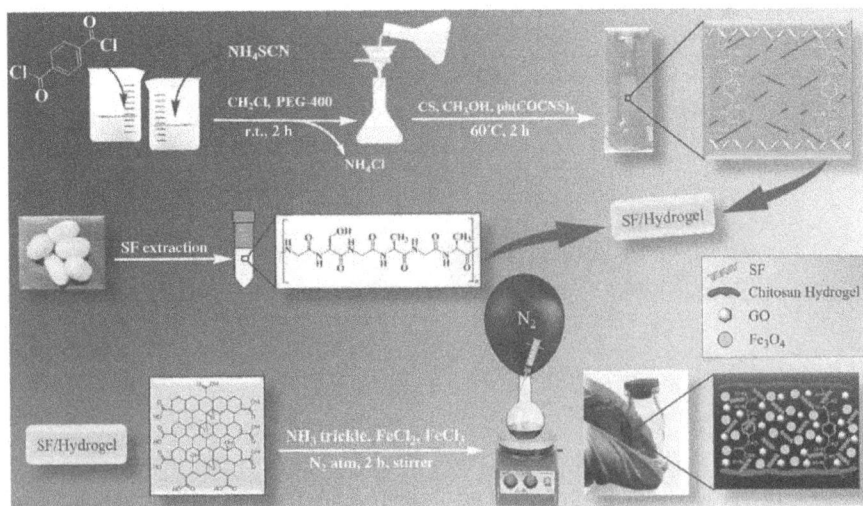

FIGURE 10.10   Preparation of CS hydrogel/SF/GO/Fe_3O_4 nanobiocomposite scaffold. [32].

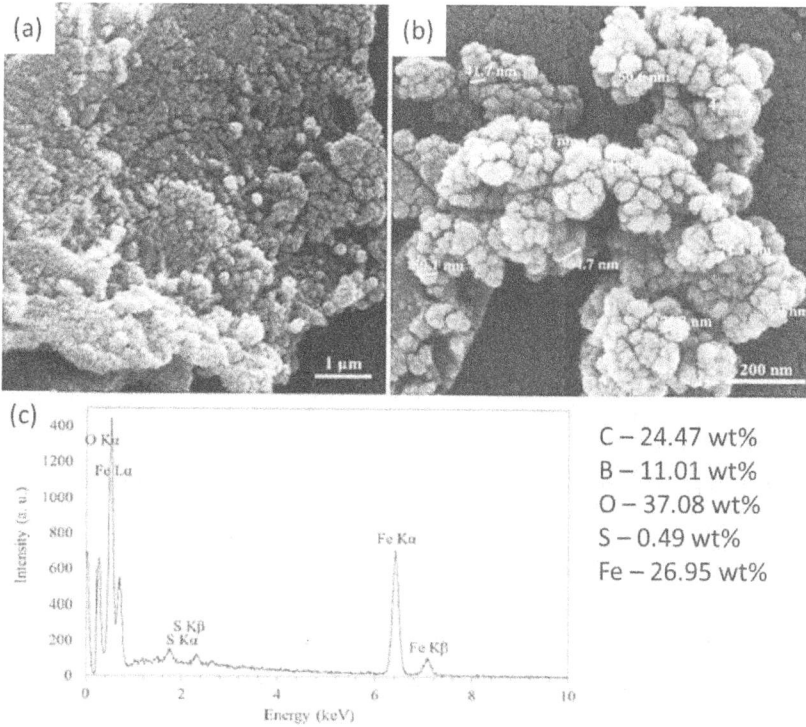

C – 24.47 wt%
B – 11.01 wt%
O – 37.08 wt%
S – 0.49 wt%
Fe – 26.95 wt%

FIGURE 10.11 The FE-SEM images of the CS hydrogel/SF/GO/Fe_3O_4 nanobiocomposite magnified (a) 1 μm and (b) 200 nm. (c) The EDX spectrum and (d) elemental mapping analysis of the CS hydrogel/SF/GO/Fe_3O_4 nanobiocomposite [32].

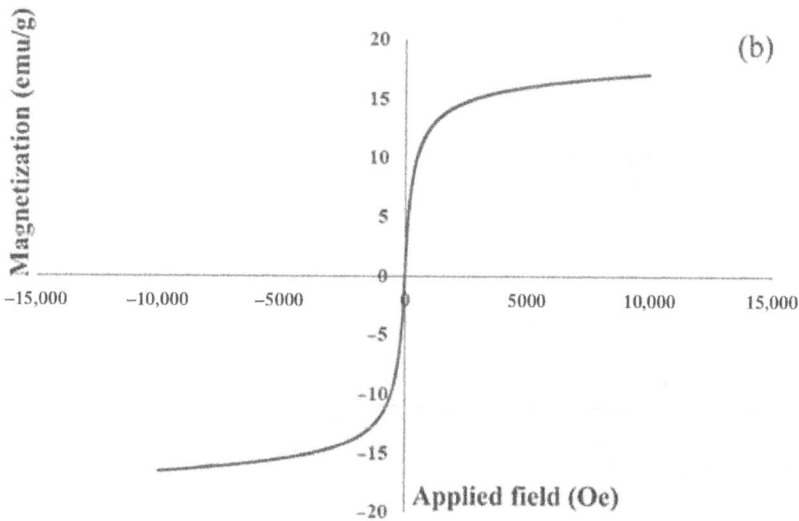

FIGURE 10.12 VSM analysis of the CS hydrogel/SF/GO/Fe_3O_4 nanobiocomposite [32].

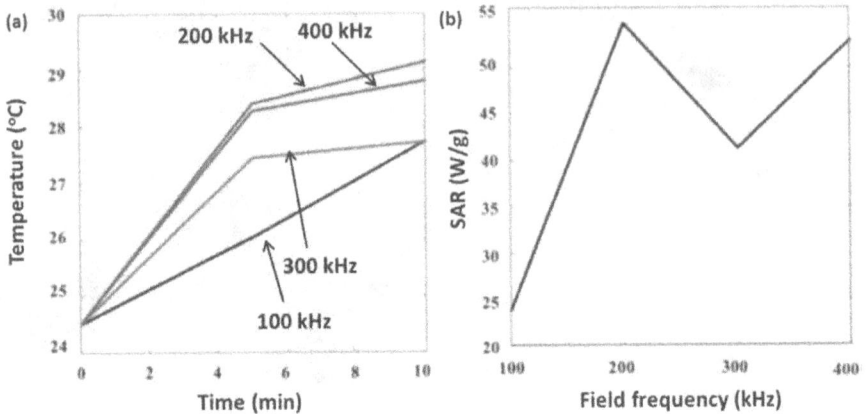

FIGURE 10.13 (a) Heating profile of CS hydrogel/SF/GO/Fe_3O_4 nanobiocomposite with a 1 mg/mL concentration in magnetic fields with various field frequencies. (b) Maximum SAR as a function of field frequency for CS hydrogel/SF/GO/Fe_3O_4 nanobiocomposite with a concentration of 1 mg/m [32].

10.8   CONCLUSION

A technique for fighting cancer called hyperthermia is based on the magnetic properties of biocomposites. Because it allows for localized drug administration, this technique has made it possible to lessen the negative effects of traditional treatments like chemotherapy. However, since issues with these biocomposites' surface properties and size lead to undesirable physiological reactions, this technological tool still requires refinement in the future. The type of biocomposite that will act more effectively will therefore depend on the process used to create these materials, the coating employed, and the features of the malignancy.

REFERENCES

1. Ramos-Castaneda, M.; Moghaddam, S.J.S.J. Lung Cancer Murine Models and Methodology for Immunopreventive Study. *Methods Mol. Biol.* 2022, *2435*. https://doi.org/10.1007/978-1-0716-2014-4_15.
2. Dhillon, P. K.; Mathur, P.; Nandakumar, A.; Fitzmaurice, C.; Kumar, G. A.; Mehrotra, R.; Shukla, D. K.; Rath, G. K.; Gupta, P. C.; Swaminathan, R.; et al. The Burden of Cancers and Their Variations Across the States of India: The Global Burden of Disease Study 1990–2016. *Lancet Oncol.*, 2018, *19* (10). https://doi.org/10.1016/S1470-2045(18)30447-9.
3. Arum, Y.; Song, Y.; Oh, J. Controlling the Optimum Dose of AMPTS Functionalized-Magnetite Nanoparticles for Hyperthermia Cancer Therapy. *Appl. Nanosci.*, 2011, *1* (4). https://doi.org/10.1007/s13204-011-0032-1.
4. Shaw, S. K.; Kailashiya, J.; Gangwar, A.; Alla, S. K.; Gupta, S. K.; Prajapat, C. L.; Meena, S. S.; Dash, D.; Maiti, P.; Prasad, N. K. γ-Fe2O3 Nanoflowers as Efficient Magnetic Hyperthermia and Photothermal Agent. *Appl. Surf. Sci.*, 2021, *560*. https://doi.org/10.1016/j.apsusc.2021.150025.

5. Ganguly, S.; Margel, S. 3D Printed Magnetic Polymer Composite Hydrogels for Hyperthermia and Magnetic Field Driven Structural Manipulation. *Prog. Polym. Sci.*, 2022. https://doi.org/10.1016/j.progpolymsci.2022.101574.
6. Borgheti-Cardoso, L. N.; Viegas, J. S. R.; Silvestrini, A. V. P.; Caron, A. L.; Praça, F. G.; Kravicz, M.; Bentley, M. V. L. B. Nanotechnology Approaches in the Current Therapy of Skin Cancer. *Adv. Drug Delivery Rev.*, 2020. https://doi.org/10.1016/j.addr.2020.02.005.
7. Mehrafzoon, S.; Hassanzadeh-Tabrizi, S. A.; Bigham, A. Synthesis of Nanoporous Baghdadite by a Modified Sol-Gel Method and Its Structural and Controlled Release Properties. *Ceram. Int.*, 2018, *44* (12). https://doi.org/10.1016/j.ceramint.2018.04.244.
8. Hassanzadeh-Tabrizi, S. A.; Bigham, A.; Rafienia, M. Surfactant-Assisted Sol-Gel Synthesis of Forsterite Nanoparticles as a Novel Drug Delivery System. *Mater. Sci. Eng. C*, 2016, *58*. https://doi.org/10.1016/j.msec.2015.09.020.
9. Hilger, I.; Kaiser, W. A. Iron Oxide-Based Nanostructures for MRI and Magnetic Hyperthermia. *Nanomedicine*, 2012. https://doi.org/10.2217/nnm.12.112.
10. Kotoulas, A.; Dendrinou-Samara, C.; Sarafidis, C.; Kehagias, T.; Arvanitidis, J.; Vourlias, G.; Angelakeris, M.; Kalogirou, O. Carbon-Encapsulated Cobalt Nanoparticles: Synthesis, Properties, and Magnetic Particle Hyperthermia Efficiency. *J. Nanoparticle Res.*, 2017, *19* (12). https://doi.org/10.1007/s11051-017-4099-9.
11. Bauer, L. M.; Situ, S. F.; Griswold, M. A.; Samia, A. C. S. High-Performance Iron Oxide Nanoparticles for Magnetic Particle Imaging-Guided Hyperthermia (HMPI). *Nanoscale*, 2016, *8* (24). https://doi.org/10.1039/c6nr01877g.
12. Dalal, M.; Greneche, J. M.; Satpati, B.; Ghzaiel, T. B.; Mazaleyrat, F.; Ningthoujam, R. S.; Chakrabarti, P. K. Microwave Absorption and the Magnetic Hyperthermia Applications of $Li_{0.3}Zn_{0.3}Co_{0.1}Fe_{2.3}O_4$ Nanoparticles in Multiwalled Carbon Nanotube Matrix. *ACS Appl. Mater. Interfaces*, 2017, *9* (46). https://doi.org/10.1021/acsami.7b12091.
13. Mazario, E.; Menéndez, N.; Herrasti, P.; Cañete, M.; Connord, V.; Carrey, J. Magnetic Hyperthermia Properties of Electrosynthesized Cobalt Ferrite Nanoparticles. *J. Phys. Chem. C*, 2013, *117* (21). https://doi.org/10.1021/jp4023025.
14. Hassanzadeh-Tabrizi, S. A. $Mg_{0.5}Ni_{0.5}Fe_2O_4$ Nanoparticles as Heating Agents for Hyperthermia Treatment. *J. Am. Ceram. Soc.*, 2019, *102* (5). https://doi.org/10.1111/jace.16160.
15. Mascolo, M. C.; Pei, Y.; Ring, T. A. Room Temperature Co-Precipitation Synthesis of Magnetite Nanoparticles in a Large Ph Window With Different Bases. *Materials (Basel).*, 2013, *6* (12). https://doi.org/10.3390/ma6125549.
16. Mascolo, M. C.; Pei, Y.; Ring, T. A. Nanoparticles in a Large PH Window With Different Bases. *Materials (Basel).*, 2013, *6*.
17. Medeiros, S. F.; Santos, A. M.; Fessi, H.; Elaissari, A. Stimuli-Responsive Magnetic Particles for Biomedical Applications. *Int. J. Pharm.*, 2011. https://doi.org/10.1016/j.ijpharm.2010.10.011.
18. Yang, G. Q.; Han, B.; Sun, Z. T.; Yan, L. M.; Wang, X. Y. Preparation and Characterization of Brown Nanometer Pigment With Spinel Structure. *Dye. Pigment.*, 2002, *55* (1). https://doi.org/10.1016/S0143-7208(02)00056-6.
19. Qin, M.; Shuai, Q.; Wu, G.; Zheng, B.; Wang, Z.; Wu, H. Zinc Ferrite Composite Material With Controllable Morphology and Its Applications. *Mater. Sci. Eng. B Solid-State Mater. Adv. Technol.*, 2017, *224*. https://doi.org/10.1016/j.mseb.2017.07.016.
20. Lee, H.; Jung, J. C.; Kim, H.; Chung, Y. M.; Kim, T. J.; Lee, S. J.; Oh, S. H.; Kim, Y. S.; Song, I. K. Effect of PH in the Preparation of $ZnFe_2O_4$ for Oxidative Dehydrogenation of N-Butene to 1,3-Butadiene: Correlation between Catalytic Performance and Surface Acidity of $ZnFe_2O_4$. *Catal. Commun.*, 2008, *9* (6). https://doi.org/10.1016/j.catcom.2007.10.023.

21. Lee, H.; Jung, J. C.; Kim, H.; Chung, Y. M.; Kim, T. J.; Lee, S. J.; Oh, S. H.; Kim, Y. S.; Song, I. K. Preparation of $ZnFe_2O_4$ Catalysts by a Co-Precipitation Method Using Aqueous Buffer Solution and Their Catalytic Activity for Oxidative Dehydrogenation of n-Butene to 1,3-Butadiene. *Catal. Letters*, 2008, *122* (3–4). https://doi.org/10.1007/s10562-007-9371-7.

22. Cabuil, V.; Dupuis, V.; Talbot, D.; Neveu, S. Ionic Magnetic Fluid Based on Cobalt Ferrite Nanoparticles: Influence of Hydrothermal Treatment on the Nanoparticle Size. *J. Magn. Magn. Mater*, 2011, *323* (10), 1238–1241. https://doi.org/10.1016/j.jmmm.2010.11.013.

23. Thirunavukkarasu, G. K.; Cherukula, K.; Lee, H.; Jeong, Y. Y.; Park, I. K.; Lee, J. Y. Magnetic Field-Inducible Drug-Eluting Nanoparticles for Image-Guided Thermo-Chemotherapy. *Biomaterials*, 2018, *180*. https://doi.org/10.1016/j.biomaterials.2018.07.028.

24. Livesey, K. L.; Ruta, S.; Anderson, N. R.; Baldomir, D.; Chantrell, R. W.; Serantes, D. Beyond the Blocking Model to Fit Nanoparticle ZFC/FC Magnetisation Curves. *Sci. Rep.*, 2018, *8* (1). https://doi.org/10.1038/s41598-018-29501-8.

25. Rezoanur Rahman, M.; Bake, A.; Jumlat Ahmed, A.; Md Kazi Nazrul Islam, S.; Wu, L.; Khakbaz, H.; FitzGerald, S.; Chalifour, A.; Livesey, K. L.; Knott, J. C.; et al. Interplay between Thermal and Magnetic Properties of Polymer Nanocomposites With Superparamagnetic Fe_3O_4 Nanoparticles. *J. Magn. Magn. Mater.*, 2023, *579*. https://doi.org/10.1016/j.jmmm.2023.170859.

26. Eivazzadeh-Keihan, R.; Maleki, A. Design and Synthesis of a New Magnetic Aromatic Organo-Silane Star Polymer With Unique Nanoplate Morphology and Hyperthermia Application. *J. Nanostructure Chem.*, 2021, *11* (4). https://doi.org/10.1007/s40097-021-00401-0.

27. Eivazzadeh-Keihan, R.; Radinekiyan, F.; Asgharnasl, S.; Maleki, A.; Bahreinizad, H. A Natural and Eco-Friendly Magnetic Nanobiocomposite Based on Activated Chitosan for Heavy Metals Adsorption and the in-Vitro Hyperthermia of Cancer Therapy. *J. Mater. Res. Technol.*, 2020, *9* (6). https://doi.org/10.1016/j.jmrt.2020.08.096.

28. Hernández, R.; Zamora-Mora, V.; Sibaja-Ballestero, M.; Vega-Baudrit, J.; López, D.; Mijangos, C. Influence of Iron Oxide Nanoparticles on the Rheological Properties of Hybrid Chitosan Ferrogels. *J. Colloid Interface Sci.*, 2009, *339* (1). https://doi.org/10.1016/j.jcis.2009.07.066.

29. Ghorpade, V. S. Preparation of Hydrogels Based on Natural Polymers via Chemical Reaction and Cross-Linking. *Hydrogels Based Nat. Polym.*, 2019. https://doi.org/10.1016/B978-0-12-816421-1.00004-5.

30. Jawad, A. H.; Mubarak, N. S. A.; Abdulhameed, A. S. Hybrid Crosslinked Chitosan-Epichlorohydrin/TiO_2 Nanocomposite for Reactive Red 120 Dye Adsorption: Kinetic, Isotherm, Thermodynamic, and Mechanism Study. *J. Polym. Environ.*, 2020, *28* (2). https://doi.org/10.1007/s10924-019-01631-8.

31. Eivazzadeh-Keihan, R.; Farrokhi-Hajiabad, F.; Aliabadi, H. A. M.; Ziabari, E. Z.; Geshani, S.; Kashtiaray, A.; Bani, M. S.; Pishva, B.; Cohan, R. A.; Maleki, A.; et al. A Novel Magnetic Nanocomposite Based on Alginate-Tannic Acid Hydrogel Embedded With Silk Fibroin With Biological Activity and Hyperthermia Application. *Int. J. Biol. Macromol.*, 2023, *224*. https://doi.org/10.1016/j.ijbiomac.2022.10.236.

32. Eivazzadeh-Keihan, R.; Pajoum, Z.; Aghamirza Moghim Aliabadi, H.; Ganjali, F.; Kashtiaray, A.; Salimi Bani, M.; Lalebeigi, F.; Ziaei Ziabari, E.; Maleki, A.; Heravi, M. M.; et al. Magnetic Chitosan-Silk Fibroin Hydrogel/Graphene Oxide Nanobiocomposite for Biological and Hyperthermia Applications. *Carbohydr. Polym.*, 2023, *300*. https://doi.org/10.1016/j.carbpol.2022.120246.

11 Magnetic Polymer Composites for Soft Robotics

Anasuya Patil, A. Venkata Badarinath, Narayana Raju Padala, Kavati Ramkrishna, Raju A., Rasapelly Ramesh Kumar, H. N. Deepakumari, and Konatham Teja Kumar Reddy

11.1 INTRODUCTION

Soft robotics, a concept initially developed as a biomimetic design approach to facilitate adaptable interactions with unstructured surroundings, has evolved into a pivotal technology that has the potential to blur the demarcation between humans, robots, and the allocation of physical tasks. According to recent research studies [1], in the realm of robotics, there is a prevailing challenge associated with conventional robots that are constructed using skeletal structures and equipped with actuators at the joints [2, 3]. These robots often encounter difficulties when attempting to execute basic tasks, such as grasping objects like fruits or eggs. The primary reason for these struggles lies in the complexity of implementing feedback algorithms that are necessary to manipulate the rigid structures of these robots. Soft robotics, in contrast to traditional robotics, offer advantages in terms of gripping objects, navigating through narrow spaces, and projecting objects. These capabilities in robots are inspired by those of natural organisms such as octopi, starfish, caterpillars, and chameleons. Soft robots achieve these functionalities by utilizing stretchable elastomers and gels instead of rigid structures. These materials possess mechanical toughness and can undergo shape and dimensional changes. In a recent study, it was found that [3]. This finding is consistent Numerous innovative applications have been proposed in various fields, showcasing the distinctiveness and novelty of soft robotics. These applications encompass a wide range of areas, such as humanoid co-robots designed specifically for elder care, wearable joint supports and soft exoskeletons aimed at assisting human motion, surgical devices like endoscopes and catheters, prosthetic body parts, and artificial organs. According to recent research, the values have been identified as a significant data point [4]. The expansion of capabilities in soft robotics is closely linked to the development of enhanced "smart functionalities" in elastomers and gels. The smart functionalities encompass a range of controllable modulations in dimension, shape, and stiffness [5]. These modulations enable the actuation of soft robotics in response to external stimuli that are not limited to pneumatic or hydraulic sources. The external stimuli can originate from various sources such as electric, magnetic, chemical, thermal, or photonic means.

DOI: 10.1201/9781003454236-11

Soft robotics relies on the utilization of soft actuators to facilitate the movements of its mechanisms [6]. Conventional robots typically utilize traditional actuators, which are driven by gearmotors and controlled by proportional-integral-derivative (PID) controllers to ensure precise movement through the use of precision linkages [7]. In contrast, soft actuators are capable of propelling the movement of robotic systems through the control of a soft material's dimensions and stiffness, mirroring the functionality of muscles in living organisms. Although soft actuators lack precise and linear control of motions, they are widely acknowledged for their compactness, lightweight nature, high energy efficiency, and mechanical compliance. Given the envisioned potential uses of soft robotics in constrained and unpredictable settings, it appears that soft actuators are highly compatible with propelling soft robotic systems [8]. The classification of soft actuators is commonly based on the materials that govern their motion. As an illustration, gels or elastomers lacking inherent functionality, such as responsiveness to non-mechanical stimuli, can incorporate compressed air or fluids to facilitate pneumatic or hydraulic actuation, respectively. Magnetic actuation is regarded as a potentially advantageous alternative to conventional actuation techniques due to the non-contact nature of its energy transfer. This characteristic enables untethered control and renders it a viable option for sterile actuation in biomedical applications [9]. The integration of magnetic imaging and actuation is significant within the medical field due to its potential to eliminate the detrimental effects of radiation associated with techniques like fluoroscopy [10]. The utilization of magnetic actuation provides a prompt reaction time. Furthermore, magnetic actuation can be employed in various environments, such as air or vacuum and conducting or non-conducting liquids, as long as the workspace allows for magnetic transparency. Additionally, magnetic actuation remains unaffected by the ionic concentration present in the surrounding medium. The utilization of magnetic actuation obviates the necessity for energy storage components integrated within the system, a desirable attribute particularly in the context of scaling down to micro- and nanoscale dimensions [11].

The feasibility of achieving magnetic actuation in soft robots has been demonstrated by incorporating magnetic particles into polymers, resulting in the development of responsive materials [12]. The magnetic characteristics of magnetic polymer composites (MPCs) are influenced by the particle type, particle morphology, and particle volume fraction, whereas the mechanical properties are determined by the polymer matrix [13]. In a previous study, Garstecki et al. employed the same methodology to construct a flexible swimmer with the ability to self-propel in water through the application of an externally applied rotating magnetic field [14]. The utilization of magnetized soft material to achieve various modes of locomotion on both solid and fluid interfaces has been demonstrated by Hu et al. and Diller et al. Additionally, Lum et al. have proposed a technique for deriving the required magnetization profiles using Fourier series [15–17]. These works employed a singular strip of material, thereby imposing limitations on the range of motion associated with transporting payloads. Recent studies have demonstrated the application of multiple limbs for the purpose of grasping. These limbs were created through the utilization of 3D printing technology to fabricate ferromagnetic domains. Additionally, a robot with multiple legs was developed, but it was operated using permanent magnets and maintained a consistent magnetic profile. This approach imposes limitations on the ability to exert control over the robot's movements [18, 19].

The application of phase transformation in the polymeric matrix or metallic constituent has been employed in the context of shape-memory polymer (SMP) actuators or shape-memory alloy (SMA) actuators, respectively. The principles of actuation for dielectric elastomer actuators (DEAs) and ionic electroactive polymers (IEAPs) involve electronic and ionic conduction, often accompanied by associated chemical reactions [20]. The former approach leverages the dielectric property and deformability of elastomers and gels, while the latter also incorporates ionic conductivity within soft materials. The utilization of composite materials with magnetic properties has facilitated the remote manipulation of magnetorheological (MR) fluids and elastomers. These actions can be triggered by stimuli from diverse sources, such as an electric field, magnetic field, pressure gradient, thermal energy, electromagnetic radiation, and chemical reactions. Sitti et al. conducted a comprehensive review, which presented a concise table outlining various soft materials and the specific stimuli that have been demonstrated to induce actuation in these materials [21].

11.2   SOFT ROBOTICS AND ITS SIGNIFICANCE

Soft-bodied robots, which consist of materials with moduli similar to those found in soft biological materials (ranging from 10^4 to 10^9 Pa), exhibit significant potential for various critical applications. This is particularly true in the field of biomedicine, where their mechanical compliance offers improved safety during operation. The utilization of soft materials serves to mitigate potential harm to biological tissues or organs during interaction [22], as it enables sustained deformation and maintains mechanical resemblance to the surrounding tissue [23, 24]. Despite the purported benefits of soft robots and the growing endeavors to employ them in biomedical applications, the field of soft robotics continues to encounter a series of significant obstacles. The majority of conventional soft robots rely either on pneumatic or hydraulic actuation, which involves the use of fluids, or on antagonistic pairs of mechanical wires, known as tendon-driven systems. However, these robots typically require extensive tethering in order to establish connections with the driving actuators and supporting hardware. The limited practicality of these devices in applications that necessitate or prioritize tether-free actuation, such as implantable devices for internal organ support or controlled drug release, as well as minimally invasive devices for therapeutic or diagnostic operations, is evident. Furthermore, the accurate modeling and control of most soft robots pose significant challenges. This is primarily due to the fact that their actuation mechanisms often involve highly nonlinear deformation or even instabilities. Additionally, these robots necessitate complex structural designs that incorporate numerous passive degrees of freedom. The inherent elusiveness of soft robots significantly limits the applicability of the conventional model-based approach used for controlling rigid robots [25]. Additionally, the downsizing of mainstream soft robots with conventional actuation mechanisms to submillimeter scales poses significant challenges due to the unfavorable fabrication methods employed in their production.

While there are many types of soft materials that can self-actuate in response to external stimuli, recent advances in the design and fabrication of magnetically responsive soft materials have allowed for the creation of soft robots that are actuated and controlled by magnets [26]. Due to a number of benefits associated with magnetic actuation, these magnetic soft robots show a lot of promise for usage in the medical field. First, biological systems can withstand remote applications of magnetic fields across a broad spectrum of actuation field strengths and frequencies without any undesired effects, because natural tissues and organs are impermeable to static and low-frequency magnetic fields [27]. Furthermore, magnetic fields are amenable to rapid and precise modulation of their amplitude, phase, and frequency [28]. Due to these factors, magnetic fields can provide a secure and efficient method of controlling untethered soft robots in tight places within the human body [29].

11.3 CLASSIFICATIONS OF SOFT MAGNETIC COMPOSITES

11.3.1 MAGNETIC MATERIALS: TYPES, PROPERTIES, AND APPLICATIONS

The phenomenon of magnetism, which is considered one of the fundamental forces in nature, has captivated human interest for countless generations [30]. The capacity of specific materials to demonstrate magnetic properties has facilitated the emergence of diverse technological innovations and practical implementations. Magnetic materials can be classified into various categories according to their distinct properties and behaviors, each exhibiting a unique array of characteristics and applications [31]. This section examines the various classifications of magnetic materials, their inherent characteristics, and their extensive utilization within a multitude of industries.

Magnetic materials can be categorized into four primary classifications (Table 11.1): namely, ferromagnetic, paramagnetic, diamagnetic, and ferrimagnetic. Ferromagnetic materials are widely recognized and widely employed magnetic materials [32]. These materials demonstrate robust permanent magnetism, indicating their ability to retain their magnetic characteristics even in the absence of an external magnetic field. Ferromagnetic materials are distinguished by their inherent magnetization, characterized by the alignment of atomic magnetic moments in a consistent orientation, resulting in the formation of magnetic domains. Frequently encountered ferromagnetic materials are iron, nickel, and cobalt. Ferromagnetic materials are widely utilized in various technological applications, including electric motors, generators, transformers, magnetic recording systems (e.g., hard drives and tapes), and magnetic locks, owing to their notable magnetic properties [33]. Paramagnetic materials exhibit a relatively low level of affinity toward external magnetic fields. Upon being subjected to a magnetic field, the atomic magnetic moments of the entities in question align themselves in accordance with the field, thereby leading to a transient state of magnetization. In contrast to ferromagnetic materials, paramagnetic materials do not exhibit the phenomenon of magnetization retention upon removal of an external magnetic field. Paramagnetic materials encompass a variety of substances, such as aluminium, platinum, and oxygen, that exhibit a propensity to become magnetized when subjected to an external magnetic field. Paramagnetic materials play a critical role in various domains, including medical diagnostics through the utilzsation of MRI (magnetic resonance imaging) machines, as well as research endeavors aimed at investigating the magnetic properties of materials [34–36]. Although their practical applications may be somewhat constrained, the significance of paramagnetic materials should not be overlooked [37]. Diamagnetic materials demonstrate a modest degree of repulsion upon exposure to an external magnetic field.

TABLE 11.1

Basic Characteristics of Magnetic Materials and Their Probable Applications

Classification	Basic Features	Applications
Ferrimagnetic	• Strong permanent magnetism • High susceptibility to external fields • Spontaneous magnetization • Hysteresis effect (remnant magnetization) • Domains align in presence of a field • Can retain magnetization after field removal	• Electric motors and generators • Transformers and inductors • Magnetic recording (hard drives, tapes) • Magnetic locks and sensors • Loudspeakers and headphones • Magnetic separation (mineral processing)
Paramagnetic	• Weak attraction to external magnetic fields • Temporary magnetization • Random orientation of magnetic moments • Approaches ferromagnetic behavior with heat • Limited practical applications	• MRI machines • Research in materials science • Electronic components testing • NMR spectroscopy
Diamagnetic	• Weak repulsion from external fields • No permanent magnetization • Electron orbits create induced currents • Weak effect often masked by stronger ones	• Levitation (Maglev trains, frogs) • Superconducting materials research • High-frequency circuits • Magnetic resonance imaging calibration
Ferrimagnetic	• Opposite spins, net magnetization • Lower magnetic moment than ferromagnetic • Magnetic compensation reduces net moment • Found in ferrites and magnetic garnets	• Memory storage (magnetic cores) • Microwave devices (gyrators, circulators) • Recording heads in tape drives • Non-volatile magnetic memory (MRAM) • Used in microwave and radar applications

The phenomenon of repulsion occurs as a result of the generation of induced currents within the material, which act in opposition to the externally applied magnetic field [38]. Diamagnetic materials exhibit a lack of inherent magnetization and do not maintain their magnetic properties once the external magnetic field is eliminated. Water, copper, and bismuth are frequently cited as typical diamagnetic materials. Although diamagnetic materials exhibit weak magnetic behavior, they are utilized in various applications involving levitation, such as Maglev trains and frogs, as well as in fundamental investigations of superconductivity [39]. Ferrimagnetic materials exhibit a net magnetization due to the presence of magnetic moments with opposite spins. In contrast to ferromagnetic materials, ferrimagnetic materials exhibit incomplete alignment of their magnetic moments. The presence of this asymmetry results in a reduced overall magnetic moment in comparison to that observed in ferromagnetic materials. Ferrimagnetic materials find utility in various applications, including memory storage, notably in magnetic cores, microwave devices such as gyrators and circulators, and non-volatile magnetic memory (MRAM) [40]. The characteristics of magnetic materials are predominantly determined by their reaction to external magnetic fields, their magnetic arrangement, and their magnetization curves. The concept of magnetic susceptibility pertains to the ability of a substance to undergo magnetization in response to an external magnetic field. Ferromagnetic substances exhibit a pronounced susceptibility toward magnetic fields, whereas diamagnetic substances display a comparatively low susceptibility [41]. Ferromagnetic materials demonstrate the occurrence of hysteresis, a phenomenon characterized by a delay in magnetization response relative to the applied magnetic field. This delay is attributed to the process of alignment and realignment of magnetic domains within the material. This particular characteristic holds significant importance in various applications, including but not limited to transformers and electromagnets. The Curie temperature refers to the specific temperature at which the magnetic properties of a ferromagnetic material are no longer present as a result of thermal influences. Once the temperature surpasses the Curie temperature, the material undergoes a transition and exhibits paramagnetic behavior [42].

The unique properties of magnetic materials make them highly versatile and applicable in a diverse range of industries. Ferromagnetic materials play a crucial role in the functioning of electric motors, generators, transformers, and inductors. The utilization of ferromagnetic materials in these devices serves to augment efficiency and mitigate energy losses. Magnetic materials play a vital role in the field of information storage, wherein ferromagnetic materials serve as the fundamental components of hard drives and magnetic tapes [43]. The capacity to maintain magnetization facilitates the storage of data in the form of binary information. Paramagnetic materials are of significant importance in the operation of MRI machines. The utilization of paramagnetic contrast agents results in a transient magnetization, which serves to improve the detectability of tissues and structures during medical imaging procedures, thereby facilitating the process of medical diagnosis. The repulsive behavior exhibited by diamagnetic materials in the presence of magnetic fields facilitates various applications, such as the implementation of Maglev trains. These trains are capable of levitating above the tracks, thereby minimizing frictional forces and enhancing overall operational efficiency. Ferrimagnetic materials find application in microwave devices such as circulators and isolators, which serve to regulate the propagation direction of electromagnetic waves within communication systems. The utilization of magnetic materials is imperative in the investigation of fundamental characteristics of matter. These findings offer valuable perspectives on the dynamics of quantum mechanical phenomena and the characteristics of electrons within diverse crystalline arrangements.

Magnetic materials exhibit a wide range of variations, characterized by unique attributes and practical uses. The comprehension and application of magnetic materials have brought about a technological revolution across various industries, ranging from the commonplace utilization of robust ferromagnetic materials in everyday devices to the extraordinary capabilities of diamagnetic materials that enable levitation. The advancement of our comprehension of material properties is anticipated to lead to further groundbreaking utilization of magnetic materials in various domains, including but not limited to electronics and medical diagnostics, thereby facilitating progress in these fields.

11.4   TECHNIQUES FOR DISPERSING MAGNETIC PARTICLES INTO POLYMERS

Composite materials such as magnetic elastomers and gels exhibit optimal properties when the magnetic particles are uniformly dispersed, thereby avoiding agglomeration. The blending technique, alternatively referred to as physical doping, involves the amalgamation of magnetic particles, un-cross-linked polymer chains, and cross-linking agents within a singular container, typically containing a solvent or non-polymerizing silicone oil. During the process of cross-linking, the magnetic particles, which lack the ability to form chemical bonds with the polymeric constituents, become physically confined within the cross-linked networks of gels or elastomers. The blending technique is widely favored due to its inherent simplicity; however, it is accompanied by certain difficulties pertaining to the management of particle dispersion and the prevention of particle agglomeration. The absence of chemical bonding between magnetic particles and the polymeric network presents a significant challenge in the bleaching process of swollen gels.

In contrast to elastomers, gels can be formed using the in situ precipitation technique. This technique uses a polymeric gel network infused with a magnetic metal precursor solution to precipitate magnetic particles. The magnetic particles created by the in situ precipitation approach have more uniform dispersion than those prepared by mixing, and it is easier to incorporate a large volume of particles. However, this approach is limited to hydrogels that are stable in alkaline environments. Some biomedical uses, such as cell encapsulation, are infeasible because of the method's requirement of an alkaline solution. In order to create Fe_3O_4 nanoparticles, Albertsson and colleagues synthesized the hydrogel from a precursor solution consisting of a hemicellulose oligomer (O-acetyl-galactoglucomannan, degree of polymerization 40) and a cross-linker (epichlorohydrin) [44]. Nanoparticles were evenly distributed in the hydrogel's cellular structure, but some agglomeration was seen; the outcome was a magnetic hydrogel with high sensitivity.

Another technique frequently used for making magnetic elastomers and gels involves grafting-onto, where chemically functional groups on the surface of the magnetic particle act as a cross-linker. The particles and polymer chains form a covalent link, increasing the magnetic composite material's structural and chemical stability. Both attapulgite (a common type of magnesium aluminum phyllosilicate) and magnetic fly ash (the magnetic fraction of fly ash was prescreened and collected) were functionalized by Jiang and Liu with acrylic surface functional groups. To that end, the authors synthesized a hydrogel composite material based on polyacrylic acid, using the nanoparticles as cross-linkers and strengthening agents [45]. To create a magnetic hydrogel, Schmidt and colleagues employed unsaturated methacrylic acid derived from silicone as a cross-linking agent to covalently connect $CoFe_2O_4$ magnetic nanoparticles with polyacrylamide (Figure 11.1) [46]. The resulting gel could be affected by a magnet. Transmission electron microscopy (TEM) imaging provided compelling evidence for the structural roles of nanoparticles as cross-linking locations. Poly(dimethylsiloxane-co-aminopropylmethysiloxane) is a silicone polymer that Evans et al. adsorbed onto the surface of Fe_2O_3 nanoparticles to stop them from sticking together [47]. In order to aid in the cross-linking of

(a)

(b)

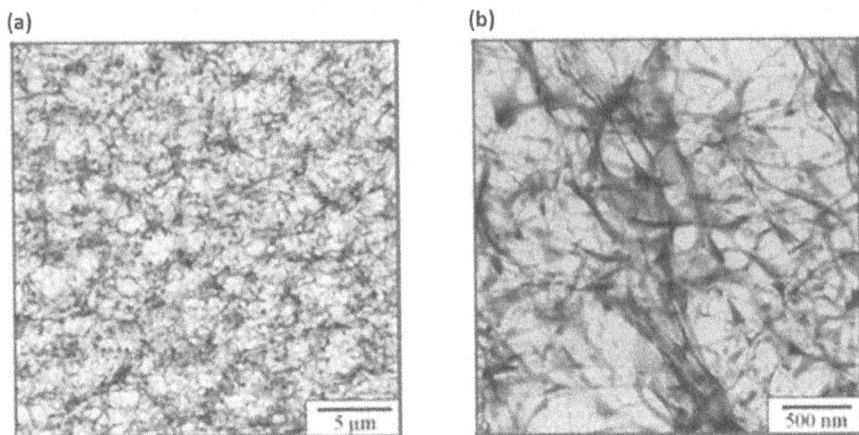

FIGURE 11.1   TEM images of a swollen, freeze-fractured, freeze-dried NP-FHG_10.8 at different resolutions. (Reproduced with permission from Ref. [49]. © 2011 American Chemical Society.)

2-hydroxy-ethyl-methacrylate (HEMA) polymer chains during the batch co-polymerization process, Stark and colleagues functionalized the surface of cobalt nanoparticles with a 4-vinylbiphenyl group [48].

Magnetic particles can have an isotropic or an anisotropic distribution. The actuation behavior may become directed if magnetic particles are ordered in such a way. In order to achieve both directionality and efficiency in magnetic actuation, it is necessary to confine the magnetic particles within organized microsized domains [49]. The soft actuator's response to an applied field can be locally localized by discretizing magnetic susceptibility by spreading magnetic particles inside organized regions.

11.5   ELASTOMER-BASED MAGNETIC SOFT ROBOTICS

The utilization of specific materials in magnetic robotics is of utmost importance as it significantly contributes to the overall functionality, control, and efficiency of these systems. The selection of these materials is based on their magnetic properties, mechanical flexibility, and adaptability to diverse applications. The following are several essential materials frequently employed in the field of magnetic robotics. The concise format of the soft robotics and their advantages are depicted in Figure 11.2.

Permanent magnets play a pivotal role as essential components in the field of magnetic robotics. The magnetic fields they produce engage in interactions with other magnetic elements within the system. Neodymium-iron-boron (NdFeB) and samarium-cobalt (SmCo) are materials frequently employed in various applications, owing to their notable magnetic potency and robustness. Soft magnetic materials play a crucial role in various applications such as actuators, sensors, and transformers. Ferromagnetic materials possess notable characteristics such as elevated magnetic permeability and reduced coercivity, rendering them highly receptive to external magnetic fields. Iron, nickel, and their respective alloys are frequently utilized in various applications due to their advantageous magnetic characteristics. Ferromagnetic elastomers are composite materials that integrate the mechanical flexibility of elastomers with the magnetic properties of ferromagnetic particles. This unique combination enables these materials to exhibit both mechanical adaptability and magnetic responsiveness. The aforementioned materials find application in the construction of soft robotic grippers and manipulators due to their ability to undergo deformation and alter their shape in response to magnetic fields. The utilization of magnetic nanoparticles, frequently incorporated within elastomers or other matrices, enables the manipulation of magnetic characteristics in a controllable manner. The manipulation of these nanoparticles through the application of external magnetic fields enables their potential utilization in drug delivery, targeted therapeutic interventions, and minimally invasive medical procedures. Flexible substrates, such as polymers, elastomers, and textiles, are commonly employed as the underlying materials for magnetic robotics applications. The robotic system is capable of flexing, elongating, and adapting to various configurations while simultaneously preserving the necessary magnetic characteristics. The utilization of microfabrication techniques enables the creation of intricate magnetic structures characterized by precise magnetization patterns. The aforementioned structures find application in micro- and nanoscale magnetic robots, wherein they are utilized for various purposes such as drug delivery, cell manipulation, and sensing. Magnetic sensors, such as Hall effect sensors or magnetoresistive sensors, are incorporated into magnetic robots in order to obtain feedback pertaining to their position, orientation, and surrounding environment. These sensors facilitate accurate control and navigation. Smart materials, such as magnetostrictive and piezoelectric materials, exhibit the ability to alter their physical shape or generate forces when subjected to magnetic fields.

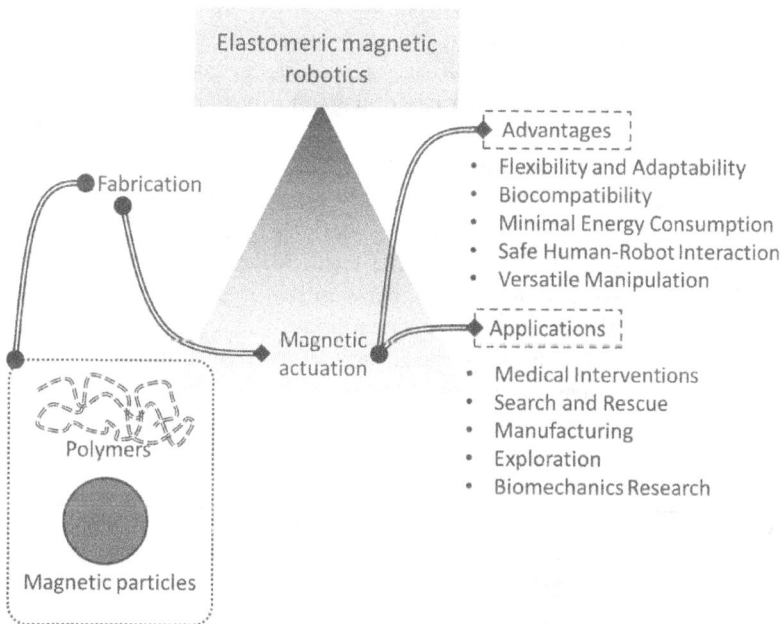

FIGURE 11.2 Diagrammatic representation of soft robotics, their mode of fabrication, advantages, and applications.

These devices find application in actuation systems, haptic feedback mechanisms, and vibration control systems. Soft ferromagnetic materials are characterized by their elevated magnetic permeability, which is accompanied by a relatively low saturation magnetization compared to that of conventional ferromagnetic materials. These materials find utility in various applications such as magnetic shielding and signal manipulation. Biocompatible materials play a crucial role in medical applications. Soft magnetic materials that possess biocompatibility and the ability to interact with external magnetic fields are utilized in the field of medical robotics for various applications such as diagnostics, drug delivery, and surgical assistance. The selection of materials is contingent upon the specific demands of the magnetic robotic system, encompassing factors such as the intended usage, the surrounding conditions, and the targeted performance attributes. The progression of magnetic robotics has prompted ongoing investigations into novel materials and material combinations, aimed at augmenting the functionalities and versatility of these robotic systems.

Zhang et al. showed a soft monolithic composite film that uses liquid crystal elastomers (LCE) as its base soft matrix and has magnetic microparticles (MMPs) built into it for remote magnetic controllability. This makes it possible for soft miniature machines to be changed in place without being attached [44]. It also has LCE that responds to environmental triggers like changes in temperature and UV light. This gives it multiple degrees of freedom (DOF) control, which makes it possible for soft machines to be responsive, multifunctional, and changeable. Machines made from the composite can change their shape on the spot to adapt to different surroundings and have more than one purpose. Small, untethered robots are getting a lot of attention in biomedical areas because they could be used in vivo for things like delivering stem cells or taking tissue samples from places where this has not been possible before. But the average speed of current soft polymer-based mini robots is still too slow for them to be useful in real life. Ijaz and colleagues made a chained magnetic microparticles–embedded elastomer-based magnetically actuated tiny walking soft robot with a high average speed (~2.7 mm/s), higher than that of traditional soft polymer–based robots of the same size. To get the most out of the average speed, they adjusted the magnetization of the robot by controlling how the magnetic microparticles were lined up [50]. It is difficult to design a robot gripper that can reliably grasp things of varying fragility, size, and form. A different team of researchers has put forward a proposition for a gripper skin that is universal, shape-adaptive, and reversible, with the ability to vary its hardness. This innovative solution offers a means of grasping objects without causing any damage to them. The magneto-rheological elastomer–based universal gripper skin is affixed to a robotic gripper in accordance with the proposed design. Upon the gripper's contact with the object, the skin promptly assumes the form of the target object. Currently, the gripper skin is consolidated through the application of a magnetic field, facilitating the enhanced ability of the gripper to securely hold the intended object [51]. Upon the release of the object (Figure 11.3), the magnetic field is eliminated, resulting in the prompt restoration of the deformed gripper skin to its initial shape. The adaptive gripper skin under consideration has been designed to effectively grasp a range of target objects, including cylinders, cuboids, and triangular prisms. The evaluation of its grasping performance is based on these specific objects.

(a)

(b)

(c)

(d)

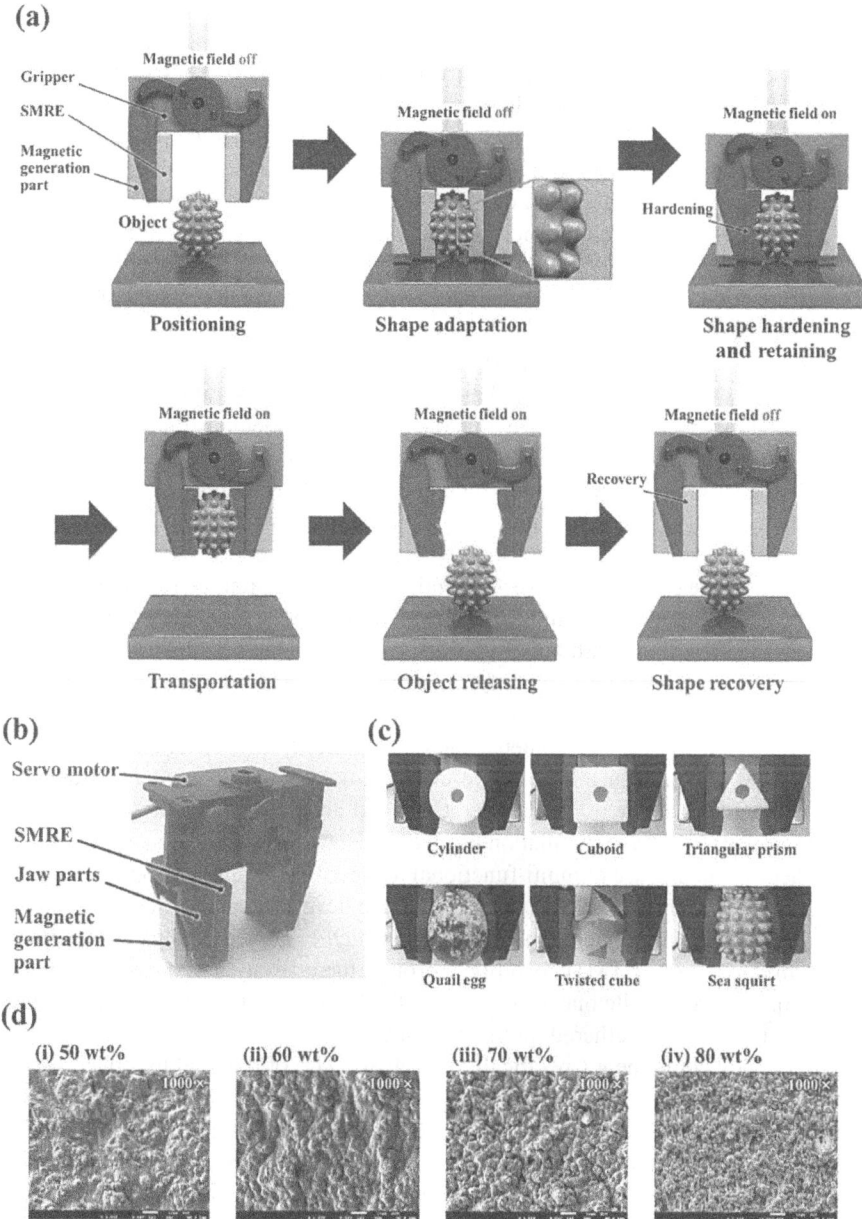

FIGURE 11.3 Proposed SMRE-based skin parallel jaw gripper. (a) Operating scenario with suggested SMRE-based parallel jaw gripper. (b) Photograph of planned parallel jaw gripper with SMRE-based skin. (c) Photos of suggested gripper holding cylinder, cuboid, triangular prism, quail egg, twisted cube, and sea squirt. (d) FESEM images (1000×) of cracked surfaces of isotropic MREs with carbonyl iron content of 50, 60, 70, and 80 wt %. (Reproduced with permission from Ref. [51]. © 2020 American Chemical Society.)

The investigation of intrinsically self-healing stretchable polymers has been the subject of extensive research in the field of soft robotics. This is primarily due to their mechanical compliance and ability to withstand damage. Nevertheless, the widespread utilization of these robotic applications in practical settings is presently impeded by several constraints, including inadequate mechanical durability, protracted recovery duration, and the necessity for external energy input. Researchers introduced a self-healing supramolecular magnetic elastomer (SHSME) that possesses a hierarchical dynamic polymer network containing a significant number of reversible bonds [52]. The SHSME material demonstrated notable mechanical strength, as indicated by its Young's modulus of 1.2 MPa, which is comparable to that of silicone rubber. Additionally, it exhibited a rapid self-healing capability, allowing it to withstand a stretch strain of 300% after autonomously repairing itself within 5 seconds at ambient temperature. This chapter presents several demonstrations of robotic systems based on the SHSME framework. These demonstrations include the rapid recovery of amphibious functions, the development of modular-assembling-prototyping soft robots with complex geometries and diverse functionalities, and the implementation of a dismembering-navigation-assembly strategy for robotic tasking in confined spaces.

There has been growing interest recently in miniature untethered robots due to their increased functionality and applicability in disruptive biomedical applications. Specifically, the robots with softer characteristics demonstrate distinct advantages in terms of their willingness to conform, adaptability, and nimbleness. Due to limited onboard capacity, these devices primarily acquire energy from the surrounding environment or physical phenomena, such as magnetic and acoustic fields, as well as patterned lights. Typically, a single device employs only one energy transmission mode (ETM) to facilitate its operations and accomplish predetermined tasks, such as movement and manipulation of objects. However, practical tasks in the real world necessitate the use of multi-functional devices that require a greater amount of energy in diverse forms. A study documented the development of a composite material consisting of liquid metal and elastomer, which exhibited dual electromagnetic torque manipulation (ETM) capabilities through the utilization of a single magnetic field [53]. This composite material was specifically designed for the purpose of constructing miniature untethered multi-functional robots. The initial ETM utilizes the low-frequency component (specifically, below 100 Hz) of the field to induce shape-morphing. In contrast, the second ETM utilizes radio-frequency induction, ranging from 20 kHz to 300 GHz, to transmit energy and power the onboard electronics. Additionally, this induction generates excess heat, thereby enabling the emergence of new capabilities. The introduction of these novel functions does not interfere with the shape-morphing capabilities achieved through utilization of the initial ETM. The provided data facilitate the incorporation of electric and thermal capabilities into small-scale flexible robots, presenting numerous possibilities for versatile miniature robots that utilize advancements in electronics to demonstrate practicality beyond just movement.

Instead of employing longitudinal "muscle" akin to that of a biological inchworm, the current generation of inchworm robots based on magnetic active elastomers (MAEs) relies on magnetic torque to propel and manipulate their flexible bodies, thereby impeding their locomotion capabilities. In one study, a novel milli-robot with micropillars, referred to as a pre-strained MAE inchworm millirobot, was introduced [54]. The pre-strained elastomer functions as a pre-load muscle, enabling the contraction of the soft body. Additionally, the micropillars serve as minuscule appendages, providing anchorage to the body during locomotion (Figure 11.4). The magnetic inchworm robot, as suggested, has a fabrication process that is straight-forward and does not necessitate the use of specialized magnetization equipment.

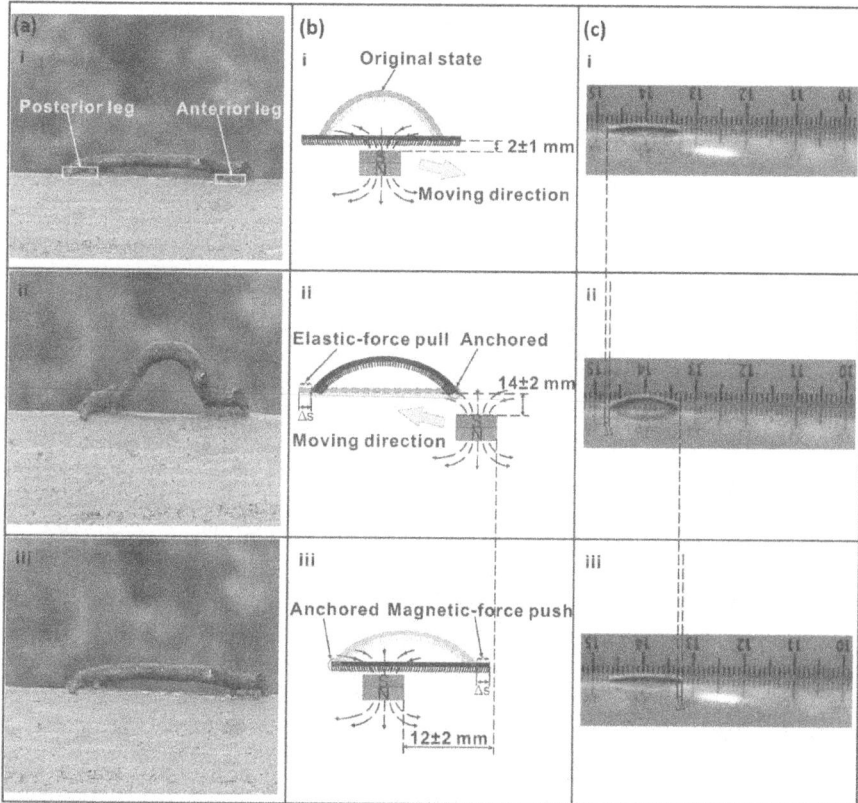

FIGURE 11.4 (a) Looping biological inchworm motility. (i) Initial inchworm posture. (ii) The inchworm anchors its anterior leg and drags its posterior leg forward. (iii) The inch-worm anchors its rear leg and pushes its anterior leg forward. (b) Magnetic inchworm robot motion breakdown schematic. Gray and black forms are undeformed and deformed, respec-tively. (i) The permanent magnet situated underneath the curved magnetic inchworm robot (~2 mm vertical distance) flattens it due to the attractive magnetic force. (ii) Move the magnet to the lower right (~14 mm vertical distance) to anchor the anterior leg, then contract the pre-load elastic muscle to draw the posterior leg forward. (iii) Move the magnet to the upper left (~12 mm horizontal distance) to anchor the posterior leg and flatten the magnetic body, push-ing the anterior leg forward. (c) Fabricated magnet Inchworm robot movement. The locomo-tion in (i–iii) matches (i–iii) in parts (a) and (b). (Reproduced with permission from Ref [54]. © 2023 Multidisciplinary Digital Publishing Institute.)

The incorporation of a pre-load muscle represents a novel addition to the magnetic inchworm robot's design, resulting in a locomotion mechanism that closely resembles that of an actual inchworm. The investigation focused on the locomotion principle and parametric design in order to achieve the desired locomotion performance.

Soft robotic grippers have been made in the past using soft pneumatic motors or smart soft materials, but they need external pressure sources, break easily, or are used on a millimeter scale, which makes them more dependent, unstable, and vulnerable. In their study, Zhang et al. suggested an EPM-MRE (electropermanent magnet–magnetorheological elastomer) actuation system that makes soft actuation by non-contact magnetic force more stable and independent. They found its basic principle and looked into parameter design by making a robot hand that uses a suction cup [55]. They made a prototype of an EPM-MRE-activated suction cup using magnetic-charge and tensile models. They also improved the structure of the EPM by adding an axisymmetric EPM with a frustum pole and a bi-silicone structure to the suction cup so that the MRE membrane would be activated evenly. The effect of different current pulses on EPM activation and the effect of contact shape and air gaps on suction force were studied. Tests showed that the suction cup could be turned on in 10 ms and could produce a maximum suction force of 9.2 N at a steady state while using no energy. They concluded that the EPM-MRE soft actuation system can be used as a robot hand. By programming the pattern of magnetization on the magnetic rubber, the magnetic soft robot can be made to change shape in a controlled way. But the magnetization patterns are usually fixed, which means that a magnetic soft robot can only do one thing. In another work, a programmable magnetic elastomer made of mixtures of liquid metal, NdFeB, and silicone was created. Through thermally aided magnetic programming, useful groups of liquid metal/NdFeB wrapped in a silicone matrix could be magnetized over and over again [56]. The researchers made a hybrid magnetic elastomer that can be controlled at 15.7, 41.5, and 47°C by using several liquid metals with melting points below 60°C. The SEM and EDS data show the micromorphology and element content of the elastomer. The complete physical property measurement system (PPMS) was used to test the magnetic properties of the elstomer. The moment–temperature (M–T) graph shows that the moment of the material jumps at the point where the metal melts. And the moment–magnetic field (M–H) graphs show that the magnetic properties of the elastomer are both hard and soft. Robots that can be changed, are soft, and can do more than one thing are helping to advance medical and rehabilitation robotics, human-machine contact, and smart home technology. A key part of making soft robots is being able to use flexible and efficient methods to allow the seamless merging of structures that can be changed. Another group of experts showed how design features and functions can be programmed into elastomeric surfaces [57]. They carefully changed elastomeric surfaces by scanning them with a laser and then penetrating them with solvent containing active particles. This made it possible to bend, fold, and add functions in a way that they could control. The usefulness of the elastomers can be taken away with a solvent retreatment, and the active particle infusion process can be done again and again. Researchers developed a platform method for making elastomeric sheets that can be programmed and used again and again by changing the specific morphology patterns and active particles.

11.6 HYDROGELBASED MAGNETIC SOFT ROBOTICS

The overall size of robots becomes an important consideration for activities in various settings. This is especially true when the goal is to have the robots function within the human vascular system. In this particular scenario, the comprehensive measurements of the robots may play a crucial role in determining the capacity of the microrobots to navigate through the vasculature. For example, in the context of delivering therapeutic agents to a tumor that is encompassed by intricate microvascular networks consisting of vessels with varying diameters ranging from a few micrometers to tens of micrometers, the utilization of micrometer-scale robots may enable a more precise approach to reaching the intended tumoral site. Consequently, rather than subjecting the entire patient's body to therapeutic agents, microrobots have the potential to enhance therapeutic effectiveness by selectively delivering higher drug dosages to specific target areas. This phenomenon mitigates the secondary toxic effects associated with the administration of the drug throughout the circulatory system. One of the primary obstacles encountered in the deployment of micrometer-scale robots is the need to optimize their functionality within the constraints of limited space. In order to achieve this objective, rather than incorporating additional components specifically designed for each function, a potentially effective approach involves utilizing the same components for multiple functions.

The composite matrices in which the magnetic filler particles are inserted into magnetic soft materials are elastic, thanks to the polymeric components. Internal mechanical stresses are generated by forces and torques acting on the individual particles in an external magnetic field, causing the soft polymer matrix to deform; the degree of deformation depends on the pliability of the material or structure and the polymeric material's response to the applied stress. The choice of polymeric substance, such as a soft elastomer or gel, to host the embedded magnetic filler depends on several factors, including the desired mechanical (viscoelastic) properties of the composite, the physicochemical properties of the chosen polymers and their compatibility with magnetic filler particles (especially when the particles are surface-treated or functionalized), the required functions and target applications, and the environmental conditions. The polymeric components of conventional magnetic soft materials are largely inert, with the active behavior coming from the magnetic particles contained inside the material in reaction to an applied field. Shape memory polymers [58] and liquid crystal elastomers [44], both of which are active polymeric components, have been used in recent research to demonstrate multi-functional magnetic soft materials that use thermal or photothermal actuation in tandem with magnetic actuation (Figure 11.5). In the sections that follow, we will talk about several typical soft polymeric materials and their properties as they relate to magnetic soft actuators.

A manufacturing paradigm shift away from the mechanical assembly of basic building blocks typical of rigid-body robots will be required for soft robots to adopt capabilities more akin to those of living organisms [60]. And 3D printing is an effective fabrication process for building complex, finely detailed, and seamless structures. Although mechanical strengthening requirements often are not met for the final printed hydrogel, pre-cross-linking the hydrogel precursors can successfully enhance the viscosity of the ensuing solution to allow printing [61]. To achieve the

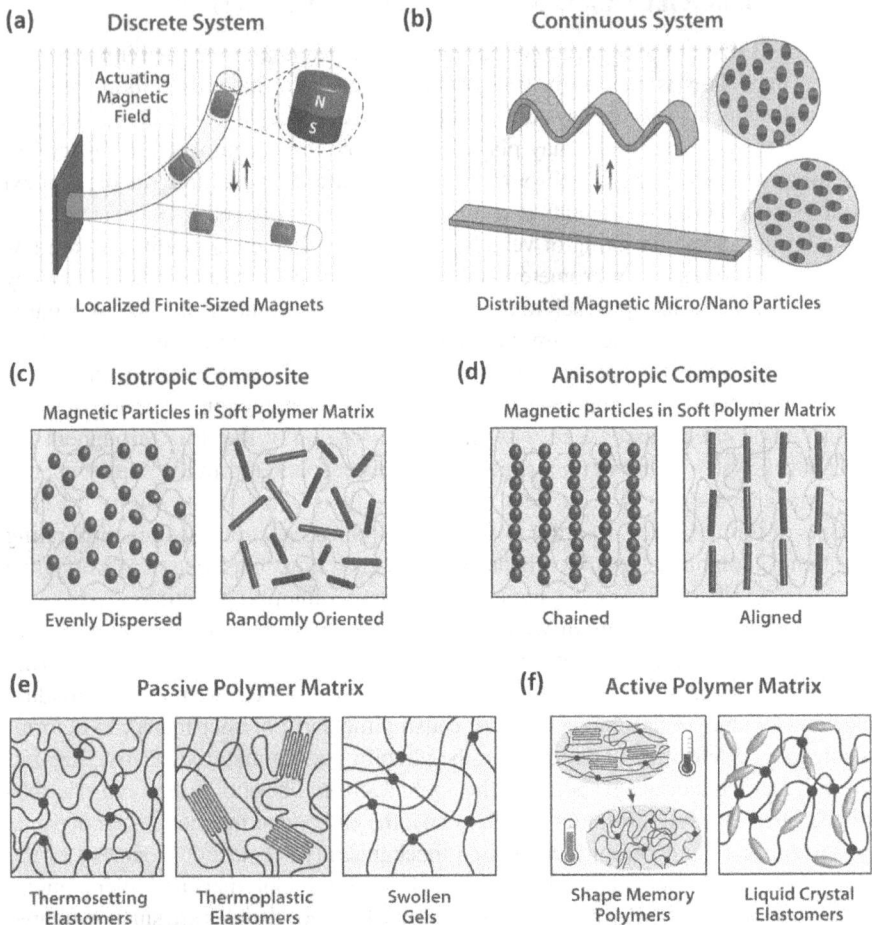

FIGURE 11.5 Magnetic soft materials and their classifications. Magnetic soft materials can be put into either (a) discrete or (b) continuous systems, based on whether the magnetic components are discrete magnets embedded in the flexible structure or micro- or nanoparticles dispersed in the soft polymer matrix. Continuous magnetic soft materials can also be divided into (c) physically isotropic composites or (d) anisotropic composites, based on how the magnetic filler particles are arranged in the polymer matrix. The polymeric part of magnetic soft materials can be either (e) passive polymeric materials, like thermosetting or thermoplastic elastomers and swollen gels, or (f) active polymeric materials, like shape memory polymers or liquid crystal elastomers, depending on whether the polymer matrices themselves respond to external stimuli to change their physical properties or produce actuation through deformation. (Reproduced with permission from Ref. [59]. © 2022 American Chemical Society.)

appropriate rheological qualities for printing, nanoclays and silica particles have also been used as non-homogeneous thickening materials in the hydrogel precursor ink [62]. However, the inclusion of such materials with properties different from those of hydrogels can undermine the characteristics associated with hydrogels. Hydrogel-based soft robots are the subject of almost no research in the literature.

The rheological tuning materials used to apply hydrogels in biomimetic soft robotics not only must retain the desirable properties specific to hydrogels, such as hydrophilicity, softness, and permeability, but also must function at low dosage to retain the targeted functions of the host hydrogels, such as stimuli-responsiveness. For soft robots to be employed in scenarios involving close interaction with the human body and biomedical applications, it is essential that the rheological modifier exhibits well-proven biocompatibility and minimal toxicity. The rapid design of hydrogel-based biomimetic soft robots with the appropriate level of architectural complexity and robotic performance is currently hindered by a lack of simple and adaptable manufacturing strategies. Utilizing biocompatible alginate as a rheological modifier, Cheng et al. used direct-ink-write (DIW) printing to produce 3D freeform constructions of chemically and physically cross-linked hydrogels (Figure 11.6) [63]. Due to alginate's naturally hydrophilic polymer network, the host hydrogels' intended functions can be maintained, and the hydrogels' mechanical toughness can be improved.

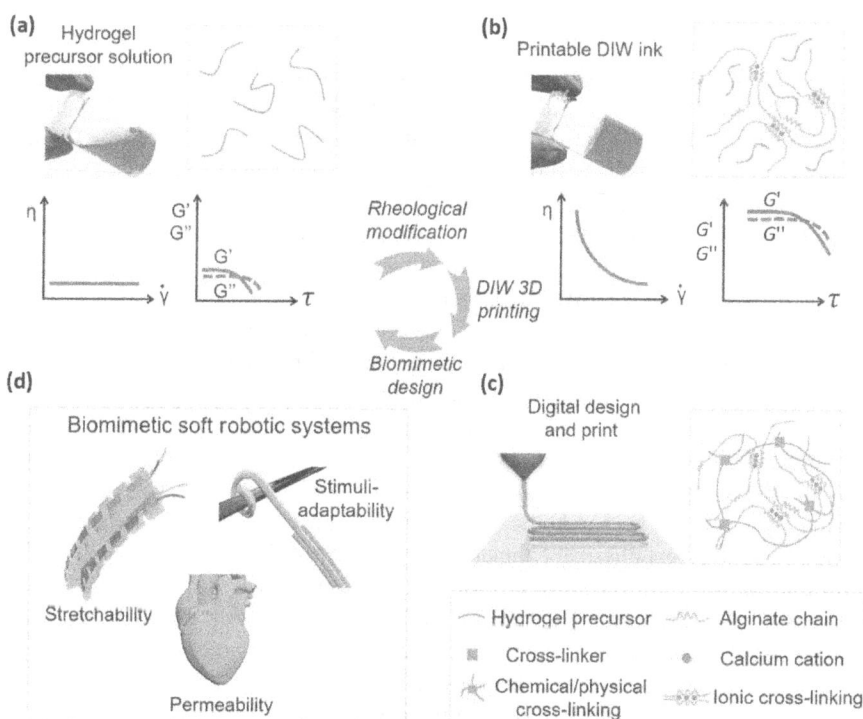

FIGURE 11.6   Schematic of DIW 3D printing of biomimetic soft robots from hydrogels. (a) A photo of a hydrogel precursor solution, in this case AAM, with a diagram of its ingredients and normal water-like rheological behavior. (b) A photo of DIW ink (AAM DIW ink is used as an example) that can be used for printing, along with a diagram of its ingredients and normal gel-like rheological behavior after rheological modification. (c) Digital creation and printing of DIW ink (schematic) and the composition diagram after the hydrogel has hardened. (d) Schematic of three biomimetic soft robotic systems: a bioengineered robotic heart, an artificial tentacle, and an artificial tendril. (Reproduced with permission from Ref. [63]. © 2019 American Chemical Society.)

A bioengineered robotic heart with beating transportation functions, an artificial tendon with phototropic motion, and other bioinspired fluidic and stimulus-activated robotic prototypes can all benefit from the incorporation of free structures and available functionalities from a diverse hydrogel family. The design method increases the geometrical flexibility, mechanical tunability, and actuation complexity of hydrogels for use in biocompatible soft robotics.

11.7 CONCLUSION

The potential for soft robotics to disrupt established industries like healthcare and manufacturing has attracted a lot of interest. Finding materials that can emulate the suppleness and adaptability of natural organisms is one of the primary difficulties in soft robotics. Recently, magnetic polymer composites have shown promise as a means of expanding soft robots' utility. The unique features of polymers and magnetic materials are combined in these composites, opening up new avenues for the development of cutting-edge, highly effective soft robotic systems.

In order to create magnetic polymer composites, magnetic particles like iron oxides or nickel-based compounds are infused into a polymer matrix. A revolutionary approach to precise control, motion, and deformation is introduced by the use of magnetic components in soft robots. To successfully navigate uncertain and complicated surroundings, soft robots require the ability to dynamically modify their shape, stiffness, and mobility in reaction to external magnetic fields. This focus on flexibility is a departure from conventional robotics, which has typically relied on inflexible components and mechanical joints.

Magnetic polymer composites' non-intrusive actuation technique is one of their main selling points. Robots made from these composites may be remotely manipulated utilizing magnetic fields, unlike traditional soft robots, which rely on pneumatic or hydraulic systems. Because of this improvement, simpler, lighter robots can be built without sacrificing functionality. Furthermore, these robots are particularly suited for applications where delicate interactions are required, such as medical treatments within the human body, because of their ability to exert forces and create motion without direct physical contact.

Magnetic polymer composites have a wide variety of possible uses in soft robotics. These composites have the potential to completely change the way minimally invasive procedures are performed by making it possible for highly flexible and biocompatible robotic equipment to be used. By precisely navigating through blood vessels or other complex anatomical systems, magnetic microbots have the potential to lessen the invasiveness of treatments and speed up the recuperation time for patients. Furthermore, the malleability of these composites could lead to the creation of wearable assistive devices that fit inconspicuously on the human body to help people who have movement limitations. Magnetic polymer composites may also have uses outside of healthcare, such as in exploration and environmental monitoring. These composites would allow soft robots to move over difficult terrain, enter small areas, and adjust to new environments. As a result, they would be useful for search-and-rescue missions in disaster zones or investigation of dangerous, off-the-grid locations, where conventional, inflexible robots would be useless.

While magnetic polymer composites show promise for applications in soft robotics, further work is needed before their full potential can be realized. Complex modeling and simulation approaches are required for designing and controlling the behavior of these composites in response to magnetic fields. It is also important to make sure that these materials are safe and biocompatible before putting them into contact with any kind of living thing.

For these reasons, it is encouraging to see so much focus on the ways in which magnetic polymer composites can alter the future of soft robotics. The use of these materials has the potential to extend the capabilities of soft robots beyond those of their rigid counterparts. New applications that make use of the magnetic polymer composites' malleability, accuracy, and non-invasiveness are sure to appear as research and development in this area continue, eventually leading to breakthroughs with far-reaching ramifications across industries and sectors.

REFERENCES

1. Kim, S., C. Laschi, and B. Trimmer, *Soft robotics: A bioinspired evolution in robotics.* Trends in Biotechnology, 2013. **31**(5): p. 287–294.
2. Mengel, T., *Creative connections: Fiction, futures studies, and leadership (for the future).* Creativity Matters, 2020. **1**: p. 72–79.
3. Whitesides, G.M., *Soft robotics.* Angewandte Chemie International Edition, 2018. **57**(16): p. 4258–4273.
4. Majidi, C., *Soft robotics: A perspective—current trends and prospects for the future.* Soft Robotics, 2014. **1**(1): p. 5–11.
5. Liu, J.A.-C., et al., *Photothermally and magnetically controlled reconfiguration of polymer composites for soft robotics.* Science Advances, 2019. **5**(8): p. eaaw2897.
6. El-Atab, N., et al., *Soft actuators for soft robotic applications: A review.* Advanced Intelligent Systems, 2020. **2**(10): p. 2000128.
7. Qiu, H., and Q. Zhang, *Feedforward-plus-proportional-integral-derivative controller for an off-road vehicle electrohydraulic steering system.* Proceedings of the Institution of Mechanical Engineers, Part D: Journal of Automobile Engineering, 2003. **217**(5): p. 375–382.
8. Tang, G., et al., *Proportional-integral-derivative controller optimization by particle swarm optimization and back propagation neural network for a parallel stabilized platform in marine operations.* Journal of Ocean Engineering and Science, 2022.
9. Kim, J., et al., *Programming magnetic anisotropy in polymeric microactuators.* Nature Materials, 2011. **10**(10): p. 747–752.
10. Sitti, M., et al., *Biomedical applications of untethered mobile milli/microrobots.* Proceedings of the IEEE, 2015. **103**(2): p. 205–224.
11. Li, H., et al., *Photopatternable NdFeB polymer micromagnets for microfluidics and microrobotics applications.* Journal of Micromechanics and Microengineering, 2013. **23**(6): p. 065002.
12. Nguyen, V.Q., A.S. Ahmed, and R.V. Ramanujan, *Morphing soft magnetic composites.* Advanced Materials, 2012. **24**(30): p. 4041–4054.

13. Gray, B.L., *A review of magnetic composite polymers applied to microfluidic devices.* Journal of The Electrochemical Society, 2014. **161**(2): p. B3173.
14. Garstecki, P., et al., *Propulsion of flexible polymer structures in a rotating magnetic field.* Journal of Physics: Condensed Matter, 2009. **21**(20): p. 204110.
15. Hu, W., et al., *Small-scale soft-bodied robot with multimodal locomotion.* Nature, 2018. **554**(7690): p. 81–85.
16. Diller, E., et al., *Continuously distributed magnetization profile for millimeter-scale elastomeric undulatory swimming.* Applied Physics Letters, 2014. **104**(17). https://doi.org/10.1063/1.4874306
17. Lum, G.Z., et al., *Shape-programmable magnetic soft matter.* Proceedings of the National Academy of Sciences, 2016. **113**(41): p. E6007–E6015.
18. Kim, Y., et al., *Printing ferromagnetic domains for untethered fast-transforming soft materials.* Nature, 2018. **558**(7709): p. 274–279.
19. Lu, H., et al., *A bioinspired multilegged soft millirobot that functions in both dry and wet conditions.* Nature Communications, 2018. **9**(1): p. 3944.
20. Hilber, W., *Stimulus-active polymer actuators for next-generation microfluidic devices.* Applied Physics A, 2016. **122**(8): p. 751.
21. Erin, O., et al., *Magnetic resonance imaging system–driven medical robotics.* Advanced Intelligent Systems, 2020. **2**(2): p. 1900110.
22. Rus, D., and M.T. Tolley, *Design, fabrication and control of soft robots.* Nature, 2015. **521**(7553): p. 467–475.
23. Sitti, M., *Miniature soft robots—road to the clinic.* Nature Reviews Materials, 2018. **3**(6): p. 74–75.
24. Laschi, C., B. Mazzolai, and M. Cianchetti, *Soft robotics: Technologies and systems pushing the boundaries of robot abilities.* Science Robotics, 2016. **1**(1): p. eaah3690.
25. Polygerinos, P., et al., *Soft robotics: Review of fluid-driven intrinsically soft devices; manufacturing, sensing, control, and applications in human-robot interaction.* Advanced Engineering Materials, 2017. **19**(12): p. 1700016.
26. Bira, N., P. Dhagat, and J.R. Davidson, *A review of magnetic elastomers and their role in soft robotics.* Frontiers in Robotics and AI, 2020. **7**: p. 588391.
27. Nelson, B.J., I.K. Kaliakatsos, and J.J. Abbott, *Microrobots for minimally invasive medicine.* Annual Review of Biomedical Engineering, 2010. **12**: p. 55–85.
28. Hines, L., et al., *Soft actuators for small-scale robotics.* Advanced Materials, 2017. **29**(13): p. 1603483.
29. Zhao, X., and Y. Kim. Soft microbots programmed by nanomagnets. *Nature*, 2019. **575**(7781), 58–59.
30. Hossain, M., et al., *Synthesis, characterization, properties and applications of two-dimensional magnetic materials.* Nano Today, 2022. **42**: p. 101338.
31. Jakubovics, J.P., Magnetism and magnetic materials. 2023, CRC Press.
32. Kalashgrani, M.Y., et al., *Recent advances in multifunctional magnetic nano platform for biomedical applications: A mini review.* Advances in Applied NanoBio-Technologies, 2022. **3**(2): p. 31–37.
33. Sajid, M., *Nanomaterials: Types, properties, recent advances, and toxicity concerns.* Current Opinion in Environmental Science & Health, 2022. **25**: p. 100319.
34. Ganguly, S., and S. Margel, *3D printed magnetic polymer composite hydrogels for hyperthermia and magnetic field driven structural manipulation.* Progress in Polymer Science, 2022. **131**: p. 101574.
35. Ganguly, S., and S. Margel, *Design of magnetic hydrogels for hyperthermia and drug delivery.* Polymers, 2021. **13**(23): p. 4259.
36. Ganguly, S., and S. Margel, *Remotely controlled magneto-regulation of therapeutics from magnetoelastic gel matrices.* Biotechnology Advances, 2020. **44**: p. 107611.

37. Mak, K.F., J. Shan, and D.C. Ralph, *Probing and controlling magnetic states in 2D layered magnetic materials.* Nature Reviews Physics, 2019. **1**(11): p. 646–661.

38. Mohapatra, J., et al., *Hard and semi-hard magnetic materials based on cobalt and cobalt alloys.* Journal of Alloys and Compounds, 2020. **824**: p. 153874.

39. Kumari, S., et al., *Recent developments on 2D magnetic materials: Challenges and opportunities.* Emergent Materials, 2021. **4**(4): p. 827–846.

40. Rezende, S.M., Magnetism, Magnetic Materials, and Devices, in Introduction to Electronic Materials and Devices. 2022, Springer. p. 345–419.

41. Gallego, S.V., et al., *Automatic calculation of symmetry-adapted tensors in magnetic and non-magnetic materials: A new tool of the bilbao crystallographic server.* Acta Crystallographica Section A: Foundations and Advances, 2019. **75**(3): p. 438–447.

42. Das, P., et al., *Tailor made magnetic nanolights: Fabrication to cancer theranostics applications.* Nanoscale Advances, 2021. **3**(24): p. 6762–6796.

43. Dumitru, I., and O.F. Caltun, *Ferrites use in magnetic recording*, in Ferrite Nanostructured Magnetic Materials. 2023, Elsevier. p. 733–745.

44. Zhang, J., et al., *Liquid crystal elastomer-based magnetic composite films for reconfigurable shape-morphing soft miniature machines.* Advanced Materials, 2021. **33**(8): p. 2006191.

45. Jiang, L., and P. Liu, *Design of magnetic attapulgite/fly ash/poly(acrylic acid) ternary nanocomposite hydrogels and performance evaluation as selective adsorbent for Pb2+ ion.* ACS Sustainable Chemistry & Engineering, 2014. **2**(7): p. 1785–1794.

46. Messing, R., et al., *Cobalt ferrite nanoparticles as multifunctional cross-linkers in PAAm ferrohydrogels.* Macromolecules, 2011. **44**(8): p. 2990–2999.

47. Evans, B.A., et al., *A highly tunable silicone-based magnetic elastomer with nanoscale homogeneity.* Journal of Magnetism and Magnetic Materials, 2012. **324**(4): p. 501–507.

48. Fuhrer, R., et al., *Crosslinking metal nanoparticles into the polymer backbone of hydrogels enables preparation of soft, magnetic field-driven actuators with muscle-like flexibility.* Small, 2009. **5**(3): p. 383–388.

49. Majidi, C., and R.J. Wood, *Tunable elastic stiffness with microconfined magnetorheological domains at low magnetic field.* Applied Physics Letters, 2010. **97**(16). https://doi.org/10.1063/1.3503969

50. Ijaz, S., et al., *Magnetically actuated miniature walking soft robot based on chained magnetic microparticles-embedded elastomer.* Sensors and Actuators A: Physical, 2020. **301**: p. 111707.

51. Choi, D.-S., et al., *Beyond human hand: Shape-adaptive and reversible magnetorheological elastomer-based robot gripper skin.* ACS Applied Materials & Interfaces, 2020. **12**(39): p. 44147–44155.

52. Cheng, Y., et al., *A fast autonomous healing magnetic elastomer for instantly recoverable, modularly programmable, and thermorecyclable soft robots.* Advanced Functional Materials, 2021. **31**(32): p. 2101825.

53. Zhang, J., et al., *Liquid metal-elastomer composites with dual-energy transmission mode for multifunctional miniature untethered magnetic robots.* Advanced Science, 2022. **9**(31): p. 2203730.

54. Wei, Y., et al., *Design of a magnetic soft inchworm millirobot based on pre-strained elastomer with micropillars.* Biomimetics, 2023. **8**(1): p. 22.

55. Zhang, P., et al., *EPM–MRE: Electropermanent Magnet–Magnetorheological elastomer for soft actuation system and its application to robotic grasping.* IEEE Robotics and Automation Letters, 2021. **6**(4): p. 8181–8188.

56. Zhao, R., et al., *Reprogrammable liquid-metal/NdFeB/silicone composite magnetic elastomer.* AIP Advances, 2023. **13**(2). https://doi.org/10.1063/9.0000470

57. Zhang, S., et al., *Programmable and reprocessable multifunctional elastomeric sheets for soft origami robots.* Science Robotics, 2021. **6**(53): p. eabd6107.
58. Ze, Q., et al., *Magnetic shape memory polymers with integrated multifunctional shape manipulation.* Advanced Materials, 2020. **32**(4): p. 1906657.
59. Kim, Y., and X. Zhao, *Magnetic soft materials and robots.* Chemical Reviews, 2022. **122**(5): p. 5317–5364.
60. Parida, K., et al., *Extremely stretchable and self-healing conductor based on thermoplastic elastomer for all-three-dimensional printed triboelectric nanogenerator.* Nature Communications, 2019. **10**(1): p. 2158.
61. Ouyang, L., et al., *A generalizable strategy for the 3D bioprinting of hydrogels from nonviscous photo-crosslinkable inks.* Advanced Materials, 2017. **29**(8): p. 1604983.
62. Hong, S., et al., *3D printing of highly stretchable and tough hydrogels into complex, cellularized structures.* Advanced Materials, 2015. **27**(27): p. 4035–4040.
63. Cheng, Y., et al., *Direct-ink-write 3D printing of hydrogels into biomimetic soft robots.* ACS Nano, 2019. **13**(11): p. 13176–13184.

12 3D and 4D Printing of Magnetic Polymer Composites for Diverse Applications

Shanmugasundaram Palani, Karthickeyan Krishnan, Ronald Darwin C., P. Balaji, Sridevi Ganesan, Mohamed Zerein Fathima, I. Somasundaram, and S. Jeganath

12.1 INTRODUCTION

The incorporation of magnetic characteristics into polymer composites has attracted considerable interest owing to the advantageous synergistic effect [1, 2]. These composite materials provide a unique combination of the adaptable characteristics of polymers and the functional attributes of magnetic materials, leading to the emergence of innovative materials exhibiting a wide range of features [3, 4]. The integration of these composite materials with 3D and 4D printing technologies gives an additional level of manipulation over the microstructure and functionality of the ultimate product. 3D printing (3DP), alternatively known as additive manufacturing (AM), rapid prototyping (RP), or solid-freeform (SFF), encompasses the procedure of amalgamating materials to fabricate objects based on 3D model data, typically employing a layer-by-layer approach [5]. This technique was initially elucidated by Charles Hull in 1986 [6].

The emerging discipline of 4D printing (4DP) technology is dedicated to the additive fabrication of shape memory structures and intelligent materials, which are capable of dynamic responses. In the context of 4DP, the structural configuration is engineered to undergo autonomous assembly in the absence of conventional driving mechanisms [3, 4]. Creative concepts can be encoded into various structures through a modeling process, thereby providing 4D-printed objects with increased design flexibility [5, 6]. Products can be printed in simple shapes, and 3DP and 4DP permits them to attain complex shapes without wasting material on support structures. This technology has facilitated the exploration of futuristic research endeavors, enabling the development of intricate designs and goods. Additionally, it has contributed to the reduction of production and delivery time, the fabrication of lighter and safer items, and the promotion of environmentally conscious practices, such as minimizing waste and embracing green manufacturing.

Additionally, the utilization of this technology can facilitate the development of new businesses and prospects [7, 8].

In the contemporary market, the usage of 3D printers is prevalent across various sectors including medical manufacturing, construction, arts and jewelery, chemical industry, food sector, aeronautics and space businesses, automotive sector, textile and fashion industry, fabric industry, electronic industry, furniture industry, and numerous others [8, 9]. The use of 3D printing technology is progressively expanding across several industries every day. Currently, there is a growing interest in the production of composite materials. This is primarily due to the potential for achieving functional structures through processes such as polymerization [10, 11] and advanced 3D-printing [12, 13]. These composite materials hold promise for various applications in fields such as microelectromechanical systems (MEMS), medicine, electronics, and the automotive, aeronautics, and energy industries [11, 14–16]. The composite material is composed of a polymeric matrix, such as polyamide, ABS, polyethylene, or polystyrene, that incorporates particles, including magnetic and metallic particles [7, 17]. The precise manipulation and non-cytotoxic properties of magnetic polymer composites have significantly contributed to the advancement of biomimetic soft robots to the next generation. The regulation of bare magnetic nanoparticles by remote control presents challenges. However, when these nanoparticles are confined within polymeric matrices, the resulting system exhibits characteristics similar to those of an integrated system resembling an artificial soft mussel. Simultaneously, these polymeric magnetic soft materials also exhibit susceptibility to external magnetic fields, whether static or oscillatory in nature. Additive manufacturing of magnetic soft materials via spatial assembly of polymeric precursors followed by actuation-like behavior is a relatively recent manufacturing process. In this review, we focus on magnetic nanoparticles entrapped in a polymeric matrix and their diverse shape-morphing applications [3].

12.2   3D AND 4D PRINTING TECHNIQUES FOR MAGNETIC POLYMER COMPOSITES

This section covers a range of 3D printing processes that are utilized in the fabrication of magnetic polymer composites. Some examples of additive manufacturing techniques are fused deposition modeling (FDM), stereolithography (SLA), and selective laser sintering (SLS). In the realm of 3D printing, the initial supplies used, commonly referred to as feedstock, can be categorized into various forms. These include filament or wire, which is utilized in FDM printing; powdered form, employed in SLS; and fluid, utilized in SLA and poly jet matrix printing [18]. Additionally, there are methods that involve the combination of a binder and a powder or other components, known as binder jetting [19].

Various technologies can be classified based on the specific method employed for processing the feedstock. The utilization of FDM is prevalent in the cost-effective printers commonly found in domestic or educational settings, such as households, schools, or universities. A polymer filament with a certain diameter is drawn into a heated end, where it becomes molten and is subsequently extruded via a nozzle. The process involves depositing the molten polymer in a sequential manner along predetermined pathways, resulting in the gradual formation of an object [20, 21]. Wire arc additive manufacturing (WAAM) is a comparable technique employed for metals [22]. SLS operates through the fusion of consecutive layers within a powder bed. This additive manufacturing technique exhibits versatility, as it can be employed for processing various materials, including polymers, metals, and ceramics [23, 24]. Additional powder-bed-based techniques include selective laser melting, direct metal laser sintering, and electron beam melting [25]. Within the realm of photopolymerization processes, SLA holds the distinction of being the most ancient and widely utilized method. This process involves the submersion of the printing bed into a reservoir containing resin, where the uppermost layer undergoes polymerization with the use of laser light. Subsequently, the printing bed is raised by a distance equivalent to one layer, and the subsequent layer is subjected to polymerization [26]. Digital light processing (DLP) employs a digital micro-mirror device within the optical pathway, facilitating simultaneous exposure of an entire layer [27]. In the context of PJM technology, the process involves the creation of items on a printing bed. This is achieved by applying a liquid polymer resin onto specific places and subsequently curing it using UV light [18]. Two-photon or multi-photon polymerization techniques allow for the development of even more precise structures. These techniques include using a highly concentrated laser to simultaneously absorb two or more photons, which restricts the area where polymerization may occur [28, 29].

Due to cutting-edge discoveries that will be useful in solving many industrial difficulties, 4DP technology has sparked the interest of engineers, designers, and scientists. Intelligent materials, 3D printing, and a well-programmed design are used. Under changing circumstances, this method develops a range of metamaterial structures [30]. Novel and cutting-edge 3D printers are mostly used in cutting-edge additive manufacturing (AM) technologies [31]. 4DP's dynamic and multi-functional products are created using technology methodologies. The addition of time-dependent components to 3DP results in 4D-printed objects. The development of dynamic and multi-functional goods in 4DP is achieved by the utilization of several technical methodologies. The incorporation of temporal elements into the process of three-dimensional printing results in the production of four-dimensional printed objects. Currently, a range of AM technologies have been utilized, which have established the groundwork for sophisticated four-dimensional printing approaches [32–34]. Four-dimensional printing is typically accomplished through the utilization of three-dimensional printing techniques. The prevailing 3DP techniques commonly employed include extrusion-based processes such as fused deposition modeling and direct ink writing (DIW), light-based vat photopolymerization (VP) processes, selective laser sintering, material jetting (MJ)/inkjet printing (IJP), and selective laser melting (SLM) [35–37]. Nevertheless, the utilization of the aforementioned 3D printing technologies [38–40] does not allow for the production of irreversible deformation in the 4D-printed structures. Furthermore, it is worth noting that human pre-programming must be implemented as a preliminary step before the model undergoes deformation. This particular approach renders the technology imprecise, demanding, and challenging to regulate [41]. One potential solution to this challenge involves the utilization of 4D-printed reversible structures. These structures are capable of undergoing rapid transformations in material properties and shape in response to external stimuli [42]. The predominant methods utilized for 4D printing are light-based and extrusion-based procedures [43]. The ability to change the shape, porosity, or magnetic responsiveness of magnetic polymer composites over time has intriguing possibilities. Soft robotics, adaptive structures, and biomedical devices are emphasized as examples of 4D printing applications.

12.3   3DP AND 4DP OF MAGNETIC POLYMER COMPOSITES FOR DIVERSE APPLICATIONS

3DP and 4DP are an advanced and expeditious prototyping method that possesses limitless potential for use in several technical domains, such as biomedicine, electronics, robotics, and the food, automotive, construction, and aerospace industries. Nevertheless, the rapid progress in technology within many engineering fields necessitates the undertaking of multi-disciplinary research [44].

12.3.1   BIOMEDICAL APPLICATIONS

3D images of tissues and organs have grown more informative and higher resolution with the advancement of computed tomography (CT) and magnetic resonance imaging (MRI) technology [45]. Using the obtained picture data, 3DP technology might be employed to create patient-specific tissues and organs with intricate 3D microarchitecture. Polymer materials commonly employed in the realm of biomedical applications for printing purposes encompass naturally derived polymers such as gelatin, alginate, and collagen, as well as synthetic polymer molecules including polyethylene glycol (PEG), poly lactic-co-glycolic acid (PLGA), and polyvinyl alcohol (PVA). In the field of biomedical applications, there exists an urgent requirement for materials that possess a specific set of desirable traits. These traits include printability and biocompatibility, as well as favorable mechanical and structural properties [46].

The importance of biocompatibility and favorable mechanical properties cannot be overstated when it comes to 3D-printed scaffolds. These factors play a crucial role in ensuring the success and effectiveness of such scaffolds in various applications. Biocompatibility refers to the ability of a material to interact with living tissues without causing any adverse reactions or harm. In the context of 3D-printed scaffolds, it is imperative that the material used be biocompatible to promote cell adhesion and proliferation. The successful fabrication of composite scaffolds with enhanced biocompatibility was accomplished by incorporating bioactive particles into a polymer matrix. The incorporation of biodegradable and biocompatible polymers within the printed scaffolds serves to preserve their toughness, while the addition of brittle bioceramic particles enhances their biocompatibility. Enhancing mechanical stability is crucial for scaffolds to effectively facilitate and sustain cellular activity. In order to enhance the structural integrity of scaffolds, a majority of composite scaffolds have been fabricated through the incorporation of tricalcium phosphate (TCP), hydroxyapatite (HA), magnetic nanoparticles, graphene, silica, and carbon nanotube into a polymer matrix [47–52].

An investigation by Zhang et al. involved the fabrication of composite scaffolds consisting of three-dimensional magnetic Fe_3O_4 nanoparticles incorporated within mesoporous bioactive glass/polycaprolactone (Fe_3O_4/MBG/PCL) materials [52]. The fabrication process employed in this study utilized 3D-printing technology. The findings indicated that the Fe_3O_4/MBG/PCL scaffolds exhibited consistent macropores of 400 μm, a substantial porosity of 60%, and impressive compressive strength ranging from 13 to 16 MPa. The introduction of magnetic Fe_3O_4 nanoparticles into MBG/PCL scaffolds did not have any impact on their ability to mineralize apatite. However, it did enhance their magnetic heating capability and substantially promote the proliferation, alkaline phosphatase (ALP) activity, osteogenesis-related gene expression, and extracellular matrix (ECM) mineralization of human bone marrow–derived mesenchymal stem cells. Furthermore, the study employed doxorubicin (DOX) as a representative anti-cancer drug to investigate the efficacy of Fe_3O_4/MBG/PCL scaffolds in facilitating controlled and prolonged drug release, thereby making them suitable for localized drug delivery therapy. Hence, the 3D-printed scaffolds composed of Fe_3O_4/MBG/PCL demonstrated promising multi-functionality, encompassing augmented osteogenic activity, localized administration of anti-cancer drugs, and magnetic hyperthermia. In addition to better mechanical qualities, the introduction of Fe_3O_4 or graphene into printed scaffolds provides unique thermal and electrical properties. Magnetic Fe_3O_4 nanoparticles on printed scaffolds can generate local heat in the presence of an alternating magnetic field, making hyperthermia therapy useful in cancer treatment [52]. Magnetic-responsive polymers are synthesised through the incorporation of magnetic nanoparticles (MNPs) such as Fe_2O_3 and NdFeB [53–56]. The utilization of magnetically activated shape memory polymers (SMPs) has occurred in various biomedical contexts, wherein they serve as a foundation for the creation of drug delivery mechanisms and the construction of biorobotic structures tailored for minimally invasive medical procedures [56]. Lin et al. successfully fabricated remotely controllable and biodegradable occluders using a 4DP technique [57]. The occluders were composed of polylactic acid (PLA) and incorporated Fe_3O_4 magnetic particles. The magnetically responsive shape memory polymer composites (SMPCs) demonstrated exceptional shape memory properties. The remarkable cytocompatibility and histocompatibility exhibited by the occluders promote the adhesion and ingrowth of granulation tissues, thereby facilitating the rapid process of endothelialization. Furthermore, the utilization of personalized shape memory occluders serves to guarantee an optimal fit and deliver ample support for defects. Hence, the utilization of 4D-printed shape memory occluders as a plausible alternative to conventional metal occlusion devices emerges as a promising avenue. Similarly, Zhang and colleagues conducted a study wherein they utilized polydimethylsiloxane (PDMS)–based magneto-responsive shape memory polymer composites made by integrating NdFeB-based magnetic nanoparticles, as illustrated in Figure 12.1a–d [58]. The origami structures demonstrated programmable transformation, notable shape memory properties, and manageable locomotion. In this study, a novel universal finite element simulation method was developed to provide design guidance for the magnetization and magnetic actuation deformation process of magnetoactive soft material (MASM) objects. The integration of 3D injection printing with the origami-based magnetization technique in this 4D printing approach holds promise for the development of actuators and robots based on MASM materials that exhibit both rigid and flexible properties. The technology used by Wan et al. combined multi-material design and triple shape memory response to enable the creation of 4D-printed structures [59]. The successful achievement of 4D printing was demonstrated in the fabrication of a multi-material structure composed of poly (D,L-lactide-co-trimethylene carbonate) (PLMC), poly (trimethylene carbonate) (PTMC), and Fe_3O_4. This innovative structure exhibits the ability to undergo several shape-changing transformations when subjected to successive stimuli of a remotely applied magnetic field and heat. These structures exhibit localized and sequential shape changes when subjected to certain stimuli such as a magnetic field and heat. The triple-SMP created using 4D printing techniques and its nanocomposite counterparts demonstrate remarkable biocompatibility. The CCK-8 assay, as demonstrated in Figure 12.1e, was used to investigate the influence of printed PLMC/PTMC on fibroblast cell proliferation within 3 days. The 3D-printed material had no effect on the survival of NIH/3T3 fibroblast cells, both dose and time dependently. The results of the LIVE/DEAD cytotoxicity experiment reported in Figure 12.1f reveal that all printed films promote cell growth. After 48 hours, almost all of the discovered cells were alive, as indicated by the green color, with only a few dead cells detected in all scaffolds. This technology signifies a significant advancement in advanced additive manufacturing for the production of biocompatible 4D-printed triple-SMP and multi-material structures. It shows considerable promise for many biomedical applications, including drug delivery and tissue engineering.

Wei and co-workers successfully demonstrated the fabrication of 4D active shape-changing architectures with custom-defined geometries [60]. These structures exhibit thermally and remotely actuated behaviors. The fabrication process involves the direct-write printing of ultraviolet (UV) cross-linking poly(lactic acid)-based inks. The findings demonstrate that the use of a UV cross-linking agent yields printed objects that exhibit remarkable shape memory characteristics. This allows for the transformation of configurations in 3D–1D–3D, 3D–2D–3D, and 3D–3D–3D formats. Even more important, the incorporation of iron oxide allows for the successful integration of 4D shape-changing devices with quick remotely operated and magnetically guidable properties. A spiral construction with a 1 mm inner diameter was placed in a plastic tube holder with a 3 mm inner diameter to demonstrate the design strategy. Remotely actuated self-expandable behavior was seen after they were placed together in a coil with an alternating magnetic field. Within 10 s, the inner diameter of the spiral structure increased from 1 to 2.7 mm (Figure 12.1g). A 4D scaffold of this type has high potential for usage as a self-expandable intravascular stent (Figure 12.1h). This study printed both SMPs and shape memory nanocomposites (SMNCs), presenting a viable technique for designing 4D active shape-changing architectures with multi-functional features. This paves the way for the advancement of 4D printing, soft robotics, flexible electronics, minimally invasive medical procedures, and other technologies.

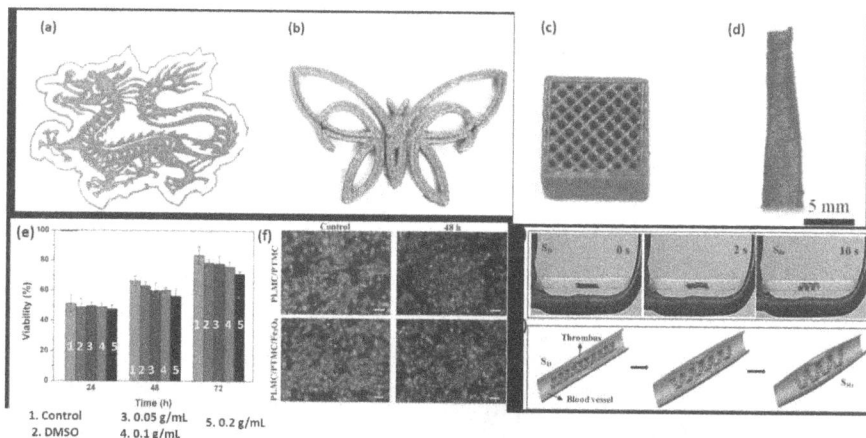

FIGURE 12.1 4D-printed magnetically driven shape memory occluders: (a) Chinese dragon, (b) butterfly, (c) porous cuboid, and (d) Shanghai Tower, China. (From [58].) (e) Viability of NIH/3T3 fibroblast cells cultured for 1–3 days in extract medium from 3D-printed PLMC/PTMC. (f) Fluorescence image of NIH/3T3 cells with 3D-printed PLMC/PTMC and PLMC/PTMC/Fe$_3$O$_4$ in 48 h. (From [59].) (g) A 30 kHz alternating magnetic field is used to activate the restricted shape recovery process. (h) The 4D scaffold's potential use as an intravascular stent. The temperature during deformation was 80°C. Under constrained conditions, the original, deformed, and recovered shapes are denoted as S$_O$, S$_D$, and S$_{Rr}$, respectively. (From [60].)

For 4D printing robots, a unique smart hydrogel made of NIPAM, Laponite nanoclay, and NdFeB magnetic particles with simultaneous temperature sensation and magnetic actuation was devised [53]. A variety of soft millirobots were fabricated, each possessing a distinct structure and functionalities. These included a catheter featuring a magnetic head with multiple segments, a robot resembling a leptasteria, and a shellfish-like robot. Notably, the latter two robots exhibited responsiveness to both magnetic and thermal fields. The locomotive capabilities of the millirobot have been experimentally confirmed by having it successfully navigate physical obstacles within a simulated human stomach environment, thereby accomplishing the task of actively transporting cargo. Furthermore, as a drug carrier, the hydrogel could not only accomplish controlled drug release but also decrease unexpected leakage to some level (Figure 12.2a,b). The combined effects of the magnetic field and thermal field elicit synergistic responses, thereby enhancing the adaptability of the robot and mitigating drug leakage during transportation. It is noteworthy that smart hydrogels exhibit remarkable biocompatibility (Figure 12.2c–e) and offer a favorable environment for cellular proliferation.

New 4D-printed composite structures that could be activated by microwave radiation were created utilizing ferromagnetic polylactic acid (PLA) and PLA [61]. The approach suggested in the study might be utilized with a wide range of SMPs and dielectric materials, offering up new opportunities for a wide range of innovative 4D printing composites. This research could have applications in robotics, deployable structures, product packaging, reducing additive manufacturing time, medicine, and

(a)

(b)

(c)

(d)

(e)

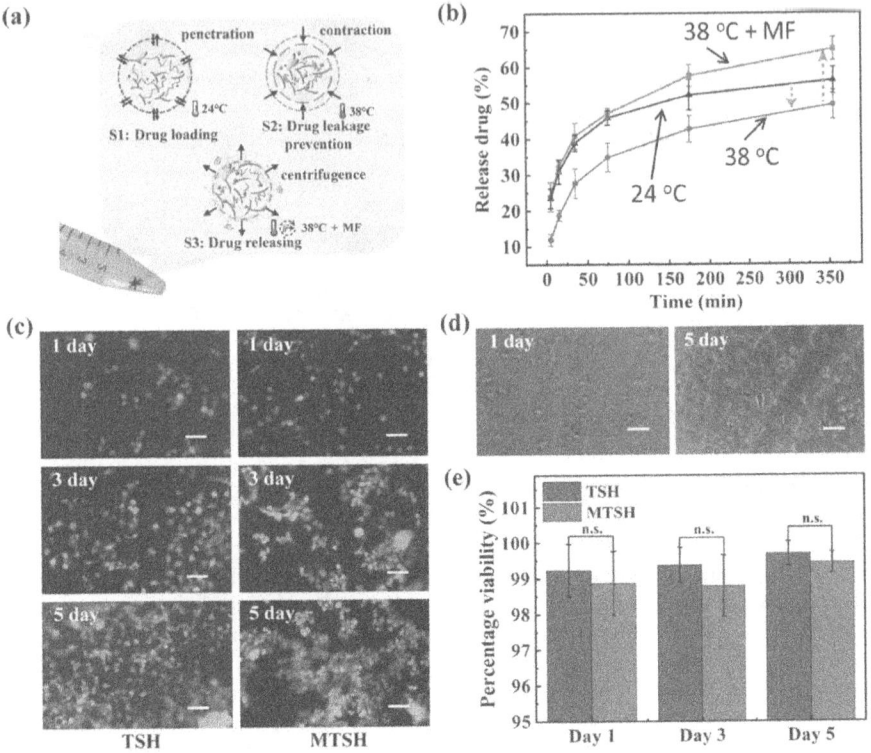

FIGURE 12.2 Hydrogel drug delivery and cell behavior. (a) Schematic showing three different drug release statuses, S1–S3. MF: magnetic field. (b) Drug release profiles under different settings. (c) MCF-7 fluorescence after 1–5 days of TSH and MTSH culture. (d) Optical microscope pictures of MCF-7 incubated on TSH after 1 and 5 days. (e) MCF-7 cell proliferation after 1, 3, and 5 days [53].

other fields. A sandwich structural soft actuator (MPDMS/MXene/PTFE) with electrothermal/magnetic coupling actuation was developed by integrating a magnetic NdFeB/polydimethylsiloxane (MPDMS) composite layer, MXene film, and PTFE tape. The soft actuator demonstrated typical magnetic biomimetic actuation, and the dragonfly actuator's programmable actuation capabilities were shown by both laboratory experiments and finite element simulations [62]. Furthermore, electrothermal/magnetic coupling actuation was employed successfully in an intelligent crawling robot to overcome the classic inchworm movement simulation.

12.3.2 OTHER APPLICATIONS

Ferrara et al. utilized magnetite in their study to fabricate composites suitable for space-compliant electrical motors [63]. In this study, a PEEK matrix was reinforced with magnetite microparticles at a weight percentage of 50 in order to fabricate filaments suitable for FDM printing. The magnetic properties associated with the aforementioned phenomenon were employed in a computational model to simulate the performance of an axial flux brushless DC electric motor. This simulation demonstrated that the motor generated a satisfactory level of torque, rendering it suitable for employment in aerospace applications with low power requirements. Furthermore, it was observed that the motor exhibited a reduction in mass of up to 50% compared to a conventional magnet.

Several applications focus on actuators [64, 65], magnetic energy harvesters [66], and other rigid parts in addition to motor parts and similar rigid objects. An additional expansive domain of utilization involves the exploitation of the magnetic characteristics exhibited by nanoparticles that are integrated within a polymeric matrix, in conjunction with the dielectric properties of the matrix. This kind of application finds relevance in various areas, such as electromagnetic shielding [67, 68] and the development of a tunable ring resonator designed for the GHz frequency range [69]. Soft magneto-active materials are also being explored for potential applications such as grippers or in soft robotics generally. Qi et al. devised a PLA/carbonyl iron particle filament with a 1:7 volume ratio of magnetic particles supported by a silane coupling agent [70]. The FDM technique was employed to fabricate the shapes, which were subsequently coated with a layer of silicone rubber, as depicted in Figure 12.3. This post-processing step facilitated the creation of specimens featuring distinct magnetic regions within a nonmagnetic matrix. The authors were able to

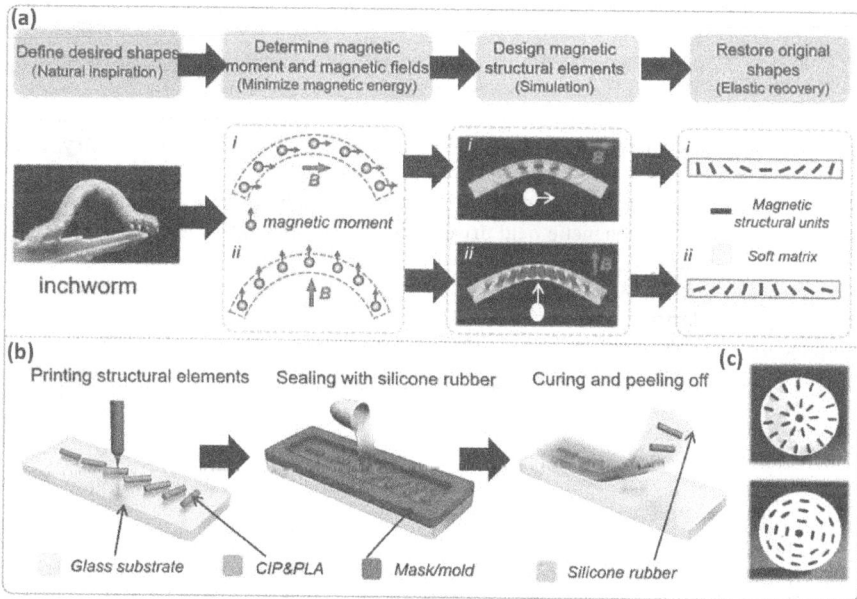

FIGURE 12.3 MASMs' shape-programming method and fabrication process. (a) A flow chart of a shape-programming method and a programming case for inchworm mimicry. COMSOL is used to simulate the deformations, and the direction of the UMF is indicated by the red arrows. (b) Pictorial depiction of 3D printing and encapsulation. The numerous magnetic structural parts, which are enclosed by silicone rubber, are manufactured via 3D printing. (c) Photographs of MASM samples including oriented magnetic structural elements. The disc samples have a radius of 10 mm. (From [70].)

manipulate the inchworm-like structures by modulating the application or cessation of an external magnetic field. Additional structures exhibiting diverse orientations of inserted magnetic 3D-printed units were subjected to manipulation through the manipulation of external magnetic fields, as visually demonstrated in various video recordings documented in reference [70].

12.4   CONCLUSION

In summary, the integration of 3D and 4D printing methodologies with magnetic polymer composites has facilitated the exploration of a vast array of potential applications. The aforementioned materials, characterized by their customized magnetic properties, possess the capacity to significantly transform multiple industries, thereby facilitating novel approaches to address enduring obstacles. As ongoing research endeavors persist and technological advancements progress, the use of these materials is projected to expand, thereby influencing a magnetic trajectory for the field of additive manufacturing. The realization of the full potential of this unique technique can be achieved with the implementation of additional developments in SMPC-based multi-material printing and the precise alignment of additives using magnetic and electric fields.

REFERENCES

1. Kalia, S., et al., Magnetic polymer nanocomposites for environmental and biomedical applications. Colloid and Polymer Science, 2014. **292**: p. 2025–2052.
2. Jazzar, A., et al., Recent advances in the synthesis and applications of magnetic polymer nanocomposites. Journal of Industrial and Engineering Chemistry, 2021. **99**: p. 1–18.
3. Ganguly, S., and S. Margel, 3D printed magnetic polymer composite hydrogels for hyperthermia and magnetic field driven structural manipulation. Progress in Polymer Science, 2022. **131**: p. 101574.
4. Ganguly, S., et al., Mussel-inspired polynorepinephrine/MXene-based magnetic nanohybrid for electromagnetic interference shielding in X-band and strain-sensing performance. Langmuir, 2022. **38**(12): p. 3936–3950.
5. Standard, A., Standard terminology for additive manufacturing technologies. 2012. ASTM International. p. 1–9.
6. Hull, C. Apparatus for production of three dimensional objects by stereolithography, US Patent **4575330**. 1986.
7. Wang, X., et al., 3D printing of polymer matrix composites: A review and prospective. Composites Part B: Engineering, 2017. **110**: p. 442–458.
8. Mouzakis, D.E., Advanced technologies in manufacturing 3D-layered structures for defense and aerospace. in Lamination-Theory and Application. 2018. IntechOpen.
9. Ehrmann, G., T. Blachowicz, and A. Ehrmann, Magnetic 3D-printed composites—Production and applications. Polymers, 2022. **14**(18): p. 3895.
10. Jackson, N., et al., Integration of thick-film permanent magnets for MEMS applications. Journal of Microelectromechanical Systems, 2016. **25**(4): p. 716–724.
11. Fu, K., et al., Progress in 3D printing of carbon materials for energy-related applications. Advanced Materials, 2017. **29**(9): p. 1603486.
12. Kokkinis, D., M. Schaffner, and A.R. Studart, Multimaterial magnetically assisted 3D printing of composite materials. Nature Communications, 2015. **6**(1): p. 8643.

13. Grant, P.S., et al., Manufacture of electrical and magnetic graded and anisotropic materials for novel manipulations of microwaves. Philosophical Transactions of the Royal Society A: Mathematical, Physical and Engineering Sciences, 2015. **373**(2049): p. 20140353.

14. Murr, L.E., Frontiers of 3D printing/additive manufacturing: From human organs to aircraft fabrication. Journal of Materials Science & Technology, 2016. **32**(10): p. 987–995.

15. Castles, F., et al., Microwave dielectric characterisation of 3D-printed BaTiO3/ABS polymer composites. Scientific Reports, 2016. **6**(1): p. 22714.

16. Zuniga, J., et al., Cyborg beast: A low-cost 3d-printed prosthetic hand for children with upper-limb differences. BMC Research Notes, 2015. **8**(1): p. 1–9.

17. Matsuzaki, R., et al., Three-dimensional printing of continuous-fiber composites by in-nozzle impregnation. Scientific Reports, 2016. **6**(1): p. 23058.

18. Koziol, T., and C. Kundera, Viscoelastic properties of cell structures manufactured using a photo-curable additive technology—PJM. Polymers, 2021. **13**(11): p. 1895.

19. Mostafaei, A., et al., Binder jet 3D printing—process parameters, materials, properties, modeling, and challenges. Progress in Materials Science, 2021. **119**: p. 100707.

20. Durgun, I., and R. Ertan, Experimental investigation of FDM process for improvement of mechanical properties and production cost. Rapid Prototyping Journal, 2014. **20**(3): p. 228–235.

21. Noorani, R., Rapid prototyping: principles and applications. 2006.

22. Bartsch, H., et al. Fatigue analysis of wire arc additive manufactured (3D printed) components with unmilled surface. in Structures. 2021. Elsevier.

23. Zhou, W.Y., et al., Selective laser sintering of porous tissue engineering scaffolds from poly (l-lactide)/carbonated hydroxyapatite nanocomposite microspheres. Journal of Materials Science: Materials in Medicine, 2008. **19**: p. 2535–2540.

24. Tan, K., et al., Selective laser sintering of biocompatible polymers for applications in tissue engineering. Bio-Medical Materials and Engineering, 2005. **15**(1-2): p. 113–124.

25. Awad, A., et al., Advances in powder bed fusion 3D printing in drug delivery and healthcare. Advanced Drug Delivery Reviews, 2021. **174**: p. 406–424.

26. Wang, Z., et al., A simple and high-resolution stereolithography-based 3D bioprinting system using visible light crosslinkable bioinks. Biofabrication, 2015. **7**(4): p. 045009.

27. Kadry, H., et al., Digital light processing (DLP) 3D-printing technology and photo-reactive polymers in fabrication of modified-release tablets. European Journal of Pharmaceutical Sciences, 2019. **135**: p. 60–67.

28. Straub, M., and M. Gu, Near-infrared photonic crystals with higher-order bandgaps generated by two-photon photopolymerization. Optics Letters, 2002. **27**(20): p. 1824–1826.

29. Kumi, G., et al., High-speed multiphoton absorption polymerization: Fabrication of microfluidic channels with arbitrary cross-sections and high aspect ratios. Lab on a Chip, 2010. **10**(8): p. 1057–1060.

30. Bodaghi, M., and W. Liao, 4D printed tunable mechanical metamaterials with shape memory operations. Smart Materials and Structures, 2019. **28**(4): p. 045019.

31. Lin, Z., K. Song, and X. Yu, A review on wire and arc additive manufacturing of titanium alloy. Journal of Manufacturing Processes, 2021. **70**: p. 24–45.

32. Arif, Z.U., M.Y. Khalid, and E. ur Rehman, Laser-aided additive manufacturing of high entropy alloys: Processes, properties, and emerging applications. Journal of Manufacturing Processes, 2022. **78**: p. 131–171.

33. Van Wijk, A., and I. van Wijk, 3D printing with biomaterials: Towards a sustainable and circular economy. 2015. IOS press.

34. Ji, A., et al., 3D printing of biomass-derived composites: Application and characterization approaches. RSC Advances, 2020. **10**(37): p. 21698–21723.

35. Cooke, S., et al., Metal additive manufacturing: Technology, metallurgy and modelling. Journal of Manufacturing Processes, 2020. **57**: p. 978–1003.

36. Oladapo, B.I., A.V. Adebiyi, and E.I. Elemure, Microstructural 4D printing investigation of ultra-sonication biocomposite polymer. Journal of King Saud University-Engineering Sciences, 2021. **33**(1): p. 54–60.

37. Altıparmak, S.C., and B. Xiao, A market assessment of additive manufacturing potential for the aerospace industry. Journal of Manufacturing Processes, 2021. **68**: p. 728–738.

38. Bodaghi, M., A. Damanpack, and W. Liao, Adaptive metamaterials by functionally graded 4D printing. Materials & Design, 2017. **135**: p. 26–36.

39. Van Hoa, S., Development of composite springs using 4D printing method. Composite Structures, 2019. **210**: p. 869–876.

40. Moradi, M., et al., The synergic effects of FDM 3D printing parameters on mechanical behaviors of bronze poly lactic acid composites. Journal of Composites Science, 2020. **4**(1): p. 17.

41. Sonatkar, J., B. Kandasubramanian, and S.O. Ismail, 4D printing: Pragmatic progression in biofabrication. European Polymer Journal, 2022. **169**: p. 111128.

42. Ren, L., et al., Programmable 4D printing of bioinspired solvent-driven morphing composites. Advanced Materials Technologies, 2021. **6**(8): p. 2001289.

43. Zolfagharian, A., et al., Control-based 4D printing: Adaptive 4D-printed systems. Applied Sciences, 2020. **10**(9): p. 3020.

44. Wang, W., Y. Liu, and J. Leng, Recent developments in shape memory polymer nanocomposites: Actuation methods and mechanisms. Coordination Chemistry Reviews, 2016. **320**: p. 38–52.

45. Meaney, J.F., and M. Goyen, Recent advances in contrast-enhanced magnetic resonance angiography. European Radiology, 2007. **17: p.** Suppl 2:B2–B6.

46. Murphy, S.V., and A. Atala, 3D bioprinting of tissues and organs. Nature Biotechnology, 2014. **32**(8): p. 773–785.

47. Jakus, A.E., et al., Three-dimensional printing of high-content graphene scaffolds for electronic and biomedical applications. ACS Nano, 2015. **9**(4): p. 4636–4648.

48. Yildirim, E.D., et al., Fabrication, characterization, and biocompatibility of single-walled carbon nanotube-reinforced alginate composite scaffolds manufactured using freeform fabrication technique. Journal of Biomedical Materials Research Part B: Applied Biomaterials, 2008. **87**(2): p. 406–414.

49. Xia, Y., et al., Selective laser sintering fabrication of nano-hydroxyapatite/poly-ε-caprolactone scaffolds for bone tissue engineering applications. International Journal of Nanomedicine, 2013. **8**: p. 4197–4213.

50. Dávila, J., et al., Fabrication of PCL/β-TCP scaffolds by 3D mini-screw extrusion printing. Journal of Applied Polymer Science, 2016. **133**(15): 43031–43040.

51. Lee, H., et al., Mineralized biomimetic collagen/alginate/silica composite scaffolds fabricated by a low-temperature bio-plotting process for hard tissue regeneration: Fabrication, characterisation and in vitro cellular activities. Journal of Materials Chemistry B, 2014. **2**(35): p. 5785–5798.

52. Zhang, J., et al., 3D-printed magnetic Fe_3O_4/MBG/PCL composite scaffolds with multifunctionality of bone regeneration, local anticancer drug delivery and hyperthermia. Journal of Materials Chemistry B, 2014. **2**(43): p. 7583–7595.

53. Hu, X., et al., Multifunctional thermo-magnetically actuated hybrid soft millirobot based on 4D printing. Composites Part B: Engineering, 2022. **228**: p. 109451.

54. Zhu, H., et al., Mechanically-guided 4D printing of magnetoresponsive soft materials across different length scale. Advanced Intelligent Systems, 2022. **4**(3): p. 2100137.

55. Huang, S., et al., 4D printing of soybean oil based shape memory polymer and its magnetic-sensitive composite via digital light processing. Polymer-Plastics Technology and Materials, 2022. **61**(9): p. 923–936.

56. Khalid, M.Y., et al., 4D printing: Technological developments in robotics applications. Sensors and Actuators A: Physical, 2022. **343**: p. 113670.
57. Lin, C., et al., 4D-printed biodegradable and remotely controllable shape memory occlusion devices. Advanced Functional Materials, 2019. **29**(51): p. 1906569.
58. Zhang, Y., et al., 4D printing of magnetoactive soft materials for on-demand magnetic actuation transformation. ACS Applied Materials & Interfaces, 2021. **13**(3): p. 4174–4184.
59. Wan, X., et al., 4D printing of multiple shape memory polymer and nanocomposites with biocompatible, programmable and selectively actuated properties. Additive Manufacturing, 2022. **53**: p. 102689.
60. Wei, H., et al., Direct-write fabrication of 4D active shape-changing structures based on a shape memory polymer and its nanocomposite. ACS Applied Materials & Interfaces, 2017. **9**(1): p. 876–883.
61. Koh, T.Y., and A. Sutradhar, Untethered selectively actuated microwave 4D printing through ferromagnetic PLA. Additive Manufacturing, 2022. **56**: p. 102866.
62. Li, W., et al., Dual-mode biomimetic soft actuator with electrothermal and magneto-responsive performance. Composites Part B: Engineering, 2022. **238**: p. 109880.
63. Ferrara, M., et al., Investigating the use of 3D printed soft magnetic PEEK-based composite for space compliant electrical motors. Journal of Applied Polymer Science, 2022. **139**(20): p. 52150.
64. Sundaram, S., et al., Topology optimization and 3D printing of multimaterial magnetic actuators and displays. Science Advances, 2019. **5**(7): p. eaaw1160.
65. Taylor, A.P., et al., Fully 3D-printed, monolithic, mini magnetic actuators for low-cost, compact systems. Journal of Microelectromechanical Systems, 2019. **28**(3): p. 481–493.
66. Burlikowski, W., et al., 3D printing of composite material for electromechanical energy harvesters. Electronics, 2022. **11**(9): p. 1458.
67. Vong, C., et al., Manufacturing of a magnetic composite flexible filament and optimization of a 3D printed wideband electromagnetic multilayer absorber in X-Ku frequency bands. Materials, 2022. **15**(9): p. 3320.
68. Duan, Y., et al., A wide-angle broadband electromagnetic absorbing metastructure using 3D printing technology. Materials & Design, 2021. **208**: p. 109900.
69. Malallah, Y., et al., RF characterization of 3-D-printed tunable resonators on a composite substrate infused with magnetic nanoparticles. IEEE Microwave and Wireless Components Letters, 2022. **32**(10): p. 1175–1178.
70. Qi, S., et al., 3D printed shape-programmable magneto-active soft matter for biomimetic applications. Composites Science and Technology, 2020. **188**: p. 107973.

13 Magnetic Polymer Composites for Agriculture Applications

Javed Iqbal, Sonia Maqbool, Nadia Bhatti, and Aziz-ur-Rehman

13.1 POLYMER AND ITS HISTORY

According to precise definitions, polymer is a material made up of molecules with long chains of a few atom species or groups connected by main, often covalent, links. Although the terms "polymer" and "macromolecular" are sometimes used interchangeably, the focus on matter in this definition emphasizes the fact that the latter term strictly refers to the molecules that make up the former. The process of creating macromolecules, known as polymerization, involves joining combined molecule after molecule of monomers through chemical processes. For instance, the polymerization of ethylene results in the production of polyethylene, a sample of which typically contains molecules with as many as 50,000 carbon atoms arranged in a chain. Polymers are distinguished from other materials by their long chain nature, which also gives birth to their distinctive features.

Since the beginning of life, polymers have been found in nature. Polymers including DNA, RNA, proteins, and polysaccharides are essential to life in both animals and plants. Man has used naturally occurring polymers from the dawn of time to make tools, clothes, decorations, living conditions, tools, weapons, writing materials, and other necessities. However, it is generally agreed that the 19th century, when significant discoveries were made about the modification of some natural polymers, is when the modern polymer industry got its start. Thomas Hancock observed in 1820 that rubber from nature becomes more fluid and simpler to combine with additives and mold when it is masticated, or repeatedly exposed to severe shear stresses. A few years afterward, in 1839, Charles Goodyear discovered that heating with sulfur increased the elastic qualities of natural rubber and reduced its tackiness. In 1844, Hancock and Goodyear were rewarded for this discovery with the invention of a process to which Hancock gave the name "vulcanization." In 1851, Charles Goodyear's brother Nelson developed the process for turning natural rubber into vulcanite, which is additionally known as durable rubber or ebonite, by vulcanizing the rubber with a lot of sulfur. In 1846, Christian Schonbein created cellulose nitrate, often known as nitrocellulose or gun cotton, and this is when it initially gained popularity. He was quick to see the material's potential as an explosive inside, and a year later, handgun cotton was being produced. However, cellulose nitrate's discovery as a rigid, elastic substance that was accessible and could be reshaped into various

 DOI: 10.1201/9781003454236-13

forms by the use of both pressure and heat was more significant for the development of the polymer industry. The first person to make use of this assortment of qualities was Alexander Parkes, who displayed items manufactured of Parkesine, a plasticized version of cellulose nitrate, in 1862. The same substance, called celluloid, was invented by John and Isaiah Hyatt in 1870. Celluloid was prepared using camphor as the activator. Celluloid had far greater commercial success than Parkesine.

Despite the fact that the synthetic polymer profession was now well established, its development was constrained by a serious lack of knowledge about the makeup of polymers. Scientists had been describing the peculiar characteristics of polymers for more than a century, and by 1920, the consensus was that they were composed of physically associated clumps of tiny molecules. Few scientists agreed with Hermann Staudinger's strongly held belief that polymers were made up of very massive molecules with a complex arrangement of simple constituent units connected by covalent bonds. In order to define polymers, Staudinger coined the term "macromolecule." He then actively went out to demonstrate the validity of his concept during the 1920s. His research on the synthesis, structure, and characteristics of polyoxymethylene and polystyrene was particularly significant, and the conclusions from his investigations left little room for questioning the viability of the macromolecular perspective. The crystallographic analyses of natural polymers published by Herman Mark and Kurt Meyer, as well as Wallace Carothers's seminal work on the preparation of polyamides and polyesters, provided more support for Staudinger's theory. Thus, by the early 1930s, most scientists had come to believe that polymers have a macromolecular structure. The study of polymers grew significantly over the next 20 years, during which time many of the fundamental concepts of the science of polymers were developed and the first works devoted to their research were published. Paul Flory's both experimental and theoretical work was important during this time, and in 1974, he was awarded the Nobel Prize in Chemistry for his lengthy and significant contributions to polymer science. Staudinger had already earned the same award for his groundbreaking work in 1953. Unsurprisingly, a lot of synthetic polymers entered the marketplace for the first time as scientific understanding of macromolecules developed. They include polystyrene, silicone compounds, nylon 6.6, polyethylene [which is poly(vinyl chloride)], poly(methyl methacrylate), poly(styrene-butadiene rubber), poly(methyl methacrylate), poly(methyl methacrylate), and polytetrafluoroethylene [1].

13.2 POLYMER APPLICATIONS

13.2.1 PHARMACEUTICAL APPLICATIONS

To manage drug distribution from diverse pharmaceutical dosage forms, a lot of effort has been expended in the pharmaceutical field during the past few decades. The creation of the first drug delivery systems to lengthen or delay medication delivery or to boost drug release for pharmaceuticals demonstrating bioavailability deficiencies was made possible by the incorporation of polymers in pharmaceutical science. Widespread use of polymers to streamline manufacturing procedures or create pharmaceutical dosage forms for a variety of administration routes has been made possible by the variety of polymers that are now readily available for use in pharmaceuticals, because of their low responses regarding illegal substances, and their safety. Additionally, the synthesis of novel polymers with distinctive features and the modification of already existing synthetic and natural polymers provide the formulator with an extensive choice of applications for improving drug administration for each individual scenario [2]. There has been a lot of interest in recent years in employing Carbopol in a variety of medicinal applications. Carbopol polymers are acrylic acid–based polymers that have been joined together by polyalkenyl ethers containing divinyl glycol. The cross-linked chemistry of acrylic acid is the foundation of the Carbopol polymer family. The items are interconnected at many levels, offering a range of performance possibilities that are functionally varied. Rheology modifiers made of Carbopol polymer compounds are efficient and effective. They offer great thickening, suspension, and stabilization advantages [3]. Polymers have been widely employed as chemical excipients in systems for drug delivery throughout the past few decades. Pharmaceutical polymers have advanced from being used as capsule gelatin shells to providing great formulation advantages, such as allowing controlled/slow discharge as well as precise focusing of therapeutic agents on the site(s) of action (the "magic bullets" concept), and as a result hold significant clinical promise. Depending on their molecular weight (MW), configuration, structure, homo/co-polymerization, and proportion of crystallinity, polymers exhibit a broad variety of physicochemical properties. So, not only as vital excipients in dosage forms for pharmaceuticals but in many other ways as well, pharmaceutical polymer compounds, which are continually changing, have been and are garnering great attention [4].

13.2.2 OPTICAL APPLICATIONS

A family of materials known as polymers is often employed in several application areas. Polymers are being proposed as an entirely novel category of substance for the creation of passive optical devices in the discipline of optical communications. A possible technique for producing structures with a changed refractive index, which is required for the guiding of light with a few wavelengths in a device with optical capabilities, is ion irradiation. After and during ion irradiation, the behavior of several polymers that meet the criteria for high transparency has been examined [5].

13.2.3 APPLICATIONS IN AGRICULTURE

In recent years, it has become clear how much soil erosion is caused by irrigation in general and by furrow erosion in particular. The most detrimental soils in the United States, covering over 1.5 million hectares, are in surface-irrigated areas in states such as Washington, Oregon, and Idaho. The soils in the area are made of ash and loess, have little cohesiveness and few persistent particles, and are deficient in biological matter and clay. Crop production in eroded regions is lower, and more inputs are needed to produce the same yield. Shear forces from overland flow produce particle separation and movement at the soil's surface. The shear stress necessary for overcoming the cohesive interactions between soil particles is finally exceeded as flow velocities rise. Sediments accumulate at the gap surface when water seeps into the soil, forming a thin seal or depositional layer. The electric conductivity of the seal's foundation soil, the Port Neuf silt loam, ranged from 0.1 to 8% of that of the seal itself. Furrow erosion can be stopped, which will maintain high infiltration while also slowing the establishment of depositional seals.

In laboratory investigations, organic polymers, namely PAM and polysaccharides, have been employed to preserve the permeability and structure of soils exposed to artificial precipitation. The application of 5020 kg ha-1 of anionic PAM to the soil surface boosted the ultimate penetration amount in soils affected by precipitation by a couple of orders of magnitude and significantly decreased runoff and interrail erosion. Studies in the lab showed that PAM solutions containing 2.5, 5.0, and 10.0 g PAM m-3 did not experience rill erosion. No soil disintegration was visible if PAM was present in the stream, even on upward slopes (30%) and at excessive flow rates with an elastic stress of 10 Pa. PAM in a tiny amount was thought to strengthen the binding forces among soil fragments at the soil's surface, which would stop rill erosion.

Investigation of materials and application techniques in the field was spurred by the success of utilizing laboratory erosion simulators to entirely stop rill erosion of a variety of soils. The idea of spraying only the surface of the soil with organic compounds rather than combining them with the cropped layer is supported by present knowledge of the significance of thin layers of organic polymers at the soil barrier in regulating rain penetration or rain and depression erosion. Because less of the polymers are used in this application method, it is more economical. This encourages the production of polymer compounds based on starch that has been grafted with PAM, since they are more efficient, simpler to use in the real world, less expensive, and biodegradable. Perhaps the most widely available and reasonably priced polymer made from natural materials on the market is starch [6].

13.3   POLYMER COMPOSITES

In order to create a material with qualities significantly superior to those of either of the underlying components, composite materials mix and sustain two or more different phases. Since ancient times, composite materials have been employed in building. Bricks have been strengthened with straw for more than two thousand years, and this technique is still in use today. Additionally, there is proof that nearly a thousand years ago in Greece, concrete beams' tension faces were strengthened with metal. Polymeric composites are multi-phase materials created by mixing additives and reinforcing filaments with polymer resins like polyester, vinyl ester, and epoxy to create a bulk compound with properties that are superior to those of the base materials used alone. Fillers are frequently employed to provide aesthetic aspects, lower bulk density, lessen cost, and add bulk to materials. In order to strengthen the polymer and enhance its mechanical qualities, such as toughness, filaments are employed. The main load-carrying components are high-strength glass, aramid, and carbon filaments, which are shielded by polymer resin and joined together to form a solid structural component. Such composites are frequently referred to as fiber composite materials. For many years, polymer composites have been used extensively in the building sector for non-critical applications including bathrooms and vanities, cladding, decorating, and finishing. Construction accounted for 35% of the world-wide market for polymer composites in 1999, ranking as the second-largest consumer in the world. Fiber composite components have recently gained popularity in the field of building for more critical uses and have proven to be a competitive and viable alternative to steel in concrete reinforcement and, to a lesser extent, new mass structures. These materials are increasingly being used for structural load-bearing applications [7].

13.4   POLYMER COMPOSITE APPLICATIONS

In high-end applications, such as race cars, auto parts, flexible optoelectronics, scaffolds, optical devices, delivery systems for pharmaceutical products, substitute medical biomaterials, spacecraft, aircraft, and tissue repair (wound dressing), advanced polymer composite materials have recently become popular. Fiber or synthetic fiber–strengthened synthetic polymer composites and synthetic and natural fiber–supplemented bio-based polymer composites are two categories of polymer composites. The substitution of ecologically friendly materials was a result of the rise in environmental awareness brought on by the inappropriate consumption of petroleum-based goods, community interest in green materials, and increasing pressure from forthcoming environmental laws by the government. According to reports, the global market for advanced polymer composites will be valued at $12.1 billion by 2020. Due to the superior benefits of natural fibers over artificial fibers in terms of cost-effectiveness, prevalent sources, biodegradability and low weight, minimal health hazards, enhanced finish on the surface of molded elements, fewer injuries during processing, and generally good mechanical properties, natural fiber polymer composites (NFPCs) have attracted a lot of attention in a variety of applications.

Lightweight natural fibers may be incorporated into polymers to create these NFPCs utilizing a broad range of fabrication techniques, including manual lay-up, pultrusion, compressive casting, plastic injection molding, vacuum sealing, spraying up, resin transfer molding, vacuuming absorption molding, filament winding, and solution casting. These procedures often entail wetting, blending, or saturating the reinforcing material of fibers and framework, which causes a chemical and thermal reaction that links the cellulose fibers and polymer framework together to form a hard structure [8]. In order to apply polymer composite technology to aircraft equipment, Dowty has been consistently developing it for the past 26 years. This technology is primarily related to the resin transfer manufacturing (RTM) production process. The main reasons composites are used in aircraft components are lighter weight, which directly affects the component's efficiency as well as the efficiency of the entire aircraft; lower cost, both initially and over time, which must take into account resistance to fatigue, impact resistance, maintainability, and repairability; and enhanced performance, which can be increased by the ability to enhance component shape, form, and stiffness. With 26 years of operational expertise on helicopters along with more than a decade on passenger-carrying aircraft, Dowty has a solid track record when it comes to the use of nanomaterials for propeller blades. Around this specific use, material optimization, design competence, and manufacturing method have been established. The potential use of composite materials in other equipment areas, such as aircraft landing mechanisms and hydraulic components, has also been regularly investigated [9].

13.5   MAGNETISM AND ITS HISTORY

Since many millennia ago, the magnetic phenomenon has been understood and used. The only material that naturally exists in a magnetized state is magnetite, which was the subject of the initial investigations into magnetism. Due to its ability to align itself in specific directions when able to rotate freely, the substance was also known as gemstone and could, in some cases, also be used to determine latitude. The ability of two pieces of lodestone to draw or even repel each other is another well-known characteristic of the material. It was discovered that magnetization was also capable of drawing iron once it was feasible to extract iron from its ores. Since there are so many magnetic materials now known, it is helpful to provide a scientific rule for what may be referred to as a magnetic substance [10].

The first magnetic substance known to man was the mineral called magnetite, which is where the history of magnetism begins. The earliest evidence of its existence is hazy, but 2500 years ago, people were aware of its ability to draw iron. Magnetite is found throughout the area of Magnesium, in what today constitutes modern Turkey and was home to the most abundant deposits in ancient times. Our term "magnet" derives from a related Greek word, which is said to have originated from the designation of this district. The Greeks also understood that if a piece of iron was contacted or rubbed with magnetite, it would turn magnetic on its own. Later, although at an unknown time, it was discovered that a suitably formed piece of magnetite would point roughly north and south if supported such that it might float on water. If previously scraped with magnetite, a pivoting iron needle would as well. Thus, the mariner's compass came into being. Because of magnetite's ability to point north, the term "lodestones," which means "way stone," was once used to describe it. William Gilbert, an Englishman, conducted the first true scientific investigation into magnetism and published his seminal work, *On the Magnet*, in 1600. After Gilbert, no fundamental discoveries were made for an additional century and a half, although there were numerous useful advancements in the production of magnets. Compound steel magnets, which were created in the 17th century and were formed of several magnetized steel strips joined together, were able to lift 28 times their weight in iron. The fact that there was just one method for creating magnets at the time—rubbing iron or steel with a lodestone—makes this feat more amazing. There was no alternative method before the first electromagnet was created in 1825 as a result of Hans Christian Oersted's groundbreaking discovery in 1820 that the flow of electricity generates a magnetic field. It may be stated that research into magnetic materials began with the development of the electromagnet, thereby making accessible far stronger fields than those created by lodestones [11].

13.6   APPLICATIONS OF MAGNETISM

The fundamentals of contemporary life, such as energy from electricity, communications, and information storage, are permeated by uses of magnetism. Magnetism contributes more than 1% of the nation's gross domestic product to the economy, but people in general are mostly ignorant of this fact [12]. In technical applications such as novel information storage, four-state logic devices, and magnetoelectric sensors, the alteration of polarization by an electromagnetic field at room temperature is crucial [13]. Inertial and gravity sensors are an important use of magnetic fluids technology. The development of all sorts of sensors required for motion measurement is possible using magnetic fluids [14]. Although permanent magnetic elements have been employed in electrical equipment for over a century, their employment in electromechanical accessible electronic devices is presently expanding quickly due to recent remarkable advancements in their characteristics and accessibility [15]. As early as the 17th century, bar magnets were being used to remove iron from the body.

The electromagnet has been used in medicine since the last decade of the 19th century. A gradient magnetic field can be used to separate the white cells [16]. Magnetic resonance imagery (MRI), a cutting-edge medical diagnostic tool, is one of the well-established uses of magnetization in medicine [17]. Additionally, magnetic fields have been employed for a variety of measures. These fields have been created by the body's own electromagnetic activity or by that of tiny quantities of material that have been intentionally injected into the body [18]. The benefits of magnetic micromachines point to more potential uses in the medical industry. The spiral-type magnetic micromachine, for instance, may swim in both a liquid and a gel. Additionally, this device has a heating feature, making it appropriate for applying localized heat to the body. At the guide-wire's tip, a small magnet is connected to provide active bending. This is a quick and practical method for placing a catheter into a blood vessel or a lung [19]. Electromagnetic particle imaging (MPI) offers several significant benefits over other methods for measuring tissue perfusion, such as nuclear methods, MRI, CT, and ultrasound. The approach enables non-invasive quantitative measures of tissue perfusion expressed as milliliters of blood per gram of tissue [20] due to the linear relationship between the marker concentrations and the MPI signal.

Due to their unusual properties, such as magnetic quantum tunneling and quantum size effects, magnetic nanoparticles have garnered a lot of attention [21]. Iron dioxide, cobalt ferrite, iron platinum, and manganese ferrite are the magnetic nanoparticles that have received the most research [22]. Magnetic nanoparticles (MNPs) have certain advantages over traditional procedures, including automation [23], a reduction in the amount of toxic chemicals required (in comparison to phenol/chloroform-based methods), and a shorter time required for analyte separation. Magnetic nanoparticles may be employed for separating any target and connected to a variety of human and automated applications because of their superparamagnetic characteristics and tiny nanodimensions [24]. Metallic ions have been eliminated from wastewater using several different treatment methods and procedures involving MNPs. In recent decades, magnetic Fe_3O_4 nanoparticles have garnered considerable interest in a variety of fields, including magnetic recording, magnetic sensing, and medical treatment. Theoretically, magnetic nanoparticles exhibit finite-size occurrences or have a large ratio of surface area to volume, both of which enhance the ability of the particles to adsorb metals. A magnetic field from the outside may also be used to separate metal-loaded magnetically adsorbent from solution with ease [25]. The magnetic field (MF), which is now a component of the surrounding atmosphere and a source of energy, influences meristem cell division and normal metabolisms. Additionally, MF has an impact on water ionization, preservation, and absorption [26].

13.7 MAGNETIC POLYMER COMPOSITES

Composite materials made of polymers that are magnetically reactive can deform or become stressed. These materials, known as agnatically active polymers, can be created physically or by chemical synthesis. In the latter scenario, magnetic components are combined with the polymeric substance to produce a composite that is magnetoactive. Polymer gels and magnetic-field-sensitive (also known as magnetorheological, or MR) elastomers are soft synthetic materials with magnetizable components contained inside them as a result of cross-linking. The resultant composites respond to the magnetic field from the outside by varying in length and have bidirectional mechanical and magnetic characteristics. Recent research and development initiatives are actively focusing on high-performance multi-functional polymer composites with unique characteristics and capabilities.

For numerous industrial and military purposes, shielding against electromagnetic radiation and active electrical barriers are of great interest. Mechanical and magnetic characteristics of magnetoactive elastic compounds were examined in a prior work. A use of such substances as components of noise cancellation systems was investigated in a subsequent study. For electromagnetic interference (EMI) applications, efforts have been undertaken to develop and characterize the electromagnetism response of steel textiles and combination steel and polymer fabrics. Magnetostrictive polymer composite materials may be woven or knitted into fibers to create highly flexible functional composites. Such materials may be used as controllers and shielding components to create a useful material for electromagnetic shielding, such as a hybrid fabric consisting of Kevlar and stainless-steel fibers.

Steel fibers in plastic resins have been investigated by Hochberg et al. for implementation in EMI applications. Additionally, efforts have been undertaken to create high-performance polymer fibers with excellent mechanical qualities. In one study, carbon fiber and polypropylene resin were combined to create composites using the vapor deposition technique. In a different study, ceramic coatings were electrophoretically deposited onto carbon and metallic fibers. A novel idea investigated and partially realized by Farshad and colleagues was the development of a new polymer composite material with comparatively potent magnetostrictive characteristics. In their article, the researchers describe the steps involved in creating this novel material as well as some of its physical characteristics [27]. Due to the growing demand for sensing capabilities in new industrial applications, including those in the automotive industry, robotics, and biomedicine, the usage of magnetic sensors has expanded tremendously. Due to their great performance, durability in hostile conditions, and cheap manufacturing costs, magnetic sensors are the best sensors for contactless position, angle, or rotational speed detection in the automobile sector. Strong permanent magnets linked to the moving mechanism, often made of rare earth minerals, create the magnetic fields that these sensors sense. The trade-off between manufacturing facilities complexities (sintering and the casting process, machining processes, and part magnetization), cost, and required magnetic performance limits the shape characteristics of these permanent magnets, limiting most applications to working with simple designs of high population density magnetic fields, such as rods or blocks.

Magnetic polymer composites (MPCs, Figure 13.1) are used to create functional magnetized components with complicated geometry, such as encoding device wheels or shafts in mind, that generate a magnetic field without the need for an additional permanent magnetic component in the sensor systems. Polymer-bonded magnets created using MPCs are a more flexible and environmentally friendly alternative to permanent magnets. The MPCs have a magnetic filler encased in a thermoplastic matrix. For injection molding, the matrix used is often a material made of polymers with a high melt flow rate, such as nylons. It is possible to utilize isotropic in shape and asymmetrical micron-sized fragments containing strontium ferrite (SrFe12O19), barium ferrite (BaFe12O19), samarium cobalt, or neodymium iron boron primary magnetic fillers. MPCs are mostly produced by injection molding because of its cost-effectiveness and the extremely tight tolerances attained, notwithstanding the latest reports of their 3D printing. Polymer-bonded magnets are more lightweight and less brittle than traditionally manufactured sintered magnets, and injection molding procedures allow for the mass manufacturing of components. They may be produced with more design flexibility in terms of form and magnetizing structure. However, because they have a lower density of magnetic material (70 to 90%) than conventional sintered permanent magnets (>95%), they exhibit inferior magnetic characteristics. The magnetic moments of the particles that are magnetic are arranged at random in the virgin arrangement, and there is no overall magnetization ($B = 0$). The moments of magnetic attraction align parallel to a magnetic field that is outside (H) when it is applied. If the field is sufficiently strong ($H > H_s$), all the moments align parallel to the field and the condition of saturation is obtained (B_s). Some of the moments of magnetic attraction remain aligned after the magnetizing field is removed ($H = 0$), and the magnetization value falls to a remaining value (B_r). Additionally, when a magnetic field is applied to anisotropic magnetic particles above the melting temperature of the polymer matrix, the particles themselves align in the field direction in addition to the magnetic moments. All the particles' easy axes line up, improving the level of saturation magnetization and lowering the saturation field.

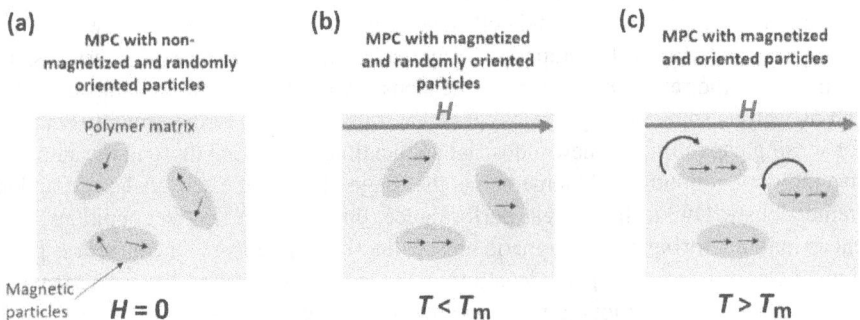

FIGURE 13.1   Orientation of the magnetic particles in the MPC under different conditions.

The magnetic characteristics of the MPC are determined by the final particle arrangement—namely, the particles' orientation and magnetization. When the polymer is injected into the molding tool under high field and high temperature conditions, both heterogeneity and momentary magnetic alignment can happen simultaneously. Alternatively, the process can happen in stages, with the particles initially being oriented during injection and then the magnetic moments being aligned by a second, stronger external field. The magnet's, MPC's, and environment's characteristics, as well as geometrical elements, all play a role in this process. The magnetic field produced by the magnetic source (such as an eternal magnet or MPC) must be accurately measured when building magnetic sensor systems in order to choose the sensor with the best sensitivity, range of movement, number of axes, positioning separation, etc. When dealing with complicated patterns or geometries, this is often done either by employing finite element technique models that can calculate the magnetic fields or by using standard models for basic magnet geometries. However, when MPCs are used as magnetic field sources, the magnetic field they produce is dependent on the MPC magnetism stage, which is established via the above-described alignment procedure. In order to develop a magnetic source that meets the application's criteria, the model must consider the non-linear behavior of the MPCs' magnetization curve. The magnetic field produced by MPCs when the casting and magnetization processes are considered cannot be accurately simulated by current FEM models [28]. For the manufacture of functional components in a wide range of applications, including the automotive, energy, and aerospace sectors, new adhesive and additive manufacturing techniques employing composite materials containing particles with magnetic properties embedded into a polymeric matrix are garnering significant interest [29]. Soft materials that are magnetically active flex or suffer mechanical stresses in response to a magnetic field from the outside. These materials are made of polymer-based elastomers and can be created by physical combination or chemical synthesis. In the latter scenario, magnetic components are combined with the polymeric substance to create a composite magnetoactive polymer [30].

13.8 NOVEL COILING BEHAVIOR IN MAGNETIC POLYMER COMPOSITES

When exposed to an external magnetic field, a flexible composite made of a polymer matrix and magnetic filler demonstrates valuable and distinctive transduction capabilities. When exposed to an external magnetic field, this kind of magnetic polymer nanocomposite (Magpol), which may be more specifically referred to as a ferrogel, magnetic gel, magnetic field–sensitive gel, magnetorheological elastomer, or magnetoactive polymer, may easily change its structure and mechanical characteristics. Controllable dampers, stiffness-tunable supports, micro fluid pumps, medication administration, cancer therapy, soft sensors with transducers, and artificial muscles are just a few of the many uses for Magpol that have been noted. Actuation is the movement of the whole composite as a result of a magnetic field from outside acting on the electromagnetic filler particles.

13.9 MAGNETIC BEHAVIOR OF IRON AND IRON OXIDE POLYMER COMPOSITES

Due to the magnetic particles' high interfacial adhesion to the matrix's polymer chains, the composite moves. Magpol has been seen to undergo deflection, elongation, and contraction during shape change. With its ability to exhibit mechanical transitions like either first- or second-dimensional phase transformations, Magpol can be used for soft actuation. As a result of its shape-changing behavior, it can be implemented for linear as well as switch and switching (i.e., on-off) applications [31]. In one study, metallic iron and nanoparticles of iron oxide were created using a gaseous inert condensation process. The coordinated oxidizing of the particle surface, which produced an Fe oxide shell-Fe core structure, passivated the particles. Spin casting of nanoparticle and polymethylmethacrylate film combinations produced nanoparticle-polymer composites. Both the nanoparticles disseminated in the composites and those compacted into pellets had their magnetic characteristics examined. Increased exchange bias and coercivity were seen in the particles. With increasing oxide shell thickness, the passing on bias was seen to rise. Because of nanoparticle loading, the magnetic attraction in the nanoscale composites was investigated. It was shown that coercivity rose as the particles were disseminated into the nanocomposite, indicating a higher anisotropy barrier. In a similar vein, the findings of the magnetic relaxation showed that, in comparison to the crushed pellets, the composites relaxed much less over the whole temperature range. This finding lends credence to the idea that weaker dipolar interactions may result in increased anisotropy barriers [32].

13.10 MAGNET FILLER–POLYMER MATRIX COMPOSITES AS ACTUATORS

In nature, actuation of the capacity of a substance to react to a change in its surroundings by mechanical means occurs often and is the foundation of living systems. Actuators may be defined as devices that can be controlled to produce work. Actuators have a broad range of performance characteristics and are based on a wide range of physical principles. The selection of a certain material for an actuator application is influenced by several performance indicators. Actuators currently do not deliver gentle, smooth motion that closely resembles human movement; this is a major unmet demand in actuation technology. Magpol can produce such actuation. The absence of contactless magnetic actuation, several actuating methods, high actuation strain, self-sensing, and rapid response are all benefits of Magpol-based systems. The capacity of actuators to "morph," or change shape, in response to the right input, such as ambient electrical, magnetic, chemical, or thermal stimuli, is an added benefit to their performance. Morphing is "efficient adjustability that could encompass microscopic, macro photography, structural and/or fluidic approaches," according to one definition. In fields like biology, structural health monitoring, defense, the aerospace industry, adaptive optics, and robotics, this "intelligence" may be quite useful. Examples of possible uses include inflatable studs for use in surgical instruments, artificial muscles, and airplanes with transforming wings to improve flight conditions. The magnetic filler may also indirectly cause morphing, as in the case of spontaneously stimulated shape memory polymers (SMPs). In SMPs, heat produced by the particles in an intermittent RF field with magnetic resonance can cause phase changes that are advantageous.

Two crucial actuation parameters, actuation stress and strain, may be compared between Magpol and other popular actuators using a property chart (Figure 13.2). Low strains that may be found in innovative materials and are frequently utilized to make actuators include piezoceramics, magnetostrictive substances, and shape memory alloys (SMAs). For instance, magnetostrictive as well as piezostrictive materials have actuation strains that are typically about 10 to 3%. Although SMAs experience substantial actuating stress and strain, their normal operating frequency is less than 1 Hz. Magpol, a material that may offer high strain values, higher dynamic characteristics, and a measuring mechanism, is therefore appealing for transducer applications. The application of electromagnetism and magnetostrictive forces, particularly magnetic forces, can present difficulties at tiny length scales, but they are essential to many of the primary actuation systems. The highest displacement and force that can be achieved in common magnetic MEMS devices are 10^{-5} m and 10^{-3} m, respectively. The establishment of Magpol technologies with high electromagnetic energy density to produce more work and force is therefore now a problem. Magpol's competitiveness has recently been greatly enhanced because of performance advances, especially when compared to other magnetically powered actuators. It is evident that the amount of activation stress, which requires significant improvement, is the fundamental drawback of Magpol. Future possibilities for raising the highest possible actuation stress include improvements in (a) the filler's magnetic properties, particularly higher saturation magnification and magnetic permeability; (b) the filler's optimal concentration; (c) the matrix's mechanical characteristics; (d) the properties

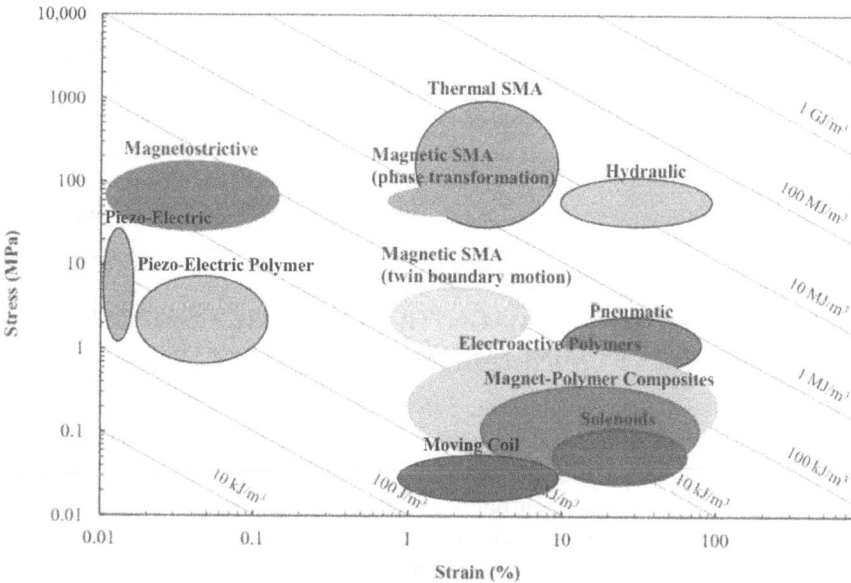

FIGURE 13.2 Typical stress and strain characteristics of actuation technologies: Magpol-based actuators compared to other actuation technologies.

of the interfacial region between the matrix and the filler; and (e) the magnetic field gradient. The maximum actuation strain among magnetically powered actuators is a key benefit of Magpol. In a recent investigation, it was discovered that the lengthened strain and wrapping strain of Magpol were, respectively, 30% and 50% more than previously thought.

13.10.1 MAGPOL COMPOSITION

In order to create actuators, Magpol combines the capabilities of an exceedingly flexible matrix of polymer with the driving force acquired by the movement of external magnetic fields on magnetic materials. Ferrogel, magnetic gel, magnetic field–sensitive gel, magnetorheological elastomers, and magnetoactive polymers are examples of materials that are similar. In most instances, the basic idea is the same. Ferrogels, for instance, are hydrogel networks that have been infused with a magnetizable solution made up of magnetic nanoparticles. The microscopic components interface with the neighboring polymer chains and change location when exposed to a magnetic field that is external. The composite changes form as a result. The polymer matrix may be selected with a lot of freedom. Matrix materials that are frequently employed include hydrogels, silicones, polyurethanes, and rubber. The materials that make up the particles of magnetic attraction can also be either mechanically soft or mechanically hard. Strong actuators and sensors may be created by connecting electric or magnetic fields with the flexibility of Magpol. The magnetic and actuation characteristics of Magpol can be influenced by its composition.

13.10.2 MAGNETIC PROPERTIES OF MAGPOL

The performance of Magpol is significantly influenced by its magnetic properties. Nguyen et al. created Magpol that incorporated silicone rubber compounds and Fe nanoparticles (size 5 m) in the elastomers. They found that the silicone samples that included 50 wt% iron had magnetic characteristics that were qualitatively comparable to those exhibited by the iron granules. Additionally, they noticed that the level of saturation ferromagnetic value of the composites increased linearly with filler content [33].

13.11 PREPARATION OF MAGNETIC POLYMERIC COMPOSITE PARTICLES BY MINI-EMULSION POLYMERIZATION

The creation of magnetic polymerized composite particles (MPCPs) for use in biomedical applications such as cell separation, immunoassays, nucleic acid purification processes and DNA separation, enzyme immobilization, electromagnetic resonance imaging, and hyperthermia is currently generating a great deal of interest. This is because MPCPs have a high sensitivity to magnetic fields from the outside and are simple to surface-modify and then functionalize with different bioactive chemicals. As an alternative to in situ chemical-based co-precipitation of the ferrous chloride and ferric chloride in order to form Fe_3O_4 underneath pore spaces of pre-formed, monodisperse, permeable polymer particles, MPCPs can also be made by coating or encapsulating electromagnetic organic and inorganic particulates, such as magnetite deposits (Fe_3O_4), with either natural or synthetic polymers. In research, a variety of MPCP synthesis techniques have been employed. Fe_3O_4 particles can be either hydrophilic or hydrophobic. Fe_3O_4 nanoparticles must be hydrophobic in order to scatter in the hydrophobic monomers when they are enclosed in a hydrophobic polymer, such as polystyrene. The most often utilized monomer is styrene (St). Because polymer formation does not take place in monomer droplets, it is exceptionally challenging for Fe_3O_4 elements to disperse into tiny particles in the case of standard emulsion formation polymerization. Mini-emulsion polymer formation should be considered the most suitable method for successfully encapsulating Fe_3O_4. For MPCPs to be acceptable for biomedical applications, the particle size distribution (PSD) should be as narrow as possible so that the MPCPs can respond to an external magnetic field as uniformly as possible. Therefore, the challenge in almost all the strategies for preparing MPCPs is to minimize or even eliminate the formation of PPPs and BMPs in production. A more difficult challenge is to obtain MPCPs with very narrow PSDs. These important issues, however, have not been reported and discussed in detail in the literature. Mini-emulsion polymerization was used to prepare hydrophobic MPCPs with St as a monomer. Advantages included no sedimentation, homogeneous size and size distribution, high and consistent magnetic content, superparamagnetic behaviors, absence of toxicity, and absence of iron leakage. However, the polymerization procedure for the sealing of Fe_3O_4 can result in three different types of particles: namely, MPCPs (with Fe_3O_4 enclosed inside), pure polymeric particles (PPPs; without Fe_3O_4 inside), and uncovered (free) magnetite deposits particles (BMPs; without an outer polymer coating). PPPs are undesirable because they will impair the magnetic conductivity of composite particles, while BMPs are undesirable because there is not enough polymerization on their exterior surfaces to allow for further functionalization. In the first step, the chemical conjugation of ferrous and ferric salts produced a water-based hematite ferrofluid with a standard size of around 10 nm. By acidifying the based-on-water magnetite ferrofluid and dispersing the acidified magnetic St, oil-based St magnetic ferrofluid was created. By using sodium dodecyl sulphate as an emulsifier, a compound called 2,2'-Azobis(2-methylpropionitrile) as a starting material, and the based-on-oil St magnetite ferrofluid, researchers were able to create MPCPs. Following the introduction of the substance's hydroxyethyl cellulose as well as polyvinylpyrrolidone as assist stabilizers, methacrylic acid was utilized as a comonomer. The enclosed conditions of magnetite were thoroughly examined with the goals of increasing the degree of magnetite encapsulation, preventing pure polymer fragments from being exposed to magnetite fragments, and producing the smallest particle size distributions. The findings demonstrated that mini-emulsion polymerization is a successful technique for encasing magnetite in a hydrophobic polymer. By choosing the right preparation conditions, it is possible to fully prevent uncovered magnetite fragments and pure polymer particles. All the MPCPs that were produced showed extraordinary magnetism and had some magnetic resistance [34].

13.12 MAGNETIC POLYMER COMPOSITES FOR AGRICULTURE APPLICATIONS

Drugs, household wastewater, acid, basic, or azo colors, and heavy metals are only a few of the toxins that are continually released into water bodies. As a result, there are significant environmental issues. Many pharmacological chemicals, including antibiotics, are only partially broken down by living organisms before being eliminated as original compounds or in the form of metabolites. Present in wastewater, in surface water, and (to a lesser degree) in drinkable and groundwater all over the world, they are frequently not completely cleaned by wastewater treatment plants. Thus, leftover antibiotics from the environment might enter the body of a person through food and water, leading to resistance to these substances and disruption of normal hormones. The antibiotics ciprofloxacin (CPX) and norfloxacin (NOX) belong to the fluoroquinolone (FQ) class, which both humans and animals utilize. Small levels of FQs in sewage from nursing homes, hospitals, and pharmaceutical industries are the main factor in severe and chronic poisoning, and they also encourage the growth of bacteria that are resistant to them. As a result, the abolition of FQs is a crucial issue. Even at very low quantities, these antibiotics are exceedingly harmful to all living things and cause health issues including diarrhea, vomiting, headaches, and nausea. Their presence has been identified in wastewater in very small amounts. Another significant category of organic contaminants is dyes. They contribute to the deterioration of the aquatic setting by reducing photosynthetic activity and reducing sunlight penetration. The triphenylmethane dye family includes the compounds methyl violet (MV) and brilliant green (BG), which are frequently used as biology stains for identifying bloody fingerprints. However, human respiratory and digestive systems might become irritated by MV and BG. Additionally, MV is an instigator that gives humans cancer and severe eye discomfort. And BG may also cause skin irritation, eye ache, and redness. These colors are present in a variety of situations because microbial enzymes only weakly destroy them. In general, wastewater includes colors as well as several heavy metals (such as Ni, Cu, and As). "Low-density chemical substances with high toxicity" is how most metals and their constituents are referred to. They are globally disseminated and non-biodegradable and have a significant impact on the environment and human health. Cu (II), for instance, severely harms human bones and kidneys. Even when present in extremely small concentrations in water, it is still detrimental to the water. Due to metals' non-biodegradability and bioaccumulation, metal ion pollution endangers people, animals, plants, and microbes. Therefore, it is crucial to rid water sources of it.

Organic dyes, heavy metals, and antibiotics have all been removed from wastewater using a variety of techniques, including coagulation, decomposition, separation using membranes, adsorption, biological treatment, complex oxidation, and electrochemical oxidation. The bulk of these techniques, however, are unsuitable and ineffectual in certain circumstances. Ion exchange and membrane separation are both relatively costly processes, and chemical precipitation produces a lot of metal-based wastes that lead to secondary contamination. Adsorption has received more attention recently for the elimination of toxicants. It is perhaps the most efficient way since it produces no more pollution and is effective. Additionally, the method is straightforward to perform, inexpensive, and free from toxicity, and it entails very simple regeneration. Many studies have reported using a variety of adsorbents to remove colors, metals, and antibiotics from aqueous solutions. Manganese-doped Fe_3O_4 nanoparticle–laden carbon–activated ultrasound-assisted cationic dye adsorption was found to be exceptionally effective. Ren and colleagues investigated the highly efficient desorption of metals such as mercury on novel, modified-pore magnetic composites with amino groups [35]. According to Zhao et al.'s research, EDTA-cross-linked cyclodextrin may be a non-toxic adsorbent that may be used to simultaneously bind cationic coloring agents and metals [36]. On Ni/Al hydroxide, Nimibofa and colleagues reported the sorption of colors and metal ions. According to earlier research, Fe_3O_4-activated charcoal/cyclodextrin/sodium alginate polymer gel beads may be a superior option to other magnetic adsorbents because of their combined advantages. Consumer-activated charcoal materials have recently been used in high-performance adsorption applications because they are affordable and easily accessible, have a large surface area and porous surface, have a high adsorption efficiency, and can be applied widely to remove any toxicants. A functionalized carbon outer shell and a magnetite inner core make up the intriguing structure of the AC-coated magnetic nanocomposite Fe_3O_4/AC. By adding cyclodextrin to the flanking chain with SA to create a novel SA derivative, the capacity for adsorption may be increased. Seven glucopyranose units make up the cyclic structure of the oligosaccharide known as beta-cyclodextrin. The cyclic arrangement of CD results in a shortened cone structure with an outward portion that is hydrophilic and an interior cavity that is hydrophobic. Given their numerous outstanding qualities, magnetic/CD/AC/cyclodextrin nanoparticles have drawn a lot of interest. For instance, they significantly improved long-term viability when mixed with sodium alginate in suspension. Finally, sodium alginate is a naturally occurring linear polysaccharide that is non-toxic, hydrocolloid, water soluble, and biodegradable. It is a common binder and gelling agent used to create a variety of composite materials. Under moderate circumstances, SA may form gels with divalent anions, such as calcium, and the surface layer of SA polymers has a significant number of functionalities, including carboxylic and hydroxyl monomers. The two-component magnetic polymer nanocomposites have the following advantages as adsorbents: (1) the inner magnetic core permits easy separation; (2) the AC-coated magnetized nanocomposite shell can prevent Fe_3O_4 development and decomposition by chemicals in a critical environment; and (3) the magnets' (Fe_3O_4)/AC nanocomposite material is based on SA, which can be in the form of straightforward-to-prepare, separable, recyclable, and non-toxic gel beads [37].

REFERENCES

1. Young, R.J. and Lovell, P.A., 2011. *Introduction to Polymers*. CRC press.
2. Amighi, K., 2001, New research on the significance of polymers in pharmaceutical formulations. *Bulletin Et Memoires De L'academie Royale De Medecine De Belgique*, *156*(6 Pt 2), pp.302–310.
3. Panzade, P. and Puranik, P.K., 2010. Carbopol polymers: A versatile polymer for pharmaceutical applications. *Research Journal of Pharmacy and Technology*, *3*(3), pp.672–675.

4. Debotton, N. and Dahan, A., 2017. Applications of polymers as pharmaceutical excipients in solid oral dosage forms. *Medicinal Research Reviews, 37*(1), pp.52–97.

5. Rück, D.M., Schulz, J. and Deusch, N., 1997. Ion irradiation induced chemical changes of polymers used for optical applications. *Nuclear Instruments and Methods in Physics Research Section B, 131*(1-4), pp.149–158.

6. Lentz, R.D., Sojka, R.E., Carter, D.L. and Shainberg, I., 1992. Preventing irrigation furrow erosion with small applications of polymers. *Soil Science Society of America Journal, 56*(6), pp.1926–1932.

7. Humphreys, M., 2003. The use of polymer composites in construction. In *Proceedings of the CIB 2003 Int'l conference on Smart and Sustainable Built Environment* (pp. 1–9). Queensland University of Technology.

8. Ilyas, R.A. and Sapuan, S.M., 2019. The preparation methods and processing of natural fibre bio-polymer composites. *Current Organic Synthesis, 16*(8), pp.1068–1070.

9. McCarthy, R.F.J., Haines, G.H. and Newley, R.A., 1994. Polymer composite applications to aerospace equipment. *Composites Manufacturing, 5*(2), pp.83–93.

10. Jakubovics, J.P., 2023. *Magnetism and Magnetic Materials.* CRC Press.

11. Cullity, B.D. and Graham, C.D., 2011. *Introduction to Magnetic Materials.* John Wiley & Sons.

12. Jacobs, I.S., 1969. Role of magnetism in technology. *Journal of Applied Physics, 40*(3), pp.917–928.

13. Wang, L., Wang, D., Cao, Q., Zheng, Y., Xuan, H., Gao, J. and Du, Y., 2012. Electric control of magnetism at room temperature. *Scientific Reports, 2*(1), pp.223.

14. Piso, M.I., 1999. Applications of magnetic fluids for inertial sensors. *Journal of Magnetism and Magnetic Materials, 201*(1-3), pp.380–384.

15. Strnat, K.J., 1990. Modern permanent magnets for applications in electro-technology. *Proceedings of the IEEE, 78*(6), pp.923–946.

16. Freeman, M.W., Arrott, A. and Watson, J.H.L., 1960. Magnetism in medicine. *Journal of Applied Physics, 31*(5), pp.S404–S405.

17. Hickey, R. and Schibeci, R.A., 1999. The attraction of magnetism. *Physics Education, 34*(6), pp.383.

18. Frei, E.H., 1969. Magnetism and medicine. *Journal of Applied Physics, 40*(3), pp.955–957.

19. Ishiyama, K., Sendoh, M. and Arai, K.I., 2002. Magnetic micromachines for medical applications. *Journal of Magnetism and Magnetic Materials, 242*, pp.41–46.

20. Borgert, J., Schmidt, J.D., Schmale, I., Rahmer, J., Bontus, C., Gleich, B., David, B., Eckart, R., Woywode, O., Weizenecker, J. and Schnorr, J., 2012. Fundamentals and applications of magnetic particle imaging. *Journal of Cardiovascular Computed Tomography, 6*(3), pp.149–153.

21. Ichiyanagi, Y., Moritake, S., Taira, S. and Setou, M., 2007. Functional magnetic nanoparticles for medical application. *Journal of Magnetism and Magnetic Materials, 310*(2), pp.2877–2879.

22. Latham, A.H. and Williams, M.E., 2008. Controlling transport and chemical functionality of magnetic nanoparticles. *Accounts of Chemical Research, 41*(3), pp.411–420.

23. Wierucka, M. and Biziuk, M., 2014. Application of magnetic nanoparticles for magnetic solid-phase extraction in preparing biological, environmental and food samples. *TrAC Trends in Analytical Chemistry, 59*, pp.50–58.

24. Marszałł, M.P., 2011. Application of magnetic nanoparticles in pharmaceutical sciences. *Pharmaceutical Research, 28*, pp.480–483.

25. Shen, Y.F., Tang, J., Nie, Z.H., Wang, Y.D., Ren, Y. and Zuo, L., 2009. Preparation and application of magnetic Fe_3O_4 nanoparticles for wastewater purification. *Separation and Purification Technology, 68*(3), pp.312–319.

26. Hozayn, M., Abdallha, M.M., A. A. Abd, E.M., El-Saady, A.A. and Darwish, M., 2016. Applications of magnetic technology in agriculture: A novel tool for improving crop productivity (1): Canola. *African Journal of Agricultural Research, 11*(5), pp.441–449.
27. Farshad, M., Clemens, F. and Le Roux, M., 2007. Magnetoactive polymer composite fibers and fabrics—processing and mechanical characterization. *Journal of Thermoplastic Composite Materials, 20*(1), pp.65–74.
28. González-Losada, P., Martins, M., Paz, E., Vinayakumar, K.B., Pereira, D., Cortez, A.R. and Aguiam, D.E., 2022. Design of arbitrary magnetic patterns on magnetic polymer composite objects: A finite element modelling tool. *Polymers, 14*(18), pp.3713.
29. Palmero, E.M., Casaleiz, D., Jiménez, N.A., Rial, J., de Vicente, J., Nieto, A., Altimira, R. and Bollero, A., 2018. Magnetic-polymer composites for bonding and 3D printing of permanent magnets. *IEEE Transactions on Magnetics, 55*(2), pp.1–4.
30. Farshad, M. and Le Roux, M., 2005. Compression properties of magnetostrictive polymer composite gels. *Polymer Testing, 24*(2), pp.163–168.
31. Nguyen, V.Q. and Ramanujan, R.V., 2010. Novel coiling behavior in magnet-polymer composites. *Macromolecular Chemistry and Physics, 211*(6), pp.618–626.
32. Baker, C., Shah, S.I. and Hasanain, S.K., 2004. Magnetic behavior of iron and iron-oxide nanoparticle/polymer composites. *Journal of Magnetism and Magnetic Materials, 280*(2–3), pp.412–418.
33. Nguyen, V.Q., Ahmed, A.S. and Ramanujan, R.V., 2012. Morphing soft magnetic composites. *Advanced Materials, 24*(30), pp.4041–4054.
34. Lu, S. and Forcada, J., 2006. Preparation and characterization of magnetic polymeric composite particles by miniemulsion polymerization. *Journal of Polymer Science Part A: Polymer Chemistry, 44*(13), pp.4187–4203.
35. Ren, A., Ding, X., Li, W., Wu, H. and Yang, H., 2017. Highly efficient adsorption of heavy metals onto novel magnetic porous composites modified with amino groups. *Journal of Chemical and Engineering Data, 62*, pp.1865–1875.
36. Zhao, F., Repo, E., Yin, D., Meng, Y., Jafari, S., Sillanpaa and M., 2015. EDTA-cross-linked beta-cyclodextrin: An environmentally friendly bifunctional adsorbent for simultaneous adsorption of metals and cationic dyes. *Environmental Science and Technology, 49*, pp.10570–10580.
37. Yadav, S., Asthana, A., Singh, A.K., Chakraborty, R., Vidya, S.S., Susan, M.A.B.H. and Carabineiro, S.A., 2021. Adsorption of cationic dyes, drugs and metal from aqueous solutions using a polymer composite of magnetic/β-cyclodextrin/activated charcoal/Na alginate: Isotherm, kinetics and regeneration studies. *Journal of Hazardous Materials, 409*, pp.124840.

14 Surface-Engineered Magnetic Nanoparticles (Iron Oxides) and Their Therapeutic Applications

Sayan Ganguly and Shlomo Margel

14.1 INTRODUCTION

The delivery of drugs is a pivotal factor in the efficacious management of diverse ailments and health conditions. The objective of drug delivery is to facilitate the precise transportation of pharmaceutical agents to the intended anatomical location within the organism, at the appropriate dosage and temporal parameters, while mitigating any potential adverse reactions [1]. The aforementioned objective can be accomplished through diverse modalities, including oral, transdermal, inhalation, and intravenous administration of pharmaceutical agents [2, 3]. The field of biomedical drug delivery encompasses a range of methodologies and techniques that are utilized to transport therapeutic agents to precise anatomical locations within the body [4]. The utilization of nanoparticles, liposomes, and microspheres, alongside targeted drug delivery systems that employ ligands or antibodies to specifically attach to particular cells or tissues, is within this field [5–8]. The administration of drugs holds noteworthy ramifications for the management of diverse medical ailments, encompassing cancer, cardiovascular disease, neurological disorders, and infectious diseases [9]. Enhancing the effectiveness of current therapies, mitigating toxicity, and presenting novel prospects for pharmaceutical advancement are potential benefits. Furthermore, the administration of drugs is of utmost importance in the advancement of custom-made medicine, which encompasses the customization of therapies to suit the specific genetic, environmental, and lifestyle attributes of individual patients [10–12]. Researchers can develop custom-made treatment plans that are more effective and have fewer side effects by providing more targeted and effective drug delivery methods [13].

The biomedical applications of iron oxide nanoparticles have been extensively researched, including their potential use in drug delivery, imaging, and cancer therapy [14–16]. The enduring biocompatibility and plausible toxicity of said nanoparticles has been a subject of apprehension, particularly in their in vivo application [17]. The utilization of biodegradable surface-engineered iron oxide nanoparticles has surfaced as a promising alternative to tackle these concerns. Iron oxide nanoparticles that are biodegradable are engineered to undergo degradation within the human body, thereby mitigating the potential hazards of prolonged accumulation and toxicity [18].

DOI: 10.1201/9781003454236-14

The process of surface engineering entails the alteration of nanoparticle surfaces through the application of biocompatible polymers or biomolecules, with the aim of enhancing their biocompatibility, stability, and targeting potential. Poly(lactic-co-glycolic acid) (PLGA)–coated iron oxide nanoparticles are a type of biodegradable surface-engineered iron oxide nanoparticles that can be cited as an example [19–21]. The polymer known as PLGA has been widely employed in drug delivery applications due to its biodegradable and biocompatible properties [22]. The application of a PLGA coating serves to safeguard iron oxide nanoparticles against degradation within the biological system, thereby enhancing their biocompatibility [23]. Several other type of materials could also have significance as coating materials for IONPs [20, 24]. An additional instance pertains to the utilization of proteins, namely albumin and transferrin, for the purpose of enveloping iron oxide nanoparticles. These proteins possess the ability to function as targeting ligands, thereby facilitating the selective binding of nanoparticles to particular cells or tissues, including but not limited to cancer cells [25]. Surface-engineered iron oxide nanoparticles that are biodegradable present various benefits in comparison to conventional iron oxide nanoparticles [26]. These materials exhibit biocompatibility and biodegradability and can be tailored to selectively target cells or tissues. In addition, these materials can be synthesized using established methodologies such as the sol-gel technique or co-precipitation, rendering them economically viable and amenable to large-scale manufacturing. The advent of biodegradable surface-engineered iron oxide nanoparticles has presented novel prospects for the advancement of safer and more efficacious biomedical applications [27]. Further investigation is required to comprehensively comprehend the enduring biocompatibility and toxicity of the aforementioned [28].

The objective of this chapter is to furnish a thorough exposition of the present status of investigation concerning the combination, interpretation, and medicinal uses of biodegradable and surface-engineered iron oxide nanoparticles. An analysis and synthesis will be conducted of the existing literature pertaining to the utilization of nanoparticles for drug delivery and other biomedical purposes. This analysis will encompass an examination of the obstacles and prospects linked with the application of these nanoparticles. Furthermore, it will emphasize the prospective avenues and possibilities for further investigation in this area. The primary objective of the chapter is to furnish readers with an enhanced comprehension of the present state and prospective possibilities of biodegradable surface-engineered iron oxide nanoparticles for biomedical utilization.

14.2 SYNTHESIS OF BIODEGRADABLE SURFACE-ENGINEERED IRON OXIDE NANOPARTICLES

The distinctive magnetic properties of iron oxide nanoparticles (IONPs) have led to their extensive use in various biomedical applications. The safe use of these substances has been a concern due to their potential toxicity and non-biodegradability. In order to address these concerns, researchers have developed biodegradable surface-engineered IONPs that are specifically designed to undergo degradation within the body into non-toxic constituents.

14.2.1 IRON OXIDE NANOPARTICLES SYNTHESIS

Various techniques can be employed to produce iron oxide nanoparticles. The co-precipitation technique involves the addition of iron salts to a base solution, such as sodium hydroxide, followed by stirring until precipitation takes place. Subsequently, the particles that ensue are subjected to a washing and drying process [29]. The approach is expeditious and facile; however, it may pose challenges in regulating the dimensions and morphology of the resultant nanoparticles. The sol-gel technique entails chemical reactions of metal alkoxides with a solvent, leading to hydrolysis and condensation and culminating in the formation of nanoparticles through annealing [30]. The procedure enables meticulous regulation of the dimensions and configuration of the nanoparticles; however, it is a multifaceted synthetic process that necessitates scrupulous management of the reaction circumstances. The process of thermal decomposition involves subjecting a precursor compound that contains iron to elevated temperatures, often in the presence of a solvent or surfactant, resulting in the formation of nanoparticles [31]. The technique yields nanoparticles of elevated purity and uniform size distribution. However, it may incur significant expenses due to the costly nature of precursor compounds and the need for high temperature conditions. The hydrothermal method entails subjecting iron salts to a hydrothermal solution at elevated pressure and temperature to initiate a reaction [29]. The nanoparticles that are obtained exhibit a significant level of crystallinity and can be produced in substantial amounts. However, the synthesis procedure is time-consuming and necessitates the use of specialized equipment [32]. The microemulsion technique involves the utilization of a blend of surfactants, water, and oil to generate micelles [33]. These micelles serve as a host for the iron precursor, which is subsequently dissolved and subjected to thermal treatment to produce nanoparticles [34–36]. The nanoparticles that are produced exhibit a uniform particle size and a narrow size distribution. However, the methodology employed can be costly due to the expenses incurred from the surfactants and the limited scalability of the process. The green synthesis approach employs biogenic and eco-friendly substances, such as bacterial cultures or botanical extracts, to fabricate nanoparticles. The approach is characterized by its cost-effectiveness and eco-friendliness; however, it may lead to reduced yields and limited ability to regulate the dimensions and morphology of the nanoparticles [37]. Table 14.1 summarizes the major synthesis methods adopted to prepare iron oxide nanoparticles. Metal ion reduction is accomplished via a system of electron shuttles, many of which are extracellular enzymes. Some bacteria, including Geothrix fermentans, Mycobacterium paratuberculosis, and Shewanella oneidensis, have been shown to secrete redox chemicals that operate as an electron shuttle between the host and substrate, allowing for the diffuse reduction of Fe^{3+} ions [38].

14.2.2 SURFACE ENGINEERING OF IRON OXIDE NANOPARTICLES

The process of affixing polymers onto IONPs can be accomplished via covalent bonding, physical adsorption, or ligand exchange. The process of covalent bonding entails the establishment of robust chemical bonds between the functional groups present on the polymer and the reactive sites located on the surface of the IONP [61]. The process of physical adsorption is dependent on non-covalent interactions, which include

TABLE 14.1

Various Methods of Iron Oxide Nanoparticles' Preparation and Their Pros and Cons

Methods	Advantages	Disadvantages	Applications	Ref.
Co-precipitation	Simple, fast, and scalable	Lack of control over size and shape	Biomedical imaging, magnetic hyperthermia	[39, 40]
Sol-gel	Precise control over size and shape	Complex synthesis process	Gas sensors, catalysis	[41–44]
Thermal decomposition	High purity and monodispersity	Expensive precursors, high temperature requirements	Magnetic data storage, ferrofluids	[45–48]
Hydrothermal	Scalable and versatile	Long synthesis times, high pressure requirements	Biomedical applications, wastewater treatment	[49–53]
Microemulsion	Narrow size distribution, uniform particle size	Expensive surfactants, limited scalability	Biomedical imaging, magnetic separation	[54, 55]
Green synthesis	Environmentally friendly, low cost	Lower yields, lack of control over size and shape	Catalysis, water treatment	[56–60]

hydrogen bonding and electrostatic forces that occur between the nanoparticle and polymer. The process of ligand exchange entails the substitution of the present surface ligands on the IONP with the intended polymer [62]. The utilization of diverse techniques offers varying degrees of stability and manipulation of the attachment process, thereby enabling the tailoring of IONPs with diverse polymers to cater to a wide range of applications, including but not limited to drug delivery, sensing, and catalysis [63].

Polymer coatings improve coated nanoparticles (NPs). Coating materials and functionality can greatly affect NP performance and application. Polymer coverings prevent NP aggregation and degradation. Coatings defend NPs from pH fluctuations, temperature changes, and oxidative damage, preserving their structural integrity [64]. Polymers coat NPs, allowing functional groups, biomolecules, and targeted ligands to adhere. This alteration increases the NPs' selectivity and affinity for selected cells or tissues, enabling targeted medication administration or diagnostics. Polymer coatings make NPs more biocompatible for biomedical purposes. Biocompatible polymers like polyethylene glycol (PEG) decrease NP–biological component contact, reducing immune reactions and enhancing circulation time [65]. Polymer coatings improve NP solubility and dispersibility in many solvents or matrices. Hydrophilic polymers like PVA or PVP improve hydrophobic NP dispersibility in aqueous solutions, enhancing their applicability. Polymer coverings can stabilize NPs in blood or cellular fluids. Coating materials and composition can prevent opsonization (i.e., immune system recognition and clearance) and increase NP circulation time, boosting bioavailability and efficacy [66]. PEGylation coats NPs, limiting immune system recognition. NPs circulate longer and accumulate in target tissues. Specific coating materials can attach ligands, antibodies, or peptides to NP surfaces. These coatings enable active targeting, allowing cells harboring receptors or antigens to selectively bind to and ingest NPs [67]. Temperature, pH, and enzymes can affect polymer coatings. These stimuli-responsive coatings enable regulated medication administration or imaging by triggering cargo release or NP behavior. Fluorescent dyes or magnetic nanoparticles on coatings allow NP visualization and tracking in biological systems. These coatings improve NP imaging and allow real-time distribution and localization monitoring. The unique physical and chemical properties of IONPs have led to their extensive application in diverse areas, including biomedicine, environmental remediation, and energy storage [68–70]. The surface characteristics of IONPs are of utmost importance in determining their stability, biocompatibility, and functional efficacy. Consequently, the engineering of the surface of IONPs is imperative in order to augment their efficacy and broaden their potential uses [71]. Various methodologies that have been employed for the surface engineering of IONPs are discussed here in brief.

The stability and biocompatibility of IONPs can be enhanced through the application of diverse organic or inorganic coatings. Silica, polyethylene glycol (PEG), chitosan, and other polymers have been employed as coatings for IONPs [70]. Functionalization of IONPs involves the attachment of diverse molecules such as antibodies, peptides, and aptamers to enable the targeting of specific cells or tissues [61]. Folic acid–conjugated IONPs have been employed to achieve specific transportation of anti-cancer agents to cancerous cells that exhibit elevated levels of folate receptors [68]. IONPs have the potential to undergo doping with various metals or elements, thereby modifying their chemical and physical characteristics [72]. Mn-doped IONPs have been employed as contrast agents in the context of magnetic resonance imaging (MRI) [73]. Surface modification of IONPs involves the introduction of different chemical groups such as carboxyl, amino, and hydroxyl groups, which facilitate their conjugation with other molecules. Carboxyl-modified IONPs have been employed in the field of targeted drug delivery and imaging [74]. A series of polymer ligands that possess multiple functions and coordination abilities has been developed [75]. These ligands were found to be highly effective in functionalizing iron oxide surfaces (Figure 14.1a), as well as other magnetic nanoparticles, and facilitating their integration into biological systems. The process of synthesizing amphiphilic polymers involves the conjugation of multiple amine-terminated dopamine anchoring groups, poly(ethylene glycol) moieties, and reactive groups onto a poly(isobutylene-alt-maleic anhydride) (PIMA) chain through nucleophilic addition. The presence of multiple dopamine groups within a single ligand significantly increases the ligand's affinity for magnetic nanoparticles through multiple coordinations. Additionally, the inclusion of hydrophilic and reactive groups promotes colloidal stability in buffer media and facilitates subsequent conjugation with target biomolecules. Transmission electron microscopy (TEM) images revealed that the nanoparticles, which were capped with dopa-PIMA or dopa-PIMA-PEG, retained their original size subsequent to surface modification, without any indication of alteration in shape or aggregation (Figure 14.1b). The process of ligand exchange with amphiphilic polymers has been observed to exhibit high efficiency, with minimal to negligible material loss (less than 10%) upon phase transfer to water.

The process of surface engineering plays a pivotal role in augmenting the characteristics of IONPs for diverse applications. IONPs exhibit distinctive

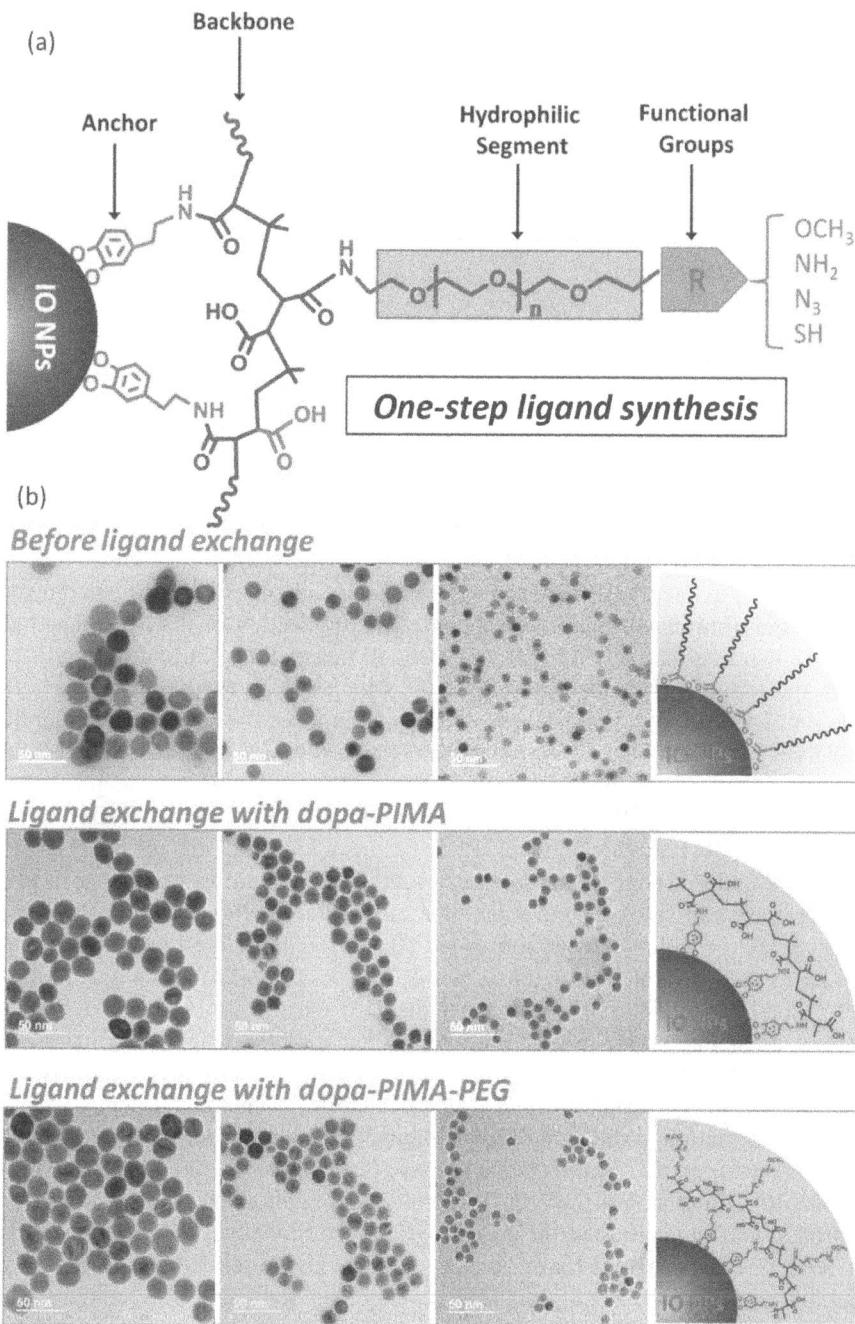

FIGURE 14.1   (a) Surface decoration of IONPs. (b) Schematic illustrations of the nanoparticles, with the corresponding surface capping ligands also shown. (Reproduced with permission from [75]. © 2014 American Chemical Society.)

physical and chemical characteristics, including elevated magnetization, bio-compatibility, and reduced toxicity, rendering them appropriate for use in biomedical and environmental contexts. The stability, biocompatibility, and functionality of IONPs are contingent upon their surface properties. Polymer coating represents a frequently employed surface engineering strategy for IONPs.

The application of polymer coatings to IONPs can yield various advantages, including but not limited to the prevention of aggregation, augmentation of sta-bility, mitigation of toxicity, prolongation of circulation time, and facilitation of targeted delivery to particular cells or tissues [76, 77]. Polymer coatings have the potential to enhance the colloidal stability of IONPs in biological fluids. This is due to the tendency of IONPs to agglomerate in the presence of different ions and biomolecules. Furthermore, the application of polymer coatings has been shown to mitigate the issue of non-specific binding of IONPs to cells and tissues, thereby enhancing their targeted delivery and minimizing their poten-tial toxicity [78].

The following are instances of polymer coatings utilized for IONPs. The utili-zation of polyethylene glycol is a frequently employed method for coating IONPs due to its favorable attributes such as minimal toxicity, remarkable biocompat-ibility, and diminished immune response. The application of PEG coatings has been shown to effectively mitigate the issue of nonspecific binding of IONPs to cells and tissues, while also promoting their longevity in the bloodstream [79]. The process of PEGylation has the potential to enhance the tumor accumulation of IONPs by means of the electron paramagnetic resonance (EPR) phenomenon [80]. Additionally, the process of PEGylation has the potential to augment the internalization of IONPs by mitigating their electrostatic hindrance from the cel-lular membrane. Polyvinyl alcohol (PVA) has been found to enhance the stability of IONPs under different physiological conditions, including variations in pH, temperature, and salt concentration. The utilization of PVA-coated IONPs has demonstrated a notable reduction in toxicity and remarkable biocompatibility, rendering them a viable option for biomedical purposes [81]. Polyvinylpyrrolidone (PVP) has been found to be effective in enhancing the colloidal stability of IONPs in biological fluids, as well as improving their cellular uptake and magnetic properties [82]. The utilization of PVP-coated IONPs has demonstrated favor-able biocompatibility and minimal toxicity, rendering them another viable option for implementation in the field of biomedicine [83–86]. Polysaccharides, namely chitosan, dextran, and hyaluronic acid, have been employed to enhance the bio-compatibility and stability of IONPs through the application of polysaccharide coatings. The application of these coatings can facilitate the precise localization of IONPs to particular cells or tissues, augment their internalization by cells, and mitigate their adverse effects [87]. Among the polysaccharides, starch is the most common and widely used polymer, used since several years ago to deco-rate iron oxide surfaces [88]. The investigation of starch-coated IONPs has been conducted for diverse purposes, including but not limited to drug transporta-tion, imaging, and remediation of the environment [89]. Starch, being a polymer that is both biodegradable and biocompatible, has the potential to enhance the stability and biocompatibility of IONPs. Starch-coated IONPs exhibit enhanced colloidal stability in biological fluids by preventing their aggregation [90]. The application of starch coatings has been observed to have the potential to mitigate the toxicity of IONPs by impeding their interaction with cells and tissues [91]. The exploitation of starch as a coating agent for IONPs is advantageous due to its widespread availability and cost-effectiveness. Starch has the potential to serve as a low-cost substitute for other polymer coatings, as it can be derived from a variety of natural sources including corn, potato, and rice. The application of starch coatings has been found to enhance the magnetic characteristics of IONPs through the mitigation of their agglomeration and the amplification of their surface area. The manipulation of starch concentration and molecular weight can regulate the dimensions and structure of IONPs that are coated with starch [89, 92–94]. According to reports, chitosan exhibits notable biodegradability, biocompatibility, and stability, while also demonstrating low toxicity and immunogenicity in cellular contexts [95]. The amine group present in chitosan is expected to enhance its solubility and hemocompatibility within the intracellular surroundings. The potential impact of chitosan coating on particle surface involves facilitating the opening of tight intracellular junctions, which can lead to an increase in cellular absorption and ultimately enhance therapeutic efficacy [96]. The Fe_3O_4/chitosan composite exhibits enhanced biomedical characteristics in comparison to uncoated Fe_3O_4 nanoparticles, rendering it a highly favorable bio-nanomaterial for a diverse range of biological applications [97]. Alginates are also used for developing polymer coated IONPs for biomedical applications. Alginates are a type of linear polysaccharide that possesses a backbone consisting of (1-4)linked β-d-mannuronic acid (M units) and α-l-guluronic acid (G units). The polymer blocks have the ability to arrange themselves into various configurations consisting of M, G, MG, and GG blocks [98]. Alginates are easily water soluble and efficient in the capping of IONPs by simple physical blending approach [99]. The magnetization and other physical features of magnetic alginate nanocomposites where biodegradability was a major concern were studied by different research groups [100, 101]. Further study was also conducted on the amalgamation of multifunctional plasmonic magnetic nanocomposites composed of iron oxide and gold. The process of thiolation was applied to sodium alginate, followed by grafting with hydrophobic amine-terminated poly butyl methacrylate ($PBMA-NH_2$) [102]. The study demonstrated that the incorporation of hydrophobically modified TSA in nanocomposites encapsulating paclitaxel (PTX) results in a significant increase of 9% in EE over that of nanocomposites coated with TSA alone.

The utilization of starch-coated IONPs has been investigated in diverse biomedical contexts, including but not limited to drug transportation and imaging. The utilization of starch coatings has the potential to facilitate the precise transportation of IONPs to particular cells or tissues through the process of conjugation with diverse biomolecules, including peptides and antibodies. The application of starch-coated IONPs is feasible for MRI owing to their elevated magnetic moment and enhanced colloidal stability [103]. Research has also been conducted on the operation of starch-coated IONPs for environmental purposes, including but not limited to the purification of water and remediation of soil. The application of starch coatings has been found to have a positive impact on the adsorption capacity of IONPs toward a range of pollutants, including heavy metals and organic contaminants.

14.2.3 BASIC CHARACTERIZATION TECHNIQUES

IONPs have been extensively employed in diverse fields, including magnetic data storage, targeted drug delivery, magnetic resonance imaging, and wastewater treatment. The characterization of IONPs with respect to particle size, shape, and distribution of nanoparticles is commonly performed using transmission electron microscopy (TEM) [104]. The methodology entails the transmission of an electron beam through a slender portion of the specimen, followed by the accumulation of the transmitted electrons to generate an image. X-ray diffraction (XRD) is a widely employed analytical method for determining the crystal structure of IONPs [105]. The methodology entails subjecting the specimen to X-ray bombardment and subsequently evaluating the diffraction pattern of the dispersed X-rays. The diffraction pattern can offer insights into the crystal structure and lattice spacing of the IONPs. Dynamic light scattering (DLS) is a method employed to ascertain the size distribution of IONPs present in a liquid suspension [106]. The methodology entails quantifying the variations in light scattering intensity that arise from the stochastic movement of the nanoparticles within the suspension. The Fourier transform infrared (FTIR) spectroscopy method is utilized for the examination of the chemical composition and functional groups of IONPs [107]. The methodology entails the quantification of the absorption of infrared radiation by the specimen, thereby furnishing insights into the chemical bonds and functional groups that are extant in the nanoparticles. Magnetic measurements are employed for the purpose of investigating the magnetic characteristics of IONPs. The methodology entails the quantification of the magnetic moment of the nanoparticles in relation to a magnetic field that is applied. The aforementioned data pertain to the magnetic anisotropy, coercivity, and saturation magnetization characteristics of the IONPs [108–110]. Particles are heterogeneous composites having a magnetic inner core of a specific core size and a coating that can alter their behavior. Hydrodynamic diameter and magnetic core size are two ways to describe nanoparticles and nanoparticle aggregates. For pinpoint accuracy, both are must-haves [111]. Nanoparticles of magnetite exhibiting precise dimensions and saturation magnetization are synthesized, and their shape has significant dominance over the concentration of the nanoparticles, as shown in Figure 14.2. This also shows the effect degradable polymer PVA could have as a non-ionic surfactant and encapsulating agent. In general, the employed characterization techniques furnish significant insights into the physical, chemical, and magnetic attributes of IONPs, which hold significance for their diverse range of applications.

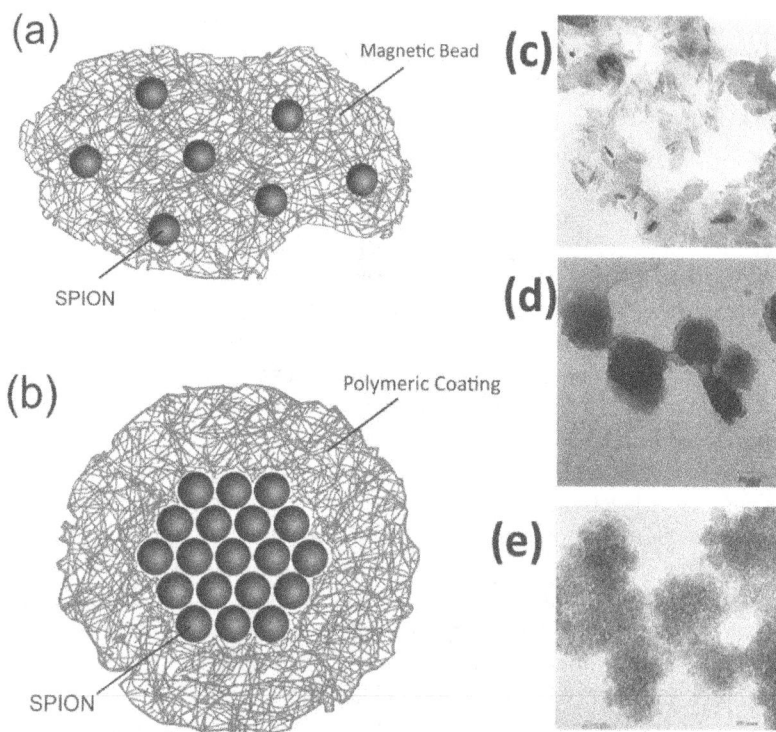

FIGURE 14.2 Formation of (a) magnetic beads and (b) CNCs due to the presence of PVA. TEM images of (c) rod nanoparticles (S5400M1.3), bar size of 80 nm, (d) CNCs (S5400M1.3), bar size of 80 nm, and (e) CNCs (S9000M1.2), bar size of 20 nm. (Reproduced with permission from [111] https://pubs.acs.org/doi/full/10.1021/jp803016n. © 2008 American Chemical Society.)

14.3 BIOMEDICAL APPLICATIONS OF BIODEGRADABLE SURFACE-ENGINEERED IRON OXIDE NANOPARTICLES

The potential applications of surface-engineered iron oxide nanoparticles that are biodegradable are in the field of biomedicine, specifically in drug delivery, magnetic hyperthermia, and magnetic resonance imaging [112]. Biocompatible polymers can be utilized to enhance stability of these entities, while targeting moieties can be functionalized to enable selective binding to afflicted cells and tissues [113]. The distinctive characteristics of these materials render them highly suitable for employment in the field of biomedicine, owing to their magnetic and optical properties, as well as their biocompatibility. Iron oxide nanoparticles have become a promising area of interest in biomedical research for the development of nanomedicines. These nanoparticles possess a diverse range of characteristics and benefits that can be utilized for various applications. The diminutive dimensions of nanoparticles, which typically fall within the range of 1–100 nanometers, represent a prominent characteristic of these entities. The ability of nanomaterials to interact with biological structures at the cellular and molecular level has opened up novel avenues in targeted drug delivery and imaging owing to their nanoscale dimensions. The biocompatibility of iron oxide nanoparticles is a critical factor that determines their applicability in the field of nanomedicine. The nanoparticles demonstrate a low level of toxicity and negligible negative impact on biological tissues, rendering them appropriate for internal administration. The crucial aspect of nanomedicines in clinical settings is their biocompatibility, which is imperative for ensuring their safety and efficacy. The magnetic properties of iron oxide nanoparticles are another noteworthy characteristic. As a result of their intrinsic magnetic properties, these nanoparticles can be controlled and directed through the application of external magnetic fields. The magnetic responsiveness of nanoparticles enables precise drug targeting to specific locations in the body, thereby improving treatment effectiveness and reducing the occurrence of systemic side effects. Iron oxide nanoparticles exhibit exceptional magnetic properties that render them highly appropriate for diverse imaging techniques, with a particular emphasis on MRI. The utilization of nanoparticles facilitates the production of contrast in MRI scans, thereby enhancing the ability to visualize and identify particular disease indicators, such as tumors or inflammatory regions. The capacity to functionalize iron oxide nanoparticles with drug molecules represents a noteworthy advantage in the field of nanomedicine. Targeted and controlled drug release can be accomplished by attaching therapeutic agents to the surface of nanoparticles or encapsulating them within the core of the nanoparticles. This methodology enhances drug stability, augments bioavailability, and facilitates targeted drug administration, thereby mitigating off-target impacts and amplifying therapeutic effectiveness.

14.3.1 Drug Delivery Applications

The distinctive characteristics of IONPs, such as their biocompatibility, magnetic properties, and high surface area-to-volume ratio, have rendered them a subject of interest in the realm of drug delivery [114]. The technique of magnetic targeting involves the functionalization of IONPs with therapeutic agents, which can subsequently be directed to the specific site of the ailment or trauma through the application of an externally generated magnetic field [106, 115]. The method referred to as magnetic targeting enables drugs to amass at the intended location while diminishing their dispersion to sound tissues, thereby mitigating adverse effects [116–119]. The utilization of magnetic targeting proves to be advantageous in the treatment of ailments, specifically cancer, as it enables the precise delivery of drugs to the tumor [120]. Magnetic targeting is predicated on the ability of magnetic nanoparticles to be precisely guided to a designated site within the body through the application of an extrinsic magnetic field. The application of an external magnetic field can direct magnetic nanoparticles to the intended location upon their injection into the bloodstream [121]. The magnetic nanoparticles are drawn by the magnetic field toward the target, where they amass and subsequently discharge the drug. The utilization of magnetic targeting presents various benefits over traditional drug delivery techniques [122]. These advantages include the capacity to administer drugs to particular tissues or organs, decreased systemic toxicity, and enhanced therapeutic effectiveness.

Extensive research has been conducted on the application of magnetic targeting in the field of cancer therapy. Magnetic nanoparticles possess the ability to be endowed with chemotherapeutic drugs or biological agents, such as siRNA or miRNA, and subsequently guided to the tumor site through the application of an external magnetic field. The utilization of magnetic targeting in cancer therapy presents various benefits such as the direct delivery of drugs to the tumor site, dose reduction, and the mitigation of chemotherapy-related side effects [123, 124]. Magnetic targeting has been investigated as a potential therapeutic approach for various medical conditions, such as cardiovascular ailments, neurological dysfunctions, and infectious pathologies. Magnetic nanoparticles have been proposed as a potential drug delivery system for the treatment of cardiovascular diseases, specifically targeting atherosclerotic plaques to mitigate the risk of restenosis. Magnetic nanoparticles have also been proposed as a potential drug delivery system for neurological disorders [125]. This approach involves the use of magnetic nanoparticles to transport drugs across the blood-brain barrier and selectively target specific regions of the brain. Magnetic nanoparticles have been proposed as a potential strategy for targeted delivery of antibiotics in the treatment of infectious diseases [126]. The aforementioned methodology also has the favorable aspect of reducing systemic toxicity and concurrently augmenting therapeutic effectiveness through the targeted transportation of the medication to the infection site.

Despite the potential benefits, magnetic targeting faces several challenges that must be addressed for effective clinical integration. Ensuring the secure delivery of magnetic nanoparticles to individuals represents a noteworthy obstacle that necessitates attention. Performing a comprehensive evaluation of the potential toxicity of magnetic nanoparticles is crucial to prevent any unfavorable impacts on the patient. An additional challenge concerns the development of magnetic targeting techniques that are efficacious in medical settings [127]. The precise regulation of the external magnetic field is imperative to ensure the targeted delivery of magnetic nanoparticles and to prevent any potential harm to non-targeted healthy tissues. In summary, magnetic targeting has emerged as a potentially efficacious approach for drug delivery, exhibiting a multitude of benefits in comparison to traditional drug delivery modalities [128]. Extensive research has been conducted on the application of magnetic targeting in cancer therapy, and its potential use in treating various other medical conditions has also been investigated. Notwithstanding its potential advantages, however, magnetic targeting encounters various obstacles that require resolution for its clinical application [129]. Additional investigation is required to enhance the magnetic targeting mechanism and assess its safety and effectiveness in real-world medical environments.

The controlled release of drugs can be achieved through the functionalization of IONPs with various biomolecules such as polymers and proteins, thereby creating an effective drug delivery system. The IONPs are utilized as a drug delivery system, whereby the release kinetics of the drug can be modulated by a range of factors such as alterations in pH or temperature or the application of an external magnetic field. Thermosensitive polymers can be utilized to functionalize IONPs for the purpose of targeted drug delivery [130]. This is achieved by inducing drug release through exposure to an external heat source. IONPs possess distinctive magnetic characteristics that render them appropriate for employment in both imaging and therapy applications [131]. IONPs have the potential to serve as contrast agents in MRI for imaging purposes. The magnetic characteristics of IONPs enable them to produce a robust \MRI signal, thereby facilitating the in vivo visualization and tracking of IONPs. IONPs have the potential to be utilized in hyperthermia therapy during therapeutic interventions [132–134].

This approach involves the application of an external magnetic field to heat the IONPs, which can subsequently lead to the destruction of cancerous cells [135].

14.4　ANTIBACTERIAL ACTIVITY OF IONPS

IONPs have been shown to have antibacterial activity through a number of different pathways. Oxidative stress in bacteria is caused when IONPs come into contact with bacterial cells and produce reactive oxygen species (ROS) such as hydroxyl radicals and superoxide ions. Bacterial DNA, proteins, and lipid membranes are all harmed by this oxidative stress, hastening their demise. Because of their tiny size and high surface area, IONPs are able to penetrate the structure of the bacterial cell membrane, causing disruption. Because of this breakdown, cellular contents can leak out and ion balances can become unstable. IONPs also prevent the correct operation of enzyme systems that are essential to bacteria's survival. IONPs prevent bacteria from surviving and multiplying by interfering with activities like energy consumption, DNA replication, and protein synthesis. IONPs have the ability to kill bacteria via numerous methods, making them useful against both antibiotic-resistant bacteria and bacterial biofilms. Taking into account considerations such as nanoparticle properties, bacterial species, and potential cytotoxicity is essential to optimize the design and application of IONPs for antibacterial purposes. The antibacterial properties of IONPs have been demonstrated against a range of bacterial strains, encompassing both Gram-positive and Gram-negative classifications. The bactericidal efficacy of IONPs is believed to stem from the production of reactive oxygen species and the interference with bacterial cell membranes. IONPs have the potential to serve as antibacterial agents in various applications, including wound healing [136–138]. The delivery of genes can be facilitated through the functionalization of IONPs with nucleic acids, including but not limited to small interfering RNA (siRNA), microRNA (miRNA), and plasmid DNA. The IONPs serve the purpose of safeguarding the nucleic acids from degradation and also aid in their cellular uptake. The magnetic characteristics of IONPs can be utilized to facilitate the localization of gene-bearing IONPs to particular cells or tissues, thereby enabling precise gene therapy.

14.4.1　Imaging Applications

The magnetic properties of IONPs have made them a popular choice for biomedical imaging applications. This section will center on the utilization of IONPs in the field of imaging and their comparative benefits vis à vis alternative imaging methodologies. Magnetic resonance imaging is a prevalent diagnostic imaging modality that employs a robust magnetic field and radio waves to produce intricate images of internal organs and tissues. IONPs can be designed to have unique magnetic properties, resulting in a distinctive "fingerprint" in the MRI signal. The utilization of magnetic resonance fingerprinting (MRF) enables precise and quantitative identification of regions labeled with IONPs, thereby augmenting the contrast and enhancing the accuracy of diagnoses [139]. Multimodal imaging approaches can be developed by integrating IONPs with other imaging agents or modalities. For example, IONPs can be coupled with fluorescent dyes or radionuclides to enable simultaneous MRI and optical imaging or MRI and positron emission tomography (PET) imaging, respectively. The integration of various imaging modalities results in an improved amalgamation of the distinct information derived from each technique [140]. IONPs have been identified as a potentially valuable contrast agent for MRI owing to their exceptional magnetic moment and superparamagnetic characteristics. The magnetic characteristics of IONPs facilitate the production of a robust contrast signal in MRI, thereby augmenting the imaging's detection sensitivity and specificity. The capacity of IONPs to selectively target tissues or organs is considered a significant benefit in the context of MRI. IONPs have the capability to be functionalized with targeting moieties, including antibodies, peptides, or aptamers, which possess the ability to selectively bind to cells or tissues of interest. The selective accumulation of IONPs in the intended tissue results in an enhanced signal-to-noise ratio and improved quality of imaging. The biocompatibility and low toxicity of IONPs represent additional benefits in the context of imaging. IONPs possess the characteristic of biodegradability, thereby enabling their elimination from the human body through natural metabolic pathways. The employment of this technique mitigates the potential for enduring toxicity linked with alternative contrast agents, such as compounds based on gadolinium. IONPs have been utilized in diverse imaging applications such as cancer imaging, cardiovascular imaging, and neuroimaging. IONPs have potential applications in cancer imaging for the purpose of detecting and monitoring the progression and metastasis of tumors. The surface-targeting ligands of IONPs exhibit a high degree of specificity toward neoplastic cells, thereby facilitating the preferential visualization of malignant tissues. IONPs have the potential to be utilized in cardiovascular imaging for the purpose of detecting atherosclerotic plaques and monitoring the advancement of the disease. The application of IONPs in neuroimaging has been observed to be effective in the identification and tracking of the advancement of neurological conditions such as Parkinson's disease and Alzheimer's disease.

Notwithstanding their numerous benefits, there exist obstacles that necessitate resolution for the clinical implementation of IONPs in medical imaging. Enhancing the imaging properties and targeting efficiency of IONPs represents a significant obstacle to their development. An additional obstacle pertains to the optimization of imaging protocols in order to guarantee that the contrast signal produced by IONPs is adequate for precise diagnosis. Overall, IONPs exhibit magnetic properties, targeting abilities, biocompatibility, and low toxicity, rendering them an advantageous contrast agent for biomedical imaging. IONPs have been utilized in various imaging applications and possess the capability to enhance the diagnosis and treatment of numerous ailments. Additional investigation is required to enhance the characteristics of IONPs and assess their safety and effectiveness in medical environments.

14.4.2 THERANOSTIC APPLICATIONS

The integration of therapeutic and diagnostic functionalities of IONPs has resulted in the emergence of theranostic platforms, which possess the ability to simultaneously diagnose and treat various ailments. IONPs have the potential to serve as a means of targeting cancer cells and administering medication, while also enabling the visualization of the tumor through magnetic resonance imaging or magnetic particle imaging (MPI). IONPs have the potential to serve as contrast agents in MRI, thereby facilitating the identification and depiction of anatomical structures that may be challenging to visualize using conventional imaging techniques. The robust magnetic characteristics exhibited by IONPs produce a contrasting signal that has the potential to distinguish between healthy and pathological tissues. In one study, iron oxide nanoparticles and gold nanoparticles (GNPs) were prepared by ultrasound-assisted and controlled seeded growth synthetic methods, respectively [141]. The findings from the experiment indicated a significant inhibition of cancer cell proliferation. Furthermore, the albumin-coated GNPs and IONPs did not exhibit any signs of toxicity. The cellular uptake and efficiency of AS1411 aptamer–functionalized nanoparticles were found to be superior to those of non-targeting nanoparticles in MCF7 breast cancer cells. This was attributed to the high affinity of the aptamer toward the overexpressed nucleolin on the cell surface of MCF7. Ligand-targeted surface-modified iron nanoparticles (INOPs) have been extensively studied in the context of brain diseases by numerous researchers [142]. MPI is a novel imaging modality that utilizes the magnetic properties of IONPs to generate accurate and intricate visual depictions of biological structures. MPI exhibits considerable potential as a viable alternative to MRI and other imaging modalities, owing to its heightened sensitivity and non-ionizing radiation properties.

The feasibility of using IONPs for hyperthermia applications has been extensively investigated. The process involves the elevation of tumor tissue temperatures beyond 42°C to trigger cell death. The distinctive magnetic characteristics exhibited by IONPs render them a highly suitable contender for employment in hyperthermia-based interventions. The present discourse aims to delve deeper into the utilization of IONPs for hyperthermia purposes. The process of inducing hyperthermia through the application of IONPs entails the precise transportation of IONPs to the affected tumor tissue, which is subsequently subjected to an alternating magnetic field (AMF) to produce thermal energy. The IONPs function as intermediaries, transforming the electromagnetic energy derived from the AMF into thermal energy, thereby inducing a rise in temperature in the tumor tissue. Cell death induced by hyperthermia is caused by various factors such as protein denaturation, lipid peroxidation, and DNA damage. Various categories of IONPs have been employed for hyperthermia purposes, such as superparamagnetic iron oxide nanoparticles (SPIONs), magnetite nanoparticles, and maghemite nanoparticles. The IONPs exhibit distinctive physical and chemical characteristics that can be adjusted to attain the most favorable hyperthermia outcomes. The selective targeting of nanoparticles to tumor tissue is a crucial factor in achieving successful hyperthermia through the use of IONPs. Several targeting strategies have been devised to attain this objective, such as the utilization of antibodies, peptides, and aptamers. The targeting moieties exhibit specificity in binding to tumor cells, thereby enabling the discriminatory transportation of IONPs to the tumor tissue. The utilization of an AMF is crucial for the efficacious implementation of hyperthermia via IONPs. By tuning the frequency and amplitude of the AMF, the optimal heating of tumor tissue can be achieved while the heating of surrounding healthy tissues is minimized. The temporal extent of the application of AMF is a significant determinant in the occurrence of cell death resulting from hyperthermia. This method has various benefits in comparison to conventional methods of hyperthermia treatment, including radio frequency ablation and microwave ablation. The mode of application of IONP-mediated hyperthermia has been observed to possess non-invasive properties, to necessitate less high temperature levels, and to exhibit enhanced capacity for localized heating. Furthermore, there exists a possibility of utilizing it in conjunction with other therapeutic modalities, such as chemotherapy and radiation therapy, to attain superior treatment results. IONPs exhibit distinctive magnetic characteristics, rendering them highly promising for hyperthermia applications. The precise localization of IONPs to the neoplastic tissue, coupled with the subsequent administration of an AMF, can elicit apoptotic cell death via hyperthermia. Using IONPs in hyperthermia presents several advantages over conventional hyperthermia techniques and holds promise for synergistic employment with other therapeutic modalities to enhance treatment efficacy.

Magnetic resonance angiography is a non-invasive medical imaging technique that utilizes MRI to generate high-resolution images of the vascular system, with a particular focus on arterial structures. This technology facilitates the identification of vascular pathologies, including but not limited to stenosis and occlusion. In contrast to non-enhanced MR angiography, contrast-enhanced MR angiography confers various benefits such as increased reproducibility and independence from blood flow. The use of contrast-enhanced magnetic resonance angiography (CE-MRA) in conjunction with standard extracellular contrast material has been widely accepted as a reliable method for vascular imaging. The clinical availability of the first blood pool contrast agent (BPA) has been reported in recent times [143]. VSOPs, which are superparamagnetic iron oxide particles with a hydrodynamic diameter of approximately 7 nm and a brief blood half-life, have been employed in T1-weighted contrast-enhanced MR angiography. The evaluation of their performance was conducted in both preclinical [144] and clinical studies [145]. Despite the formulation's favorable pharmacokinetic properties for vascular imaging and its high tolerability, it was not subjected to further clinical trials. The study demonstrated that the utilization of iron oxide nanoparticles in magnetic resonance angiography yields better results than computed tomography angiography in identifying endoleaks. Utilizing MRA instead of CT confers an additional benefit in that patients are not subjected to ionizing radiation, thereby enabling longitudinal monitoring in a more advantageous manner [146]. In addition to ferumoxytol, clinical trials have assessed the efficacy of ferumoxtran and feruglose (Clariscan) for vascular imaging purposes. Feruglose was utilized in the context of coronary angiography, facilitating the identification of coronary artery stenosis and the assessment of coronary artery bypass efficacy [147].

14.5   BIOCOMPATIBILITY AND TOXICITY OF BIODEGRADABLE SURFACE-ENGINEERED IONPS

The potential biocompatibility and biodegradability of surface-engineered iron oxide nanoparticles that are biodegradable have garnered considerable attention in the field of biomedical applications. Nevertheless, it is crucial to assess their toxicity prior to their clinical application. Broadly speaking, biocompatibility pertains to the capacity of a substance to coexist harmoniously with organic tissues, devoid of any deleterious consequences. The biocompatibility of biodegradable IONPs can be determined by their ability to undergo degradation and elimination from the body without inducing any adverse effects on the adjacent tissues. The biocompatibility of these particles can be further improved through surface engineering with biocompatible polymers or coatings.

14.5.1 IN VITRO STUDIES

In vitro investigations are a crucial methodology for assessing the biocompatibility and toxicity of surface-engineered IONPs that are biodegradable. These investigations entail subjecting cells or tissues to the particles and assessing their reaction. To assess biocompatibility, researchers can evaluate the impact of particles on cell viability, proliferation, and morphology. If the particles exhibit biocompatibility, their impact on cellular morphology ought to be negligible and they should not induce substantial cell death. The assessment of the inflammatory response of cells toward particles can be conducted by researchers through the quantification of the release of inflammatory cytokines. To assess the toxicity of particles, researchers can evaluate their impact on diverse cellular processes, including cell viability, proliferation, and apoptosis. In addition, the ROS produced by the particles and their genotoxicity potential can be quantified. In vitro investigations can aid researchers in refining the surface engineering of particles to augment their biocompatibility and mitigate their toxicity. One potential avenue of investigation involves the assessment of various surface coatings or modifications and their impact on the biocompatibility and toxicity of the particles. Surface-engineered graphene oxide (GO) nanosheets were developed by More et al. through covalent linking with amine-functionalized IONPs, resulting in the formation of GOIOIs [148]. The model drug gefitinib (Gf) was entrapped within exfoliated GO sheets (GOIGF) through π-π* stacking, followed by functionalization with IONPs. The study conducted an in vitro assay to evaluate the cytotoxicity of cells on MDA-MB-231 breast cancer adenocarcinoma cell lines. The results of the cell cytotoxicity assay indicated that a concentration of 30 ppm containing 64% of the drug brought about a 61.18% inhibition of cell growth. In comparison, 100% of the pure drug resulted in only a 56% inhibition. The combination of human serum albumin (HSA) and IONPs has been documented as a means of producing mono-, bi-, and tri-modal IO/HSA core-shell NPs for diagnostic purposes [26, 149]. This approach has been shown to enhance the accuracy and precision of signals obtained from diagnostic machines. The cytotoxicity of IONPs was examined through the implementation of a 48-hour XTT cell viability assay. The study involved the incubation of HCT116 (human colon adenocarcinoma), MCF7 (human breast cancer), and A172 (human glioblastoma) cells (10^5 cells) with tri-modal IO/HSA NPs at a final concentration of ~1 mg/mL. Lipid-based formulations have emerged as a popular nanotechnology platform in various biomedical applications due to their favorable characteristics such as biocompatibility, reduced toxicity, ease of surface modifications, and efficient loading of drugs, biologics, and nanoparticles. The utilization of magnetic nanoparticles, specifically iron oxide, in conjunction with lipid-based formulations has demonstrated significant benefits in various applications, such as magnet-mediated drug delivery and image-guided therapy. Kim et al. presented a method for synthesizing drug-loaded magnetic multi-micelle aggregates (MaMAs) in a single step (Figure 14.3) [150]. This technique involves the controlled flow infusion of a mixture of iron oxide nanoparticles and lipids into an aqueous solution of drugs while subjecting the mixture to ultrasonication.

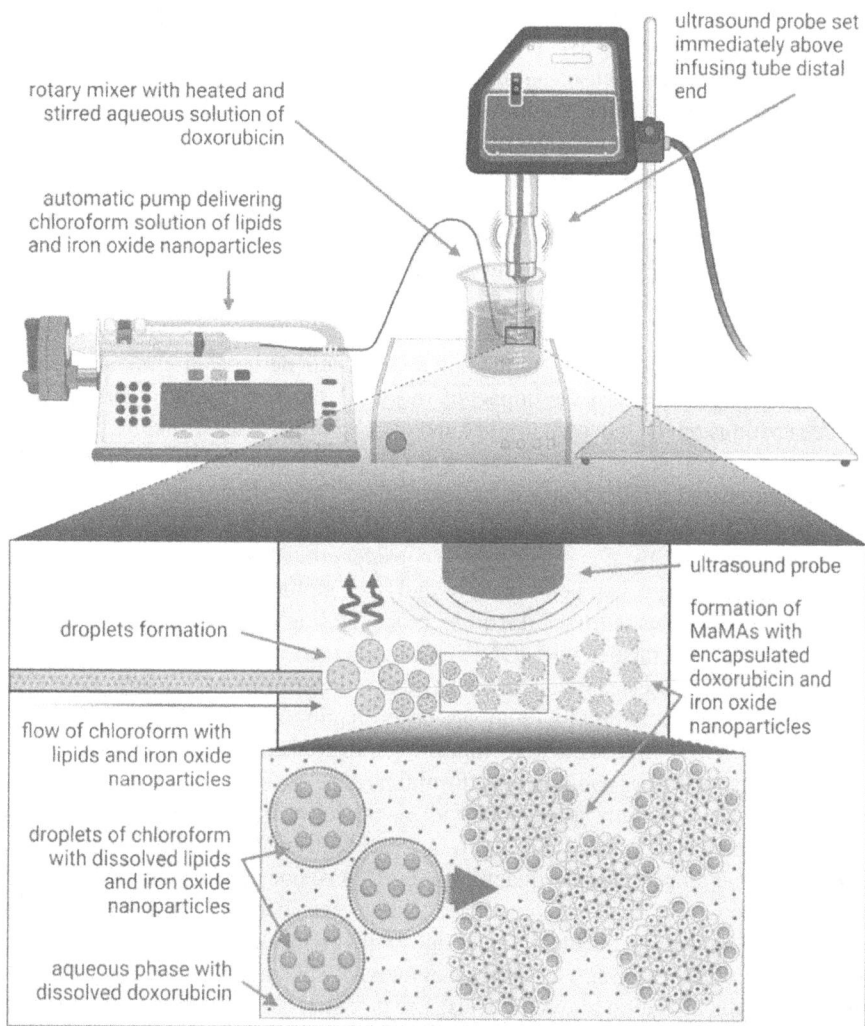

FIGURE 14.3 Drug-loaded magnetic lipid-based formulations synthesized in one pot: the lab and formation process (top). Controlled flow infusion of an IONPs/lipid mixture into an ultrasonicated doxorubicin solution produced magnetic MaMAs [150].

14.5.2 IN VIVO STUDIES

Conducting in vivo investigations is a crucial stage in the assessment of the biocompatibility and toxicity of surface-engineered IONPs that are biodegradable [151]. Such studies can furnish valuable insights into the conduct of these particles within a living organism. The utilization of in vivo studies can aid in the assessment of both the safety and the efficacy of particles in relation to their intended biomedical application [152]. The standard procedure for in vivo investigations entails the introduction of particles into animal models, followed by the observation of their impact on diverse physiological parameters [153]. The selection of an appropriate animal model holds significant importance in ensuring the relevance and validity of outcomes pertaining to the intended clinical application of particles. In order to assess biocompatibility, scholars may observe physiological indicators, including body mass, blood chemistry, and organ histopathology [154]. The measurement of inflammatory cytokine levels in the blood is a viable method for evaluating the immune response of animal models. To assess the toxicity of particles, researchers can evaluate their impact on diverse physiological processes, including but not limited to oxidative stress, apoptosis, and organ toxicity [155]. The researchers can additionally assess the enduring impacts of the particles on the animal models. The utilization of in vivo studies can aid in the enhancement of particle formulations to suit their designated clinical application. Researchers can assess the impact of various dosages, methods of delivery, and surface coatings on the biocompatibility and toxicity of particles. Conducting in vivo studies is a crucial stage in the assessment of the biocompatibility and toxicity of surface-engineered IONPs that are biodegradable. These studies can provide valuable insights that can then be used to enhance the suitability of the IONPs for use in biomedical applications. It is imperative to consider ethical issues and guarantee that the animal models are handled ethically and in compliance with regulations. In a recent study, researchers utilized doxorubicin-loaded single-walled carbon nanotubes (SWCNTs) that were labeled with IONPs and subsequently conjugated with endoglin/CD105 antibodies. This approach facilitated active targeting and provided a theranostic strategy that incorporated bioluminescence imaging (BLI) and MRI [156]. The utilization of IONP-SWCNT nanocomposites as drug delivery vehicles for doxorubicin administration has the potential to mitigate the toxicity and superfluous side effects that are commonly associated with nonselective biodistribution. The study utilized synthesized IONPs-SWCNTs to conjugate with endoglin/CD105 antibody via EDC-NHS coupling. Colloidal quantum dots (CQDs) conjugated with IONPs have been employed in the field of biomedicine and are considered promising contenders for advanced optical imaging techniques due to their remarkable luminescent properties and photostability [157]. The intrinsic cytotoxicity of magnetic CQDs is primarily attributed to the liberation of free radicals. The potential hazards associated with bare IONPs in clinical translation can be attributed to the release of free radicals into the bloodstream. As a result, the significance of surface coatings cannot be overstated [158]. The anionic biopolymer known as γ-PGA possesses natural properties that render it suitable for use as both a carbon and a nitrogen precursor. Additionally, it exhibits favorable attributes such as good water solubility and biocompatibility when incorporated into inorganic nanocomposites.

14.6 DEGRADATION PATHWAYS FOR IONPS IN BIOLOGICAL SYSTEMS

In order to fully comprehend the biological identity of IONPs, it is crucial to have a thorough understanding of the intricate molecular and functional machinery that is fueled by labile iron and ROS. This knowledge highlights the significance of comprehending IONP degradation. The degradation byproducts of IONP have the potential to alter cell responses by acting as an external source for intracellular iron and ROS. This can have significant implications for the overall impact of IONPs on biological systems. In this review article, we will be discussing the biodegradation of IONPs and how this process contributes to their biological identity. This section aims to provide a comprehensive overview of the topic. In the field of nanoparticle research, the degradation of IONPs is a topic of great interest. There are several factors that can influence the degradation process, but three main factors stand out: the protein corona, endocytosis routes, and cellular degradation machinery. These factors have been extensively studied and are known to play a crucial role in determining the fate of IONPs in biological systems. The perception of IONPs in a biological system is determined by various factors, which in turn influence the response of the biological system to the IONPs (Figure 14.2). Long-term biodegradation investigations have found that inoculated IONPs lose their magnetic characteristics, which correlates with a rise in iron metabolic routes typical of a degradation process [159]. The methods by which IONPs are internalized by cells and how this affects their breakdown and subsequent release of iron into the cytoplasm determine how much iron can be used by the cell. Nanomaterial degradation is caused by acid hydrolases and cathepsins, among other degrading enzymes, in the acidic, highly ionic environment of phagolysosomes (pH 4.5–5.5) [160]. (See Table 14.2.)

TABLE 14.2

Different types of degradation pathways and their mechanism (for IONPs)

Degradation Pathway Type	Mechanism	Ref.
Dissolution	• IONPs can undergo dissolution, where the iron oxide core dissociates into iron ions (Fe^{2+} or Fe^{3+}) in the presence of acidic or reducing conditions. • The released iron ions can then interact with biological molecules.	[161, 162]
Oxidative Dissolution	• Under oxidative conditions, IONPs can dissolve via a process called oxidative dissolution. • ROS, such as hydrogen peroxide (H_2O_2), can react with the iron oxide core, leading to the release of iron ions.	[161, 163, 164]
Surface Erosion	• IONPs may undergo surface erosion, where the outer layer of the nanoparticles gradually degrades, resulting in the release of iron ions. • This process can be influenced by pH, temperature, and the presence of biomolecules.	[165]
Biodegradation	• Biological entities, such as enzymes, can enzymatically degrade IONPs. • For instance, certain enzymes like peroxidases or reductases can catalyze the degradation of IONPs by interacting with the nanoparticle surface or reducing the iron oxide core.	[166]
Aggregation and Clearance	• IONPs can aggregate in biological environments due to various factors, such as protein corona formation. • Aggregated IONPs may then be recognized and cleared by the mononuclear phagocyte system (MPS), primarily in the liver and spleen. • Clearance may occur through endocytosis, followed by lysosomal degradation.	[167]

(Continued)

TABLE 14.2 (*Continued*)
Different types of degradation pathways and their mechanism (for IONPs)

Degradation Pathway Type	Mechanism	Ref.
Enzymatic Biodegradation	• IONPs can be enzymatically degraded by various enzymes present in biological systems. • Enzymes like catalase, peroxidase, or reductase can interact with the nanoparticle surface or the iron oxide core, leading to its degradation. • Enzymatic biodegradation is influenced by factors such as enzyme specificity and concentration.	[168, 169]
Intracellular Lysosomal Degradation	• Internalized IONPs may be trafficked to lysosomes within cells. • Lysosomes contain enzymes capable of breaking down the nanoparticle structure, resulting in the release of iron ions. • The degradation process in lysosomes involves acidic conditions and the action of hydrolytic enzymes.	[170]
Reactive Oxygen Species–Mediated Degradation	• In the presence of ROS, such as hydrogen peroxide (H_2O_2) or superoxide radicals ($O_2^{\cdot-}$), IONPs can undergo oxidative degradation. • ROS can interact with the iron oxide core, inducing chemical reactions that lead to the release of iron ions. • This pathway is often associated with inflammation and oxidative stress.	[171–173]
Surface Erosion and Dissolution	• IONPs may undergo surface erosion or dissolution, where the outer layer of the nanoparticles degrades over time. • This process can be influenced by factors such as pH, temperature, and the composition of the surrounding biological fluid. • Surface erosion or dissolution can result in the release of iron ions into the biological system.	[174]
Clearance via Mononuclear Phagocyte System (MPS)	• In biological systems, IONPs can undergo opsonization, where proteins adsorb onto the nanoparticle surface, forming a protein corona. • Opsonized IONPs can be recognized and cleared by the MPS, primarily in the liver and spleen. • Clearance may involve endocytosis, followed by degradation within lysosomes.	[175–177]

14.7   FUTURE PERSPECTIVES AND CONCLUSIONS

Iron oxide nanoparticles that are surface engineered to be biodegradable exhibit significant promise for various biomedical applications, including drug delivery. Notwithstanding, there exist several obstacles and prospective avenues that necessitate attention. It is imperative to conduct a comprehensive investigation into the biocompatibility and toxicity of these nanoparticles to guarantee their efficacy and safety in vivo. Furthermore, it is imperative that they exhibit stability and undergo controlled degradation to avoid any potential buildup within the organism. Achieving targeted delivery to cells or tissues is a significant obstacle that can be overcome by employing surface-engineering techniques involving ligands that possess the ability to recognize specific receptors. The successful application of drug delivery necessitates the consideration of effective drug loading and release properties. In addition, the establishment of scalable techniques for the synthesis and manufacture of said nanoparticles is imperative to facilitate their extensive application. The assurance of safety and efficacy in biomedical applications necessitates regulatory approval. Prospective avenues of research include the advancement of multifaceted nanoparticles and the combination of artificial intelligence and machine learning methodologies to enhance their characteristics and utilization in the domain of biomedical investigation.

To summarize, the utilization of biodegradable surface-engineered iron oxide nanoparticles exhibits promising prospects for transforming drug delivery and other biomedical applications. Notwithstanding, there exist a few obstacles that must be surmounted, such as issues pertaining to biocompatibility and toxicity, stability and degradation, targeted delivery, drug loading and release, scale-up and production, and regulatory approval. To tackle these obstacles, it is imperative to conduct further research and development to enhance the characteristics of these nanoparticles and their utilization in the field of biomedical research. Prospective avenues of research involve the advancement of multifaceted nanoparticles and the combination of artificial intelligence and machine learning methodologies to expedite their progression and enhance their efficacy. Notwithstanding the obstacles, the utilization of biodegradable surface-engineered iron oxide nanoparticles exhibits significant potential for enhancing human health and propelling the domain of biomedical investigation.

REFERENCES

1. Vangijzegem, T., D. Stanicki, and S. Laurent, *Magnetic iron oxide nanoparticles for drug delivery: Applications and characteristics.* Expert Opinion on Drug Delivery, 2019. **16**(1): p. 69–78.
2. Chertok, B., et al., *Iron oxide nanoparticles as a drug delivery vehicle for MRI monitored magnetic targeting of brain tumors.* Biomaterials, 2008. **29**(4): p. 487–496.
3. Ma, X., et al., *A functionalized graphene oxide-iron oxide nanocomposite for magnetically targeted drug delivery, photothermal therapy, and magnetic resonance imaging.* Nano Research, 2012. **5**(3): p. 199–212.
4. Stanicki, D., et al., *An update on the applications and characteristics of magnetic iron oxide nanoparticles for drug delivery.* Expert Opinion on Drug Delivery, 2022. **19**(3): p. 321–335.
5. Soetaert, F., et al., *Cancer therapy with iron oxide nanoparticles: Agents of thermal and immune therapies.* Advanced Drug Delivery Reviews, 2020. **163–164**: p. 65–83.
6. Heneweer, C., S.E. Gendy, and O. Peñate-Medina, *Liposomes and inorganic nanoparticles for drug delivery and cancer imaging.* Therapeutic Delivery, 2012. **3**(5): p. 645–656.
7. Pattni, B.S., V.V. Chupin, and V.P. Torchilin, *New developments in liposomal drug delivery.* Chemical Reviews, 2015. **115**(19): p. 10938–10966.
8. Amstad, E., et al., *Triggered release from liposomes through magnetic actuation of iron oxide nanoparticle containing membranes.* Nano Letters, 2011. **11**(4): p. 1664–1670.

9. Pedersen, B.K., and B. Saltin, *Exercise as medicine – Evidence for prescribing exercise as therapy in 26 different chronic diseases.* Scandinavian Journal of Medicine & Science in Sports, 2015. **25**(S3): p. 1–72.

10. Thu, V.T.A., et al., *Advancing personalized medicine for tuberculosis through the application of immune profiling.* Frontiers in Cellular and Infection Microbiology, 2023. **13**: p. 68.

11. Chatzopoulou, F., et al., *Dissecting miRNA–gene networks to map clinical utility roads of pharmacogenomics-guided therapeutic decisions in cardiovascular precision medicine.* Cells, 2022. **11**(4): p. 607.

12. Huang, Q., et al., *Passively-targeted mitochondrial tungsten-based nanodots for efficient acute kidney injury treatment.* Bioactive Materials, 2023. **21**: p. 381–393.

13. Devi, K.V., and R.S. Pai, *Antiretrovirals: Need for an effective drug delivery.* Indian Journal of Pharmaceutical Sciences, 2006. **68**(1): p. 1.

14. Sun, Q., et al., *A self-sustaining antioxidant strategy for effective treatment of myocardial infarction.* Advanced Science, 2023. **10**(5): p. 2204999.

15. Huang, Q., et al., *Selenium nanodots (SENDs) as antioxidants and antioxidant-prodrugs to rescue islet β cells in type 2 diabetes mellitus by restoring mitophagy and alleviating endoplasmic reticulum stress.* Advanced Science. **n/a**(n/a): p. 2300880.

16. Clark, K.E., et al., *Sequestration and cyanobacterial diet preferences in the opisthobranch molluscs dolabrifera nicaraguana and stylocheilus rickettsi.* Frontiers in Marine Science, 2021. **8**: p. 766282.

17. Abdelnour, S.A., et al., *Nanominerals: Fabrication methods, benefits and hazards, and their applications in ruminants with special reference to selenium and zinc nanoparticles.* Animals, 2021. **11**(7): p. 1916.

18. White, E., et al., *Role of autophagy in suppression of inflammation and cancer.* Current Opinion in Cell Biology, 2010. **22**(2): p. 212–217.

19. Pandey, A., D.S. Jain, and S. Chakraborty, *Poly lactic-co-glycolic acid (PLGA) copolymer and its pharmaceutical application.* Handbook of Polymers for Pharmaceutical Technologies, 2015. **2**: p. 151–172.

20. Liu, Y., et al., *Comprehensive insights into the multi-antioxidative mechanisms of melanin nanoparticles and their application to protect brain from injury in ischemic stroke.* Journal of the American Chemical Society, 2017. **139**(2): p. 856–862.

21. Huang, Q., et al., *Oral Metal-free melanin nanozymes for natural and durable targeted treatment of inflammatory bowel disease (IBD).* Small, 2023. **19**(19): p. 2207350.

22. Kurlyandskaya, G.V., et al., *Water-based suspensions of iron oxide nanoparticles with electrostatic or steric stabilization by chitosan: Fabrication, characterization and biocompatibility.* Sensors, 2017. **17**(11): p. 2605.

23. Gaspar, A.S., et al., *Biocompatible and high-magnetically responsive iron oxide nanoparticles for protein loading.* Journal of Physics and Chemistry of Solids, 2019. **134**: p. 273–285.

24. Huang, Q., et al., *Oral metal-free melanin nanozymes for natural and durable targeted treatment of inflammatory bowel disease (IBD).* Small, 2023. **19**: p. 2207350.

25. Levy, I., et al., *Bioactive magnetic near infra-red fluorescent core-shell iron oxide/human serum albumin nanoparticles for controlled release of growth factors for augmentation of human mesenchymal stem cell growth and differentiation.* Journal of Nanobiotechnology, 2015. **13**(1): p. 1–14.

26. Ganguly, S., I. Grinberg, and S. Margel, *Layer by layer controlled synthesis at room temperature of tri-modal (MRI, fluorescence and CT) core/shell superparamagnetic IO/human serum albumin nanoparticles for diagnostic applications.* Polymers for Advanced Technologies, 2021. **32**(10): p. 3909–3921.

27. Zhang, P., et al., *A green biochar/iron oxide composite for methylene blue removal.* Journal of Hazardous Materials, 2020. **384**: p. 121286.

28. Schwaminger, S.P., et al., *Magnetically induced aggregation of iron oxide nanoparticles for carrier flotation strategies.* ACS Applied Materials & Interfaces, 2021. **13**(17): p. 20830–20844.

29. Qiao, K., et al., *Application of magnetic adsorbents based on iron oxide nanoparticles for oil spill remediation: A review.* Journal of the Taiwan Institute of Chemical Engineers, 2019. **97**: p. 227–236.

30. Singh, P., et al., *Systematic review on applicability of magnetic iron oxides–integrated photocatalysts for degradation of organic pollutants in water.* Materials Today Chemistry, 2019. **14**: p. 100186.

31. Samrot, A.V., et al., *A review on synthesis, characterization and potential biological applications of superparamagnetic iron oxide nanoparticles.* Current Research in Green and Sustainable Chemistry, 2021. **4**: p. 100042.

32. Ling, D., and T. Hyeon, *Chemical design of biocompatible iron oxide nanoparticles for medical applications.* Small, 2013. **9**(9–10): p. 1450–1466.

33. Gharieh, A., S. Khoee, and A.R. Mahdavian, *Emulsion and miniemulsion techniques in preparation of polymer nanoparticles with versatile characteristics.* Advances in Colloid and Interface Science, 2019. **269**: p. 152–186.

34. Tufani, A., A. Qureshi, and J.H. Niazi, *Iron oxide nanoparticles based magnetic luminescent quantum dots (MQDs) synthesis and biomedical/biological applications: A review.* Materials Science and Engineering: C, 2021. **118**: p. 111545.

35. Kazemi, F., et al., *Oil-in-water emulsion separation by PVC membranes embedded with GO-ZnO nanoparticles.* Journal of Environmental Chemical Engineering, 2021. **9**(1): p. 104992.

36. Yousuf, M.A., et al., *Magnetic and electrical properties of yttrium substituted manganese ferrite nanoparticles prepared via micro-emulsion route.* Results in Physics, 2020. **16**: p. 102973.

37. Deng, C., et al., *Exclusively catalytic oxidation of toluene to benzaldehyde in an O/W emulsion stabilized by hexadecylphosphate acid terminated mixed-oxide nanoparticles.* Chinese Journal of Catalysis, 2020. **41**(2): p. 341–349.

38. Sundaram, P.A., R. Augustine, and M. Kannan, *Extracellular biosynthesis of iron oxide nanoparticles by bacillus subtilis strains isolated from rhizosphere soil.* Biotechnology and Bioprocess Engineering, 2012. **17**: p. 835–840.

39. Riaz, S., M. Bashir, and S. Naseem, *Iron oxide nanoparticles prepared by modified co-precipitation method.* IEEE Transactions on Magnetics, 2013. **50**(1): p. 1–4.

40. Slimani, S., et al., *Spinel iron oxide by the co-precipitation method: Effect of the reaction atmosphere.* Applied Sciences, 2021. **11**(12): p. 5433.

41. Kayani, Z.N., et al., *Synthesis of iron oxide nanoparticles by sol–gel technique and their characterization.* IEEE Transactions on Magnetics, 2014. **50**(8): p. 1–4.

42. Da Costa, G., et al., *Synthesis and characterization of some iron oxides by sol-gel method.* Journal of Solid State Chemistry, 1994. **113**(2): p. 405–412.

43. Ennas, G., et al., *Characterization of iron oxide nanoparticles in an Fe2O3– SiO2 composite prepared by a sol– gel method.* Chemistry of Materials, 1998. **10**(2): p. 495–502.

44. Duhan, S., and S. Devi, *Synthesis and structural characterization of iron oxide-silica nanocomposites prepared by the sol gel method.* International Journal of Electronics Engineering, 2010. **2**(1): p. 89–92.8.

45. Ansari, S.A.M.K., et al., *Magnetic iron oxide nanoparticles: Synthesis, characterization and functionalization for biomedical applications in the Central nervous system.* Materials, 2019. **12**(3): p. 465.

46. Varanda, L.C., et al., *Size and shape-controlled nanomaterials based on modified polyol and thermal decomposition approaches. A brief review.* Anais da Academia Brasileira de Ciências, 2019. **91**: p. e20181180.

47. Khalilzadeh, M.A., et al., *Green synthesis of magnetic nanocomposite with iron oxide deposited on cellulose nanocrystals with copper (Fe3O4@ CNC/Cu): Investigation of catalytic activity for the development of a venlafaxine electrochemical sensor.* Industrial & Engineering Chemistry Research, 2020. **59**(10): p. 4219–4228.

48. Zhao, S., et al., *Multifunctional magnetic iron oxide nanoparticles: An advanced platform for cancer theranostics.* Theranostics, 2020. **10**(14): p. 6278.

49. Ozel, F., H. Kockar, and O. Karaagac, *Growth of iron oxide nanoparticles by hydrothermal process: Effect of reaction parameters on the nanoparticle size.* Journal of Superconductivity and Novel Magnetism, 2015. **28**: p. 823–829.

50. Joshi, M.K., et al., *One-pot synthesis of Ag-iron oxide/reduced graphene oxide nanocomposite via hydrothermal treatment.* Colloids and Surfaces A: Physicochemical and Engineering Aspects, 2014. **446**: p. 102–108.

51. Lorkit, P., M. Panapoy, and B. Ksapabutr, *Iron oxide-based supercapacitor from ferratrane precursor via sol–gel–hydrothermal process.* Energy Procedia, 2014. **56**: p. 466–473.

52. Behdadfar, B., et al., *Synthesis of high intrinsic loss power aqueous ferrofluids of iron oxide nanoparticles by citric acid-assisted hydrothermal-reduction route.* Journal of Solid State Chemistry, 2012. **187**: p. 20–26.

53. Ferraz, L.C., et al., *Vertically oriented iron oxide films produced by hydrothermal process: Effect of thermal treatment on the physical chemical properties.* ACS Applied Materials & Interfaces, 2012. **4**(10): p. 5515–5523.

54. Chin, A.B., and I.I. Yaacob, *Synthesis and characterization of magnetic iron oxide nanoparticles via w/o microemulsion and Massart's procedure.* Journal of Materials Processing Technology, 2007. **191**(1–3): p. 235–237.

55. Santra, S., et al., *Synthesis and characterization of silica-coated iron oxide nanoparticles in microemulsion: The effect of nonionic surfactants.* Langmuir, 2001. **17**(10): p. 2900–2906.

56. Martínez-Cabanas, M., et al., *Green synthesis of iron oxide nanoparticles. Development of magnetic hybrid materials for efficient As (V) removal.* Chemical Engineering Journal, 2016. **301**: p. 83–91.

57. Arularasu, M., J. Devakumar, and T. Rajendran, *An innovative approach for green synthesis of iron oxide nanoparticles: Characterization and its photocatalytic activity.* Polyhedron, 2018. **156**: p. 279–290.

58. Jamzad, M., and M. Kamari Bidkorpeh, *Green synthesis of iron oxide nanoparticles by the aqueous extract of Laurus nobilis L. Leaves and evaluation of the antimicrobial activity.* Journal of Nanostructure in Chemistry, 2020. **10**: p. 193–201.

59. Ehrampoush, M.H., et al., *Cadmium removal from aqueous solution by green synthesis iron oxide nanoparticles with tangerine peel extract.* Journal of Environmental Health Science and Engineering, 2015. **13**: p. 1–7.

60. Abdullah, J.A.A., et al., *Green synthesis and characterization of iron oxide nanoparticles by pheonix dactylifera leaf extract and evaluation of their antioxidant activity.* Sustainable Chemistry and Pharmacy, 2020. **17**: p. 100280.

61. Díez-Pascual, A.M., *Surface engineering of nanomaterials with polymers, biomolecules, and small ligands for nanomedicine.* Materials, 2022. **15**(9): p. 3251.

62. Al-Qodami, B.A., et al., *Surface engineering of nanotubular ferric oxyhydroxide "goethite" on platinum anodes for durable formic acid fuel cells.* International Journal of Hydrogen Energy, 2022. **47**(1): p. 264–275.

63. Ren, S., et al., *Interface-confined surface engineering via photoelectrochemical etching toward solar neutral water splitting.* ACS Catalysis, 2022. **12**(3): p. 1686–1696.

64. Tabassum, H., et al., *Surface engineering of Cu catalysts for electrochemical reduction of CO2 to value-added multi-carbon products.* Chem Catalysis, 2022. **2**(7), p. 1561–1593.

65. Rajaji, U., et al., *Surface engineering of 3D spinel Zn3V2O8 wrapped on sulfur doped graphitic nitride composites: Investigation on the dual role of electrocatalyst for simultaneous detection of antibiotic drugs in biological fluids.* Composites Part B: Engineering, 2022. **242**: p. 110017.

66. Gamucci, O., et al., *Biomedical nanoparticles: Overview of their surface immune-compatibility.* Coatings, 2014. **4**(1): p. 139–159.

67. Barreto, J.A., et al., *Nanomaterials: Applications in cancer imaging and therapy.* Advanced Materials, 2011. **23**(12): p. H18–H40.

68. Kievit, F.M., and M. Zhang, *Surface engineering of iron oxide nanoparticles for targeted cancer therapy.* Accounts of Chemical Research, 2011. **44**(10): p. 853–862.

69. Gupta, A.K., and M. Gupta, *Synthesis and surface engineering of iron oxide nanoparticles for biomedical applications.* Biomaterials, 2005. **26**(18): p. 3995–4021.

70. Gupta, A.K., et al., *Recent advances on surface engineering of magnetic iron oxide nanoparticles and their biomedical applications.* Nanomedice, 2007. **2**: p. 23–29.

71. Malvindi, M.A., et al., *Toxicity assessment of silica coated iron oxide nanoparticles and biocompatibility improvement by surface engineering.* PLOS One, 2014. **9**(1): p. e85835.

72. Gong, Y.-K. and F.M. Winnik, *Strategies in biomimetic surface engineering of nanoparticles for biomedical applications.* Nanoscale, 2012. **4**(2): p. 360–368.

73. Xiao, S., et al., *Synthesis of PEG-coated, ultrasmall, manganese-doped iron oxide nanoparticles with high relaxivity for T1/T2 dual-contrast magnetic resonance imaging.* International Journal of Nanomedicine, 2019. **14**: p. 8499–8507.

74. Zhu, M., et al., *Iron oxide nanoparticles aggravate hepatic steatosis and liver injury in nonalcoholic fatty liver disease through BMP-SMAD-mediated hepatic iron overload.* Nanotoxicology, 2021. **15**(6): p. 761–778.

75. Wang, W., et al., *Design of a multi-dopamine-modified polymer ligand optimally suited for interfacing magnetic nanoparticles with biological systems.* Langmuir, 2014. **30**(21): p. 6197–6208.

76. Arias, L.S., et al., *Iron oxide nanoparticles for biomedical applications: A perspective on synthesis, drugs, antimicrobial activity, and toxicity.* Antibiotics, 2018. **7**(2): p. 46.

77. Feng, Q., et al., *Uptake, distribution, clearance, and toxicity of iron oxide nanoparticles with different sizes and coatings.* Scientific Reports, 2018. **8**(1): p. 1–13.

78. Toropova, Y.G., et al., *In vitro toxicity of FemOn, FemOn-SiO2 composite, and SiO2-FemOn core-shell magnetic nanoparticles.* International Journal of Nanomedicine, 2017. **12**: p. 593.

79. Kornberg, T.G., et al., *Potential toxicity and underlying mechanisms associated with pulmonary exposure to iron oxide nanoparticles: Conflicting literature and unclear risk.* Nanomaterials, 2017. **7**(10): p. 307.

80. Al-Shalabi, R., et al., *The antimicrobial and the antiproliferative effect of human triple negative breast cancer cells using the greenly synthesized iron oxide nanoparticles.* Journal of Drug Delivery Science and Technology, 2022. **75**: p. 103642.

81. Che Mohamed Hussein, S.N., et al., *Colloidal stability of CA, SDS and PVA coated iron oxide nanoparticles (IONPs): Effect of molar ratio and salinity.* Polymers, 2022. **14**(21): p. 4787.

82. Noqta, O.A., et al., *Recent advances in iron oxide nanoparticles (IONPs): Synthesis and surface modification for biomedical applications.* Journal of Superconductivity and Novel Magnetism, 2019. **32**: p. 779–795.

83. Sugumaran, P.J., et al., *GO-functionalized large magnetic iron oxide nanoparticles with enhanced colloidal stability and hyperthermia performance.* ACS Applied Materials & Interfaces, 2019. **11**(25): p. 22703–22713.

84. Cao, X., et al., *Contribution, composition, and structure of eps by in vivo exposure to elucidate the mechanisms of nanoparticle-enhanced bioremediation to metals.* Environmental Science & Technology, 2022. **56**(2): p. 896–906.

85. Xie, J., et al., *PET/NIRF/MRI triple functional iron oxide nanoparticles*. Biomaterials, 2010. **31**(11): p. 3016–3022.

86. Hohnholt, M.C., et al., *Handling of iron oxide and silver nanoparticles by astrocytes*. Neurochemical Research, 2013. **38**: p. 227–239.

87. Zamora-Perez, P., et al., *Hyperspectral-enhanced dark field microscopy for single and collective nanoparticle characterization in biological environments*. Materials, 2018. **11**(2): p. 243.

88. Casset, A., et al., *Macrophage functionality and homeostasis in response to oligoethyleneglycol-coated IONPs: Impact of a dendritic architecture*. International Journal of Pharmaceutics, 2019. **556**: p. 287–300.

89. Arami, H., et al., *In vivo delivery, pharmacokinetics, biodistribution and toxicity of iron oxide nanoparticles*. Chemical Society Reviews, 2015. **44**(23): p. 8576–8607.

90. Vo, T.M.T., et al., *Rice starch coated iron oxide nanoparticles: A theranostic probe for photoacoustic imaging-guided photothermal cancer therapy*. International Journal of Biological Macromolecules, 2021. **183**: p. 55–67.

91. Tancredi, P., et al., *Polymer-assisted size control of water-dispersible iron oxide nanoparticles in range between 15 and 100 nm*. Colloids and Surfaces A: Physicochemical and Engineering Aspects, 2015. **464**: p. 46–51.

92. Abd Elrahman, A.A., and F.R. Mansour, *Targeted magnetic iron oxide nanoparticles: Preparation, functionalization and biomedical application*. Journal of Drug Delivery Science and Technology, 2019. **52**: p. 702–712.

93. Aboushoushah, S., et al., *Toxicity and biodistribution assessment of curcumin-coated iron oxide nanoparticles: Multidose administration*. Life Sciences, 2021. **277**: p. 119625.

94. Zulauf, G.D., et al. *Targeting of systemically-delivered magnetic nanoparticle hyperthermia using a noninvasive, static, external magnetic field*. in *Energy-Based Treatment of Tissue and Assessment VII*. 2013. SPIE.

95. Alkhader, E., N. Billa, and C.J. Roberts, *Mucoadhesive chitosan-pectinate nanoparticles for the delivery of curcumin to the colon*. AAPS PharmSciTech, 2017. **18**(4): p. 1009–1018.

96. Ramasamy, T., et al., *Chitosan-based polyelectrolyte complexes as potential nanoparticulate carriers: Physicochemical and biological characterization*. Pharmaceutical Research, 2014. **31**(5): p. 1302–1314.

97. Bharathi, D., et al., *Synthesis and characterization of chitosan/iron oxide nanocomposite for biomedical applications*. International Journal of Biological Macromolecules, 2019. **132**: p. 880–887.

98. Ganguly, S., et al., *Synthesis of polydopamine-coated halloysite nanotube-based hydrogel for controlled release of a calcium channel blocker*. RSC Advances, 2016. **6**(107): p. 105350–105362.

99. Mahmoud, M.E., et al., *A sustainable nanocomposite for removal of heavy metals from water based on crosslinked sodium alginate with iron oxide waste material from steel industry*. Journal of Environmental Chemical Engineering, 2020. **8**(4): p. 104015.

100. Kloster, G.A., et al., *Alginate based nanocomposites with magnetic properties*. Composites Part A: Applied Science and Manufacturing, 2020. **135**: p. 105936.

101. Lilhare, S., et al., *Calcium alginate beads with entrapped iron oxide magnetic nanoparticles functionalized with methionine—A versatile adsorbent for arsenic removal*. Nanomaterials, 2021. **11**(5): p. 1345.

102. Arora, V., et al., *Hydrophobically modified sodium alginate conjugated plasmonic magnetic nanocomposites for drug delivery & magnetic resonance imaging*. Materials Today Communications, 2020. **25**: p. 101470.

103. Petters, C., et al., *Uptake and metabolism of iron oxide nanoparticles in brain cells*. Neurochemical Research, 2014. **39**: p. 1648–1660.

104. Fahmy, H.M., et al., *Review of green methods of iron nanoparticles synthesis and applications.* BioNanoScience, 2018. **8**: p. 491–503.
105. Saqib, S., et al., *Synthesis, characterization and use of iron oxide nano particles for antibacterial activity.* Microscopy Research and Technique, 2019. **82**(4): p. 415–420.
106. Ganguly, S., et al., *Photopolymerized thin coating of polypyrrole/graphene nanofiber/iron oxide onto nonpolar plastic for flexible electromagnetic radiation shielding, strain sensing, and non-contact heating applications.* Advanced Materials Interfaces, 2021. **8**(23): p. 2101255.
107. Drozdz, A., et al., *FTIR microspectroscopy revealed biochemical changes in liver and kidneys as a result of exposure to low dose of iron oxide nanoparticles.* Spectrochimica Acta Part A: Molecular and Biomolecular Spectroscopy, 2020. **236**: p. 118355.
108. Bonvin, D., et al., *Controlling structural and magnetic properties of IONPs by aqueous synthesis for improved hyperthermia.* RSC Advances, 2017. **7**(22): p. 13159–13170.
109. Wu, W., et al., *Recent progress on magnetic iron oxide nanoparticles: Synthesis, surface functional strategies and biomedical applications.* Science and Technology of Advanced Materials, 2015. **16**(2): p. 023501.
110. Cruz, N., et al., *A novel hybrid nanosystem integrating cytotoxic and magnetic properties as a tool to potentiate melanoma therapy.* Nanomaterials, 2020. **10**(4): p. 693.
111. Mahmoudi, M., et al., *Optimal design and characterization of superparamagnetic iron oxide nanoparticles coated with polyvinyl alcohol for targeted delivery and imaging.* The Journal of Physical Chemistry B, 2008. **112**(46): p. 14470–14481.
112. Rayhan, M.A., et al., *Biopolymer and biomaterial conjugated iron oxide nanomaterials as prostate cancer theranostic agents: A comprehensive review.* Symmetry, 2021. **13**(6): p. 974.
113. Frazar, E.M., et al., *Multifunctional temperature-responsive polymers as advanced biomaterials and beyond.* Journal of Applied Polymer Science, 2020. **137**(25): p. 48770.
114. Ahmadi, M., *Iron oxide nanoparticles for delivery purposes,* in *Nanoengineered Biomaterials for Advanced Drug Delivery.* 2020, Elsevier. p. 373–393.
115. Ganguly, S., and S. Margel, *Remotely controlled magneto-regulation of therapeutics from magnetoelastic gel matrices.* Biotechnology Advances, 2020. **44**: p. 107611.
116. Das, P., et al., *Tailor made magnetic nanolights: Fabrication to cancer theranostics applications.* Nanoscale Advances, 2021. **3**(24): p. 6762–6796.
117. Oka, C., et al., *Core–shell composite particles composed of biodegradable polymer particles and magnetic iron oxide nanoparticles for targeted drug delivery.* Journal of Magnetism and Magnetic Materials, 2015. **381**: p. 278–284.
118. Luong, T.T., et al., *Magnetothermal release of payload from iron oxide/silica drug delivery agents.* Journal of Magnetism and Magnetic Materials, 2016. **416**: p. 194–199.
119. Turiel-Fernández, D., et al., *Ultrasmall iron oxide nanoparticles cisplatin (IV) prodrug nanoconjugate: ICP-MS based strategies to evaluate the formation and drug delivery capabilities in single cells.* Analytica Chimica Acta, 2021. **1159**: p. 338356.
120. Ganguly, S., and S. Margel, *Design of magnetic hydrogels for hyperthermia and drug delivery.* Polymers, 2021. **13**(23): p. 4259.
121. Wahajuddin, N., and S. Arora, *Superparamagnetic iron oxide nanoparticles: Magnetic nanoplatforms as drug carriers.* International Journal of Nanomedicine, 2012. **7**: p. 3445–3471.
122. Turan, O., et al., *Delivery of drugs into brain tumors using multicomponent silica nanoparticles.* Nanoscale, 2019. **11**(24): p. 11910–11921.
123. Masserini, M., *Nanoparticles for brain drug delivery.* International Scholarly Research Notices, 2013. **2013**: 1–18.

124. Jafarzadeh, S., et al., *The effect of hematocrit and nanoparticles diameter on hemo-dynamic parameters and drug delivery in abdominal aortic aneurysm with consider-ation of blood pulsatile flow.* Computer Methods and Programs in Biomedicine, 2020. **195**: p. 105545.

125. Patel, T., et al., *Polymeric nanoparticles for drug delivery to the central nervous sys-tem.* Advanced Drug Delivery Reviews, 2012. **64**(7): p. 701–705.

126. Na, H.B., et al., *Development of a T1 contrast agent for magnetic resonance imaging using MnO nanoparticles.* Angewandte Chemie, 2007. **119**(28): p. 5493–5497.

127. Weiss, N., et al., *The blood-brain barrier in brain homeostasis and neurological diseases.* Biochimica et Biophysica Acta (BBA)-Biomembranes, 2009. **1788**(4): p. 842–857.

128. Azarmi, M., et al., *Transcellular brain drug delivery: A review on recent advance-ments.* International Journal of Pharmaceutics, 2020. **586**: p. 119582.

129. Thomsen, L.B., M.S. Thomsen, and T. Moos, *Targeted drug delivery to the brain using magnetic nanoparticles.* Therapeutic Delivery, 2015. **6**(10): p. 1145–1155.

130. Talelli, M., et al., *Superparamagnetic iron oxide nanoparticles encapsulated in biode-gradable thermosensitive polymeric micelles: Toward a targeted nanomedicine suit-able for image-guided drug delivery.* Langmuir, 2009. **25**(4): p. 2060–2067.

131. Tai, L.-A., et al., *Thermosensitive liposomes entrapping iron oxide nanoparticles for controllable drug release.* Nanotechnology, 2009. **20**(13): p. 135101.

132. Lorenzato, C., et al., *MRI contrast variation of thermosensitive magnetoliposomes triggered by focused ultrasound: A tool for image-guided local drug delivery.* Contrast media & Molecular Imaging, 2013. **8**(2): p. 185–192.

133. Hemery, G., et al., *Thermosensitive polymer-grafted iron oxide nanoparticles stud-ied by in situ dynamic light backscattering under magnetic hyperthermia.* Journal of Physics D: Applied Physics, 2015. **48**(49): p. 494001.

134. Lu, Y.-J., et al., *Thermosensitive magnetic liposomes for alternating magnetic field-inducible drug delivery in dual targeted brain tumor chemotherapy.* Chemical Engineering Journal, 2019. **373**: p. 720–733.

135. Hilger, I., and W.A. Kaiser, *Iron oxide-based nanostructures for MRI and magnetic hyperthermia.* Nanomedicine, 2012. **7**(9): p. 1443–1459.

136. Gudkov, S.V., et al., *Do iron oxide nanoparticles have significant antibacterial proper-ties?* Antibiotics, 2021. **10**(7): p. 884.

137. Guo, J., et al., *Iron oxide nanoparticles with photothermal performance and enhanced nanozyme activity for bacteria-infected wound therapy.* Regenerative Biomaterials, 2022. **9**: p. 1–21.

138. Wang, G., et al., *Induction of bone remodeling by raloxifene-doped iron oxide func-tionalized with hydroxyapatite to accelerate fracture healing.* Journal of Biomedical Nanotechnology, 2021. **17**(5): p. 932–941.

139. Fernández-Barahona, I., et al., *Iron oxide nanoparticles: An alternative for positive contrast in magnetic resonance imaging.* Inorganics, 2020. **8**(4): p. 28.

140. Tada, Y., and P.C. Yang, *Iron oxide labeling and tracking of extracellular vesicles.* Magnetochemistry, 2019. **5**(4): p. 60.

141. Baneshi, M., et al., *A novel theranostic system of AS1411 aptamer-functionalized albu-min nanoparticles loaded on iron oxide and gold nanoparticles for doxorubicin deliv-ery.* International Journal of Pharmaceutics, 2019. **564**: p. 145–152.

142. Wu, Y., et al., *Surface modification of iron oxide-based magnetic nanoparticles for cerebral theranostics: Application and prospection.* Nanomaterials, 2020. **10**(8): p. 1441.

143. Bremerich, J., D. Bilecen, and P. Reimer, *MR angiography with blood pool contrast agents.* European Radiology, 2007. **17**(12): p. 3017–3024.

144. Schnorr, J., et al., *Comparison of the iron oxide-based blood-pool contrast medium VSOP-C184 with gadopentetate dimeglumine for first-pass magnetic resonance angiography of the aorta and renal arteries in pigs.* Investigative Radiology, 2004. **39**(9): p. 546–553.

145. Taupitz, M., et al., *Phase I clinical evaluation of citrate-coated monocrystalline very small superparamagnetic iron oxide particles as a new contrast medium for magnetic resonance imaging.* Investigative Radiology, 2004. **39**(7): p. 394–405.

146. Sakuma, H., *Coronary CT versus MR angiography: The role of MR angiography.* Radiology, 2011. **258**(2): p. 340–349.

147. Klein, C., et al., *Improvement of image quality of non-invasive coronary artery imaging with magnetic resonance by the use of the intravascular contrast agent Clariscan™ (NC100150 injection) in patients with coronary artery disease.* Journal of Magnetic Resonance Imaging, 2003. **17**(6): p. 656–662.

148. More, M.P., and P.K. Deshmukh, *Development of amine-functionalized superparamagnetic iron oxide nanoparticles anchored graphene nanosheets as a possible theranostic agent in cancer metastasis.* Drug Delivery and Translational Research, 2020. **10**(4): p. 862–877.

149. Tzameret A, et al., *In Vivo MRI assessment of bioactive magnetic iron oxide/human serum albumin nanoparticle delivery into the posterior segment of the eye in a rat model of retinal degeneration.* Journal of Nanobiotechnology, 2019;**17**: p. 3. Available at: http://dx.doi.org/10.1186/s12951-018-0438-y.

150. Kim, C.S., et al., *One-pot, one-step synthesis of drug-loaded magnetic multimicelle aggregates.* Bioconjugate Chemistry, 2022. **33**(5): p. 969–981.

151. Tade, R.S., et al., *Graphene quantum dots (GQDs) nanoarchitectonics for theranostic application in lung cancer.* Journal of Drug Targeting, 2022. **30**(3): p. 269–286.

152. Kankala, R.K., et al., *Cardiac tissue engineering on the nanoscale.* ACS Biomaterials Science & Engineering, 2018. **4**(3): p. 800–818.

153. Laurent, S., and M. Mahmoudi, *Superparamagnetic iron oxide nanoparticles: Promises for diagnosis and treatment of cancer.* International Journal of Molecular Epidemiology and Genetics, 2011. **2**(4): p. 367.

154. Sivasankarapillai, V.S., et al., *Cancer theranostic applications of MXene nanomaterials: Recent updates.* Nano-Structures & Nano-Objects, 2020. **22**: p. 100457.

155. Patil, S., et al., *The development of functional non-viral vectors for gene delivery.* International Journal of Molecular Sciences, 2019. **20**(21): p. 5491.

156. Al Faraj, A., A.P. Shaik, and A.S. Shaik, *Magnetic single-walled carbon nanotubes as efficient drug delivery nanocarriers in breast cancer murine model: Noninvasive monitoring using diffusion-weighted magnetic resonance imaging as sensitive imaging biomarker.* International Journal of Nanomedicine, 2015. **10**: p. 157–168.

157. Das, B., et al., *Carbon nanodots from date molasses: New nanolights for the in vitro scavenging of reactive oxygen species.* Journal of Materials Chemistry B, 2014. **2**(39): p. 6839–6847.

158. Liu, X., et al., *Nitrogen-doped carbon quantum dot stabilized magnetic iron oxide nanoprobe for fluorescence, magnetic resonance, and computed tomography triple-modal in vivo bioimaging.* Advanced Functional Materials, 2016. **26**(47): p. 8694–8706.

159. Mazuel, F., et al., *Massive intracellular biodegradation of iron oxide nanoparticles evidenced magnetically at single-endosome and tissue levels.* ACS Nano, 2016. **10**(8): p. 7627–7638.

160. Xu, H., and D. Ren, *Lysosomal physiology.* Annual Review of Physiology, 2015. **77**: p. 57–80.

161. Sun, S., et al., *Pyrite-activated persulfate oxidation and biological denitrification for effluent of biological landfill leachate treatment system.* Journal of Environmental Management, 2022. **304**: p. 114290.

162. Ganiyu, S.O., S. Sable, and M.G. El-Din, *Advanced oxidation processes for the degradation of dissolved organics in produced water: A review of process performance, degradation kinetics and pathway.* Chemical Engineering Journal, 2022. **429**: p. 132492.

163. Dong, H., et al., *A critical review of mineral–microbe interaction and co-evolution: Mechanisms and applications.* National Science Review, 2022. **9**(10): p. nwac128.

164. Fitch, A., P. Balderas-Hernandez, and J.G. Ibanez, *Electrochemical technologies combined with physical, biological, and chemical processes for the treatment of pollutants and wastes: A review.* Journal of Environmental Chemical Engineering, 2022. **10**(3): p. 107810.

165. Tajvar, S., A. Hadjizadeh, and S.S. Samandari, *Scaffold degradation in bone tissue engineering: An overview.* International Biodeterioration & Biodegradation, 2023. **180**: p. 105599.

166. Bagherifard, S., et al., *Accelerated biodegradation and improved mechanical performance of pure iron through surface grain refinement.* Acta Biomaterialia, 2019. **98**: p. 88–102.

167. Ezealigo, U.S., et al., *Iron oxide nanoparticles in biological systems: Antibacterial and toxicology perspective.* JCIS Open, 2021. **4**: p. 100027.

168. Wang, L., et al., *Fe-loaded biochar facilitates simultaneous bisphenol a biodegradation and efficient nitrate reduction: Physicochemical properties and biological mechanism.* Journal of Cleaner Production, 2022. **372**: p. 133814.

169. Tang, K.H.D., et al., *Immobilized enzyme/microorganism complexes for degradation of microplastics: A review of recent advances, feasibility and future prospects.* Science of the Total Environment, 2022: p. 154868.

170. Vercellino, S., et al., *Biological interactions of ferromagnetic iron oxide–carbon nanohybrids with alveolar epithelial cells.* Biomaterials Science, 2022. **10**(13): p. 3514–3526.

171. Ding, Y., et al., *Reactive oxygen species-upregulating nanomedicines towards enhanced cancer therapy.* Biomaterials Science, 2023. **11**(4), 1182–1214.

172. Rahman, M.A., et al., *Recent advances in cellular signaling interplay between redox metabolism and autophagy modulation in cancer: An overview of molecular mechanisms and therapeutic interventions.* Antioxidants, 2023. **12**(2): p. 428.

173. Zhang, Q., et al., *Engineering a synergistic antioxidant inhibition nanoplatform to enhance oxidative damage in tumor treatment.* Acta Biomaterialia, 2023. 10.1016/j.actbio.2022.12.067

174. Lartigue, L., et al., *Biodegradation of iron oxide nanocubes: High-Resolution in situ monitoring.* Acs Nano, 2013. **7**(5): p. 3939–3952.

175. Nowak-Jary, J., and B. Machnicka, *Pharmacokinetics of magnetic iron oxide nanoparticles for medical applications.* Journal of Nanobiotechnology, 2022. **20**(1): p. 1–30.

176. Sathyamoorthy, N., and M.D. Dhanaraju, *Shielding therapeutic drug carriers from the mononuclear phagocyte system: A review.* Critical Reviews™ in Therapeutic Drug Carrier Systems, 2016. **33**(6). 10.1615/CritRevTherDrugCarrierSyst.2016012303. PMID: 27992308

177. Chrishtop, V.V., et al., *Organ-specific toxicity of magnetic iron oxide-based nanoparticles.* Nanotoxicology, 2021. **15**(2): p. 167–204.

15 Ceramic Fillers, Fibers, and Acrylics

Rahul Kumar, Ravindra Kumar Pandey,
Shiv Shankar Shukla, and Bina Gidwani

15.1 INTRODUCTION

Composites are distinct materials with varying physical and chemical properties, essential in daily life and various applications [1]. Engineering performance
of composites depends on the material, components, and interaction. Fiber-
reinforced polymer (FRP) is a composite material formed by a polymer matrix
with fibers. The matrix provides superior mechanical and thermal properties,
while reinforcement provides dispersed non-continuous phase. Fibers include
glass, carbon, asbestos, paper, and wood, while polymers include vinyl ester,
epoxy, polyester thermosetting plastic, and phenol formaldehyde resins [2].
Engineering composites include cement, concrete, metal matrix composites
(MMCs), ceramic, and polymer matrix composites, whereas biomedical composites include fiber- and particle-reinforced materials [3]. Continuous matrix
or base phase and continuous reinforcing phase or filler materials make up
composite materials. In order to improve qualities, these phases work together
macroscopically. Composite materials have benefits including stronger fatigue
strength, superior surface finishes, improved corrosion resistance, higher tensile
strength, and lighter weight. They are widely used in biomedical fields for such
applications as artificial body parts, dentistry, and regeneration medicine. With
the matrix phase being created from natural or synthetic polymers and the reinforcing elements being made from natural fibers, polymer composite materials
exhibit good biocompatibility and can be employed in a variety of applications
[4–6]. Natural fibers called "bio-fibers" are derived from biological sources, such
as wood, regenerated cellulose, and crops. These composite materials are cheap,
lightweight, biodegradable, recyclable, and biocompatible. Non-wood fibers like
cellulose and lignin have greater physical and mechanical qualities than wood
fibers, which have lower cellulose crystallinity. They have been employed since
ancient civilizations and are used in industrial applications. Combining multiple
materials, composite materials are able to maximize certain features and exhibit
superior qualities to single components. Researchers are striving to create and
construct composite materials that challenge conventional materials like metals,
alloys, and ceramic materials by having new mechanical behaviors, physical and
chemical features, and processing technology. Most of the fundamental characteristics needed for body implants are present in composite materials, including

DOI: 10.1201/9781003454236-15

a high specific elastic modulus, a strong strength to weight ratio, and excellent toughness to prevent crack propagation [7–10].

The fundamental function of ceramic fillers is to strengthen composite materials by more uniformly dispersing stress and enhancing strength, stiffness, and durability. They can also increase composites' resilience to wear, corrosion, and high temperatures, making them appropriate for use in harsh situations. Alumina, silicon carbide, and titanium dioxide are common ceramic fillers. Ceramic fillers are often blended with matrix material throughout the production process by means of procedures such as mixing, compounding, or extrusion. Adjusting elements such as filler content, particle size, and distribution can alter the qualities of the final composite. Excess filler content, on the other hand, might cause issues such as increased viscosity during processing [11, 12].

Ceramic fillers: Ceramic fillers are used in a variety of sectors, including aerospace, automotive, electronics, and construction. They can improve the structural integrity of aircraft components in aerospace. Ceramic-filled polymers in electronics can provide increased thermal conductivity for heat dissipation in electrical devices [13]. Ceramic fillers are used in the automobile industry to create lightweight yet sturdy parts that increase fuel efficiency and safety. Ceramic fillers are essential in the creation of new composite materials with improved characteristics. Because of their capacity to adjust mechanical, thermal, and electrical properties, they are critical in creating materials that can resist diverse and demanding circumstances in a variety of sectors. Ceramic fillers are solid particles composed of ceramic materials that are used to improve the mechanical, thermal, electrical, and other characteristics of composite materials. These fillers are frequently used to improve the performance of composites made from polymer, metal, or ceramic matrices [14].

Fiber filler: A fiber filler, also known as a fiber reinforcement or fiber additive, is a compound that incorporates fibers into various goods to improve their qualities. These fibers are generally constructed of glass, carbon, agamid, or natural fibers such as cellulose and hemp. The inclusion of fibers can enhance the mechanical, thermal, and other performance aspects of the host material dramatically. Fiber fillers are used in a variety of sectors, including construction, automotive, aerospace, and textiles. Fiber fillers are frequently used in the construction sector to increase the strength, durability, and fracture resistance of concrete and cementitious materials. The fibers assist in more uniformly distributing stresses throughout the material, lowering the possibility of cracks occurring and improving the overall structural integrity of the product. Fiber-reinforced concrete is widely employed in the construction of buildings, bridges, pavements, and other infrastructure projects. In the automotive sector, fiber fillers are utilized in the manufacturing of composite materials used for parts like bumpers, panels, and interior components. These fibers contribute to the lightweighting of vehicles, leading to improved fuel efficiency and reduced emissions. Additionally, they enhance impact resistance and overall safety, making vehicles more robust in collisions [15]. Advanced fiber fillers, notably carbon and glass fibers, are frequently used in aerospace applications. These materials are used to create parts for engines, wings, and fuselages of airplanes. Fiber-reinforced materials' superior strength-to-weight ratio is crucial for the aerospace industry because it enables the creation of lightweight yet robust structures that can resist the abrasive environment of flight. Fiber fillers are also useful in the textile sector [16]. Cotton, wool, and hemp have been used in textiles for generations, but synthetic fibers have gained popularity in recent decades due to their enhanced performance and adaptability. Polyester and nylon fibers are frequently mixed with natural fibers to improve the durability, flexibility, and moisture-wicking qualities of garments and other fabrics. Fiber fillers have also found use in the medical profession in a variety of applications. To facilitate tissue integration and healing, biocompatible fibers are used in medical implants and devices. These fibers may be programmed to deteriorate over time, enabling the body's natural healing mechanisms to take control while initially providing structural support. Fiber fillers significantly improve properties in various industries, including concrete, composites, textiles, and medical devices, contributing to safer, more efficient, and more sustainable products [17–19].

Ceramic Fiber Fillers: Fillers made from ceramic fibers are used to improve the mechanical, thermal, and electrical characteristics of ceramic composites. Ceramic composites are a type of advanced material composed of high-performance ceramic fibers embedded in a ceramic matrix. These fillers are critical to the overall performance and usefulness of the final composite material. This chapter will go through the definition, kinds, qualities, uses, and benefits of ceramic fiber fillers in depth. Ceramic fiber fillers are tiny, finely split particles of ceramic materials that are added to a composite matrix composed mostly of ceramics, polymers, or metals to change the characteristics of the material. These fillers frequently have unique features that are used to improve certain composite attributes like as strength, heat resistance, and electrical conductivity. Ceramic fiber fillers have different qualities depending on their composition and structure. These fillers, in general, have high melting points, great thermal stability, low thermal expansion coefficients, and strong chemical resistance. They are useful additions to ceramic composites due to their hardness, stiffness, and resistance to wear and corrosion [20, 21].

Acrylics: Acrylics are a versatile and popular type of paint that utilizes acrylic polymer emulsion as its primary binder. Since their inception in the mid-20th century, acrylics have gained immense popularity due to their ability to replicate various textures, offer a wide array of colors, and provide a quick-drying alternative to traditional oil paints. The name "acrylic" is derived from the chemical name of the resin used to create the polymer, which is composed of acrylic acid and methacrylic acid. Acrylic paints have a water-solubility when wet and become water-resistant when dried, allowing artists to manipulate the paint's consistency for various styles, from realistic portraiture to abstract expressionism. The rapid drying time of acrylics is a standout feature, allowing artists to work more efficiently, layer colors without waiting for extended drying times, and make alterations or corrections more easily. Acrylics can be applied to various surfaces, including canvas, paper, wood, and fabrics, and are compatible with various additives and media. They can mimic the appearance of other media, such as watercolor or oil, while retaining their own distinct qualities. The color range of acrylic paints has expanded significantly, offering artists an extensive palette. Advancements in acrylic paint technology have led to the creation of acrylic gouache, a hybrid medium that combines the opacity of traditional gouache with the versatility of acrylics. Overall, acrylics have revolutionized the world of painting with their fast-drying properties, wide range of colors, and adaptability to various styles and techniques [22, 23].

15.2 ADVANTAGES AND DISADVANTAGES OF CERAMIC FILLERS, FIBERS, AND ACRYLICS (CFFA)

15.2.1 Advantages of Ceramic Fillers

Strength: Ceramic fillers may greatly improve the mechanical characteristics of composite materials, resulting in high strength and stiffness.

Thermal Stability: Ceramics offer exceptional resilience to high temperatures, making them ideal for applications requiring intense heat or thermal cycling.

Chemical Resistance: Ceramic fillers are frequently resistant to corrosive chemicals, acids, and bases, which adds durability to the composite material.

Electrical Insulation: Ceramics are often good electrical insulators, which can be useful in situations needing electrical insulation.

Low Thermal Expansion: Ceramics' low coefficient of thermal expansion can assist in decreasing dimensional changes caused by temperature swings.

Abrasion Resistance: Ceramic-filled composites can provide better resistance to wear and abrasion, making them appropriate for high-wear applications. (See Refs. [24, 25].)

15.2.2 Disadvantages of Ceramic Fillers

Brittleness: Because ceramics are inherently fragile, they are prone to cracking or fracturing when subjected to impact or bending.

Processing Difficulty: Due to their hardness, ceramic fillers can be difficult to process into composite materials, necessitating specialized processes and equipment.

Cost: Ceramic materials might be more expensive than other filler alternatives, affecting the overall cost of the composite product.

Limited Design Flexibility: Ceramics are generally difficult to form and mold, restricting design options compared to those of other fillers.

Poor Shock Resistance: Ceramics' brittleness can result in low shock resistance, making them unsuitable for applications where impact resistance is required.

15.2.3 Advantages of Ceramic Fibers

Reinforcement: Fibers, such as glass or carbon fibers, can considerably improve a material's strength and mechanical qualities when integrated.

Weight: Fiber-reinforced materials may maintain high strength while being lightweight, which is advantageous in situations where weight is critical.

Flexibility: These materials can be flexible and resistant to bending or deformation depending on the fiber type and arrangement.

Corrosion Resistance: Certain fibers, such as glass fibers, are corrosion resistant, giving resilience to the material. (See Refs. [26, 27].)

15.2.4 DISADVANTAGES OF CERAMIC FIBERS

Anisotropy: Fiber orientation can result in anisotropic qualities, which mean the material's properties change with direction, which might be undesirable in some instances.

Complexity of Processing: Incorporating fibers into a matrix material necessitates precise processing to achieve uniform distribution and good bonding.

Cost: High-performance fibers can be costly, thus raising the material's overall cost.

Environmental Impact: The manufacture and disposal of some fiber types, such as carbon fibers, raises environmental problems.

15.2.5 ADVANTAGES OF CERAMIC ACRYLICS

Weather Resistance: Acrylics are recognized for their exceptional weather resilience, which makes them appropriate for outdoor applications.

Transparency: Acrylics can be translucent, making them perfect for applications requiring clarity or visibility.

Processing: Acrylic materials are generally easy to produce, which allows for efficient production and molding.

Color Retention: Acrylics have the ability to hold their color over time, keeping their look even after prolonged exposure to sunshine. (See Ref. [28].)

15.2.6 DISADVANTAGES OF CERAMIC ACRYLICS

Scratch Sensitivity: Acrylic materials are more susceptible to scratching than other materials.

Impact Resistance: While acrylics are robust, they are not as impact resistant as other materials like polycarbonate.

Thermal Instability: Acrylics have poorer heat resistance than certain other materials, which limits their usage in high-temperature applications.

Flammability: Acrylics are more flammable than certain other polymers, which might be a problem in certain settings.

15.3 METHODS OF PREPARATIONS

15.3.1 CERAMIC FILLERS, FIBERS, AND ACRYLICS

15.3.1.1 Sol-Gel Technique

The sol-gel technique entails the hydrolysis and condensation of precursor materials to make a sol, which is subsequently gelatinized to form a ceramic substance. It's a flexible approach that allows for exact composition and morphological control.

The sol-gel method is a versatile technique for preparing ceramic materials, including fillers, by converting precursor solutions into a gel-like intermediate state and then solid ceramic material through controlled drying and heating processes. The process involves dissolved metal alkoxides in a solvent, hydrolysis reactions, polymerization and gelation, aging and ripening, drying, and calcination. The gel's structure evolves, and the particles within it grow and rearrange, affecting the final properties. Controlled drying is essential to maintain the integrity of the gel's structure. The resulting ceramic material is characterized using techniques like X-ray diffraction (XRD) and scanning electron microscopy (SEM) and can be used as a filler in various applications, such as composite materials and coatings. The sol-gel method allows for customization of parameters, making it highly versatile for various applications [29, 30].

15.3.1.2 Solid-State Reaction

In the solid-state reaction method, solid powders of different reactants are combined and heated at high temperatures to initiate a chemical reaction that results in the formation of the desired ceramic phase. The solid-state reaction method is a widely used process for producing ceramic fillers with tailored characteristics for specific applications. It involves selecting appropriate raw materials with the desired chemical composition, weighing and mixing them, sample compaction, sintering, cooling, and characterizing the final product. The success of this method depends on factors like raw material purity, mixing process, sintering conditions, and desired properties. This method is suitable for producing ceramic fillers with tailored characteristics for ceramics, composites, and advanced materials [31].

15.3.1.3 Hydrothermal Synthesis

In hydrothermal synthesis, high pressure and increased temperatures in an aqueous solution are utilized under hydrothermal settings to stimulate the formation of ceramic crystals with regulated size and shape. The hydrothermal synthesis process involves preparing precursor materials, which react under hydrothermal conditions to form the desired ceramic filler. These materials are typically in the form of soluble salts, oxides, or other compounds. The precursor materials are mixed with a water-based solvent and placed in a high-pressure vessel called an autoclave. The autoclave is heated to the desired temperature and pressure, and the process is controlled by temperature, pressure, reaction time, and precursor concentrations. The hydrothermal environment promotes crystal growth, influencing the size, shape, and properties of ceramic particles. After the reaction time, the autoclave is cooled to prevent unwanted phases or structural defects. The resulting ceramic particles are collected from the solution and may undergo additional washing, drying, and calcinations steps to enhance their properties. This process is widely used in industries like materials science, electronics, and catalysis for producing high-performance ceramic materials tailored to specific applications [32].

15.3.1.4 Spray Pyrolysis

A precursor solution is sprayed over a heated substrate in the spray pyrolysis approach. The solvent evaporates, leaving a deposit, which is then heated to produce

a ceramic coating. Spray pyrolysis is a technique used to produce thin films, powders, or nanoparticles of various materials, including ceramic materials. It entails
controlled decomposition of precursor solutions in the form of fine droplets. The
process involves dissolving precursor compounds in a solvent, atomization, droplet
drying, thermal decomposition, particle collection, and characterization using techniques like X-ray diffraction, SEM, and transmission electron microscopy (TEM).
The process parameters, such as precursor concentration, atomization rate, furnace
temperature, and gas atmosphere, need to be optimized to achieve desired ceramic
filler properties. Variations in the spray pyrolysis process may exist depending on the
material, equipment, and intended application [33].

15.3.1.5   Electrospinning

A polymer solution containing ceramic precursors is electrostatically spun into
nanofibers in the electrospinning approach. These nanofibers may be heat-treated to
become ceramic fibers. Electrospinning is a technique used to create nanofibers from
various materials, including ceramics. The process involves setting up a syringe with
a ceramic precursor solution, creating an electric field between the spinneret and the
collector, and forming a liquid jet. This jet forms and solidifies the ceramic material, creating nanofibers. Fiber collection is then performed on a grounded collector,
with morphology and alignment controlled by parameters like voltage, flow rate, and
collector type. Post-processing steps may include heat treatment to remove solvents
and organic components and crystallization and ceramic phase formation. Ceramic
nanofibers have various applications, including filtration, catalysis, sensors, and reinforcement in composites. The success of the electrospinning process depends on the
chosen precursor materials, solvent, voltage, flow rate, and other parameters [34–36].

15.3.1.6   Chemical Vapor Deposition (CVD)

Chemical vapor deposition (CVD) is the chemical reaction of gaseous precursors on
a substrate that results in the formation of a ceramic material. It is used to deposit
thin ceramic coatings onto diverse surfaces. CVD is a technique used to produce
ceramic materials by depositing thin films or coatings onto a substrate. It involves
the chemical reaction of vapor-phase precursor compounds to create solid materials
on a substrate surface. The process involves substrate preparation, precursor introduction, thermal decomposition, chemical reaction, film growth, and film thickness
control. The process can be energy-driven through methods like resistive heating,
induction heating, and microwave heating. Byproducts and waste are generated during the process, which can be carried away by the carrier gas. CVD offers precise
control over film composition, thickness, and uniformity, making it widely used in
semiconductors, coatings, and advanced materials like ceramic fillers for composites. However, CVD can be complex due to the need for precise control over temperature, pressure, precursor flow, and reaction kinetics [37].

15.3.1.7   Spark Plasma Sintering (SPS)

Spark plasma sintering (SPS) is a fast consolidation technology in which a powder
compact is subjected to pulsed electric current and pressure, enabling rapid den-
sification of ceramic materials. SPS is a consolidation technique used to prepare

ceramic materials with improved properties, particularly for producing dense and high-quality ceramics. The process involves the application of pulsed direct current and uniaxial pressure simultaneously to rapidly heat and consolidate ceramic powders. The equipment used in SPS includes a graphite die, punches, a power supply unit, and a vacuum or controlled atmosphere chamber. The process involves loading ceramic powder into the die, assembly, heating, pressure application, and sintering. SPS achieves high heating rates, reducing the time required for sintering and resulting in fine-grained microstructures with improved mechanical properties. It is suitable for various ceramic materials, including oxides, non-oxides, and composites. SPS is used in various industries, including aerospace, electronics, energy, and biomedical applications, producing high-performance components like cutting tools, biomedical implants, and electrical insulators [38].

15.3.1.8 Template-Assisted Methods

Template-assisted methods include the use of templates or molds to shape the ceramic precursor material into the required structure prior to sintering. Replica molding and sacrificial tinplating are two examples. Template-assisted methods are used to synthesize ceramic fillers with controlled size, shape, and properties, which are used in various industries like catalysis, electronics, and materials science. The process involves selecting a template material, coating the template, impregnation and adsorption, template removal, heat treatment (annealing), and template-derived ceramic filler. The resulting ceramic fillers are characterized using techniques like SEM and XRD to ensure they meet desired specifications. These fillers can be used in various applications, such as reinforcement in composite materials, catalytic supports, and electronic components. The success of the template-assisted method relies on careful selection of template material, precursor solution, heat treatment conditions, and characterization techniques [39].

15.3.1.9 Melt Infiltration

In melt infiltration, molten metal is infused into a porous ceramic preform. After solidification, a composite material with increased mechanical characteristics is created. Melt infiltration is a method for preparing ceramic matrix composites by infiltrating a molten metal into a porous ceramic preform. The process involves preparing a porous ceramic preform, selecting an infiltrant, preheating the preform, infiltrating the metal, solidifying and cooling, post-processing, and quality control. Melt infiltration offers tailorable properties, improved mechanical strength, and enhanced thermal properties compared to traditional ceramics. However, careful control of temperature, atmosphere, and processing parameters is necessary to achieve successful infiltration and desired composite properties [40].

15.3.1.10 Slip Casting

Slip casting is the process of pouring slurry of ceramic powder and a liquid binder into a mold. The liquid is evacuated, leaving a solid green substance that may be treated and sintered further. Slip casting is a popular method for creating ceramic objects, involving a mixture of clay, water, and other additives. The process involves finely ground clay particles, controlled viscosity, deflocculation, mold preparation, pouring, casting time, drying, de-molding, finishing, and glazing. The clay particles form a layer on the mold's inner surface, determining the final thickness of the ceramic object. The mold is left to sit for a specific period, absorbing water and forming a thinner slip. The excess slip is poured out, leaving the clay layer on the mold to dry. The green ware is then removed from the mold, requiring additional drying and firing stages to become a finished product. After de-molding, rough edges or imperfections are smoothed out or refined using tools or sandpaper. Bisque firing is then used to transform the green ware into porous bisque ware, which can be dipped, sprayed, or brushed with glaze. The final firing is at a higher temperature to fuse the glaze particles into a glassy coating, making the ceramic body strong, durable, and waterproof [41].

15.3.1.11 Extrusion

Extrusion is the process of forcing ceramic pastes through a die to generate continuous shapes. After that, the forms are dried and fused to create ceramic goods. The extrusion method is a widely used technique for preparing ceramic fillers. It involves selecting appropriate raw materials, formulating the composition, mixing, de-airing, and extrusion. The mixture is fed into an extrusion machine, which uses a rotating screw to force the material through a specially designed die. The die design determines the shape and size of the extruded ceramic, with various shapes available. The extruded pieces are then dried to remove excess moisture, fired in a kiln at high temperatures, and cooled down to prevent cracks or warping. The final ceramic fillers may undergo finishing processes like glazing, polishing, or surface treatments. Quality control measures are taken throughout the process to ensure the final product meets desired specifications and properties. This technique is widely used in industries like ceramics, construction, and manufacturing for its precision control over shape and size, high production rates, and ability to create intricate cross-sectional designs [42].

15.3.1.12 Powder-Liquid Mixing

The most frequent approach is powder-liquid mixing. A liquid acrylic resin is blended with ceramic powder (often a combination of glass ceramics or zirconia). The slurry is then put in a mold or onto a prepared tooth structure to polymerize and solidify. The powder-liquid mixing process for preparing ceramic acrylic involves gathering ingredients, measuring the required amounts, choosing a mixing container, starting at low speed, gradually adding ceramic powder, mixing until a homogeneous mixture is achieved, scrapping and re-mixing, adjusting viscosity, performing quality checks, de-gassing (optional), resting for a short period, and conducting a final quality inspection. The process may vary depending on the type of ceramic powder and binder used and the intended application. The final quality inspection ensures the mixture meets the required specifications for the intended application. The process can be applied to surfaces using casting, molding, or coating methods, following appropriate curing and drying procedures [43].

15.3.1.13 Layering Technique

Multiple layers of ceramic-filled acrylic are applied in the layering technique. Each layer is independently polymerized, providing for complete control over the final product's color and translucency. This method is frequently utilized to create very cosmetic dental restorations. The layering technique for ceramic acrylic preparation involves layering different shades of ceramic material on a tooth-shaped acrylic base. This process involves selecting the right shade, applying an opaque layer, applying a dentine layer, and applying a translucent layer. Dental technicians use stains and characterization materials to mimic natural features. The layered ceramic-acrylic restoration is fired in a dental furnace at high temperatures, fusing the ceramic layers together, and then refined and polished. After firing, the restoration is tried in the patient's mouth to assess fit, appearance, and bite. Once approved, the restoration is permanently cemented onto the tooth structure using dental adhesive. This technique requires skill and expertise and aims to create restorations that blend seamlessly with the patient's natural dentition, ensuring a natural appearance and functionality [44].

15.3.1.14 Computer-Aided Design/Computer-Aided Manufacturing (CAD/CAM) Technology

Computer-aided design/computer-aided manufacturing (CAD/CAM) technology enables the accurate manufacture of dental restorations. The ceramic acrylic material may be milled from a pre-sintered block before being sintered to its ultimate hardness in a high-temperature furnace. Ceramic acrylic is a hybrid material that combines the properties of ceramics and acrylics, resulting in a material with aesthetic appeal, strength, and flexibility. The CAM process involves computer software, precision machinery, and modern materials science. The design is created using CAD software, which dictates the dimensions, aesthetics, and functionality of the desired part. CAM software converts the design into machine instructions, setting parameters like tool paths, speeds, and feed rates based on the material properties. Most ceramic acrylic components are milled using cellulose nanocrystals (CNCs) machines, which precisely cut and shape the material based on the CAM software. Sintering is required for some ceramic components, while finishing involves polishing, painting, or surface treatments. Quality control measures ensure the part meets specifications and is free from defects. Ceramic acrylic can be used in various applications, including dental prosthetics, jewelry, and automotive components [45].

15.4 MATERIAL USED FOR PREPARATION OF CERAMIC FILLERS, FIBERS, AND ACRYLICS

Ceramic fillers are substances that are added to ceramics or other composite materials to improve their qualities. They are mixed with the underlying material to increase properties like as strength, hardness, and thermal stability. There are several types of ceramic fillers used, each with its own set of properties.

> **Silica (SiO_2):** Silica is typical ceramic filler that improves the mechanical and thermal characteristics of ceramics. It is frequently used to improve hardness and resistance to thermal shock. There are many benefits to using

silica (silicon dioxide, SiO_2) as a filler or nanofiller in polymer matrix composites, including improved biocompatibility and bioactivity. These substances promote biological processes for bone tissue applications, including calcium phosphate synthesis, porosity, and bioresorption rate. SiO_2 particles' metabolism and dissolution must be understood in order to fully comprehend their properties. SiO_2 nanoparticles in polyesters have benefits over other nanofillers and show characteristics that make them perfect grafting and scaffolding materials. During the manufacturing of a nanocomposite, covalent connections can be created between macromolecular chains and fillers, enhancing stability, adhesion, and mechanical properties. The establishment of covalent connections between the hydroxyl end groups of polymer PBSu and the surface silanol groups from SiO_2 nanoparticles, leading to a substantial improvement in mechanical properties, has been confirmed by nuclear magnetic resonance spectra (13C-NMR) research. The physicochemical properties of nano-SiO_2 in different forms, such as nanotubes and nanoparticles, also change, affecting the hydrophilic character of the material. The fact that silica included in PLA/SiO_2 nanocomposites degrades more quickly than pure PLA suggests that silica speeds up PLA's biodegradation in the nanocomposites. Silica may enhance water and enzyme molecules' attacks on the ester groups of PLA chains, lowering the polymer's molecular weight and causing the materials to disintegrate. Silica nanoparticles added to the polymer matrix encourage osteoblast-like cells to connect with surrounding tissue, boosting cell survival. While ortho-silicate acid $[Si(OH)_4]$ induces type I collagen synthesis in human osteoblastic cells and cell differentiation, silicon can also play a role in bone formation and mineralization. The results are consistent with reports of silica materials utilized for bone regeneration, even though no discernible effects of SiO_2 incorporation on cell behavior could be seen. Mesoporous silica has the capacity to encapsulate pharmaceuticals and biological agents due to its regular, parallel pores, high specific surface area, low density, and good biocompatibility. The usage of these particles in pharmaceutical applications, especially in drug delivery systems, is widespread. The release properties of the composite may be impacted by the drug's incorporation into the silica xerogel or the polymer matrix during the sol-gel process. Core-shell architectures enhance regulated drug release and control of drug release by creating strong connections between silica and polymer. Mesoporous silica scaffolds may be produced by electrospinning, conventional methods, or rapid prototyping. The ability of cells to retain osteoblast-like cells and express osteoprotegerin genes is improved by adding silica particles to hydrogels like CS or alginate. Nanosilica improves biological characteristics, biomineralization potential, and stiffness in polymer matrix composites without reducing mechanical strength. Potential uses of this strategy include biomineralization and cells that resemble bones [46–50].

Calcium Carbonate: Common filler used in polyolefins (POs) to replace resin in applications is calcium carbonate ($CaCO_3$). High-density polyethylene

(HDPE) bottle and packaging manufacturers utilize it to cut the price of raw resin. Increasing stiffness, hardness, and dimensional stability with fine $CaCO_3$ particles is possible, but $CaCO_3$ may also reduce tensile and impact strength qualities. $CaCO_3$ makes up 95–98% of commercial calcium carbonate products; the remaining percentage is made up of metal oxides. In comparison to untreated POs, treatment of $CaCO_3$ particles enables greater dispersion, simpler processing, and stronger impact forces. Stearic acids derived from fatty acids, maleic anhydride–grafted polymers, and silane-based compounds are suitable for coating $CaCO_3$ particles [51].

Alumina (Al_2O_3): A common ceramic filler is alumina. It boosts mechanical strength, wear resistance, and electrical insulation. Building materials, refractory materials, electrical and heat insulators, and electrical insulation all depend on the ceramic metal oxide alumina. It is a common choice for catalysts, catalyst support, and adsorbents due to its high strength, corrosion resistance, chemical stability, low thermal conductivity, and strong electrical insulation. To create alumina and corundum, preparation entails heating tri-hydroxides, gibbsite, and bayerite to temperatures above 1000°C. These transition alumina are ideal for adsorbents, catalysts, catalyst support, and filtration membranes due to their high porosity, surface areas, and acidic characteristics. Several procedures, including melt growth techniques, extrusion, electro-spraying, and electrospinning, can be used to construct them into alumina fibers. The high surface areas of transition alumina are due to the presence of irregular pores, mainly microspores with diameters less than 2 nm [52, 53].

Zirconia (ZrO_2): Zirconia fillers are recognized for their exceptional strength and hardness, making them ideal for applications requiring long-term endurance. Zirconia (ZrO_2) has three crystal structures: cubic (c-ZrO_2), tetragonal (t-ZrO_2), and monoclinic (m-ZrO_2). Synthesis of t-ZrO_2 at moderate temperatures prevents expansion. Nano-ZrO_2 powders have advanced applications in refractories, abrasives, ceramic pigments, oxygen sensors, and catalytic materials. Wet chemical synthesis approaches are used, but physical methods cannot meet nano size requirements and gas chemical methods are too expensive for practical production. High-purity ZrO_2 is a biomaterial with excellent natural white color, toughness, strength, and corrosion resistance, making it suitable for dental ceramics and biocompatible implant materials. However, processing and formation are challenging due to its high strength. Nano-ZrO_2 nanopowders offer various applications, such as filling, coating, and sintering raw materials. They considerably increase shear bond strength, flexural strength, and fracture toughness. They can also act as a porous covering to increase biocompatibility. Dental ceramics, implants, membranes, basements, and tissue engineering all use nano-ZrO_2 biomaterials to improve osseo-integration by surface treatment and bioactivation [54].

Talc: A common PO filler, talc is a naturally occurring hydrated magnesium silicate mineral with variable effects on polyethylene (PE) and polypropylene

(PP) characteristics. Because of the high aspect ratio of its plate-like particles, it improves modulus and decreases impact strength. Talc comes in particle sizes ranging from 1 to 100 m and is utilized at up to 50% loading. For high-impact automotive talc polyolefin (TPO) applications, fine talcs are essential. The bulk density of fine lamellar talc increases with talc grades that "compact" the bulk talc, minimizing trapped air and enhancing processing handling and feeding. Compacted talc has a bulk density that is nearly three times that of typical fine talcs, enabling melt processing of 30%-talc PP to be increased by 300%. There are "ultrafine" grades of talc with higher aspect ratios that can be used in car bumpers, body panels, dashboards, and interior trim to enhance the characteristics of PP and PP-based TPOs. These talc grades enhance the talc's layered structure by lessening the impacts of sub-2-m talc particles' stiffening and reinforcement. Rio Tinto Minerals and IMIFabi Spa have developed improved milling techniques that result in delaminated talc that keeps its platy structure and characteristics down to 1-m diameters. This compacted talc enables lighter PP and TPO parts while maintaining the same qualities as typical fine talc compounds at larger loadings. The benefits of each filler are combined and their shortcomings are covered up by mixing talc and calcium carbonate into one substance [55].

Titanium Dioxide (TiO_2): Titanium dioxide (TiO_2) is used in ceramics to provide color and opacity while also improving mechanical qualities. Rutile, anatase, brookite, and TiO_2 (B) are the four different structural types of TiO_2, which is the metal's natural oxide. TiO_2 (II) with the -PbO_2 structure, TiO_2 (H) with hollandite, baddeleyite with ZrO_2, and cotunnite with $PdCl_2$ are examples of high-pressure forms. The chemical industry primarily produces rutile and anatase as microcrystalline minerals. Different morphologies of TiO_2, such as nanoparticles, nanowires, nanotubes, and mesoporous structures, can be created. Hydrothermal/solvothermal, sonochemical, electrochemical, sol-gel, microwave field, and vapor phase are a few examples of synthetic techniques. As particle size reduces, TiO_2 specific surface area grows, affecting agglomeration behavior and interfacial characteristics. Polymer-TiO_2 nanocomposites can be improved through research that aims to reduce agglomeration and enhance distribution [56, 57].

Calcium Phosphate: Due to calcium phosphate's chemical makeup and biocompatibility, calcium phosphate ceramics are frequently employed as alternatives to bone grafts. However, because of their fragility, their therapeutic applications are restricted to the skeleton's non–load bearing areas. Due to their bioactivity, biodegradability, and biological responsiveness, calcium phosphates are utilized in bone repair. The Ca/P ratio, crystallinity, and phase composition affect bioactivity, degradation behavior, and osteo-conductivity. Tri-calcium phosphates are more soluble while synthetic HAp is more stable. Depending on the weight ratio of the stable/degradable phases, biphasic calcium phosphate (BCP) displays intermediate characteristics. Substituted CaP ceramics are a new development in the manufacture of calcium phosphates, particularly HAp. When compared to their parent materials, calcium phosphate ceramics are less soluble and bioactive, which results in altered biological reactions. While some are osteo-inductive, the majority are osteo-conductive. Ceramics such as HAp, TCP, and BCP are frequently employed in biological applications. The development of calcium phosphate–containing composite materials, especially polymer matrix composites, is being pursued. Biphasic calcium phosphate ceramics are two-phase materials that combine the osteo-inductivity of -TCP with the low solubility and osteo-conductive nature of HAp. By adjusting the HAp/-TCP ratio, it is possible to alter the bioactivity, rate of resorption, and mechanical characteristics of these ceramics. Biphasic calcium phosphate ceramics can be altered by substituting ions, including Sr, that accelerate the proliferation and differentiation of MG63 and HOS osteoblast-like cells. Additionally boosting biodegradation and bioactivity, magnesium substitution raises osteo-conductivity both in vitro and in vivo. The biodegradable synthetic polymers and copolymers PCL, PLLA, PLGA, collagen, and gelatin-pectin can all be modified using BCP ceramics. Compressive strength and Young's modulus of PLLA-based composites have improved with modifications. Higher compressive strength can also be achieved by directly grafting L-lactide onto the surface of BCP micro particles. A collagen/BCP composite (OsteonTM II collagen, available commercially) demonstrated good osteo-conductive characteristics with delayed rate of resorption, maintaining bone tissue regeneration space. It offers a consistent release profile, making it a good choice for a carrier of recombinant human BMP-2. In vitro and in vivo experiments revealed that the gelatin-pectin/BCP nanocomposites scaffold was effective at promoting cell proliferation and the production of new bone [58–69].

Wollastonite: Wollastonite is calcium and silicon oxide mineral used as filler in reinforcing materials, increasing tensile and flexural strength, dimensional stability, and mold shrinkage. The size of its needle-like particles ranges from 5 to 20 m. In place of talc or glass fiber, the non-hazardous mineral fiber wollastonite has gained popularity as filler. Although it has characteristics that are comparable to those of glass fibers, its fiber-like shape can harm processing machinery. Automotive panels, housings, door handles, and appliances can benefit from the usage of wollastonite to reinforce PE and PP. It has more tensile, flexural, and impact strength and equivalent mechanical characteristics to 15% talc-filled PP. Wollastonite offers the same heat deflection temperature (HDT), flex modulus, and tensile strength as 20% glass fiber alone and can also be used as a partial replacement for glass fiber. Improved dispersion, processing, impact resistance, surface gloss, dimensional stability, and scratch resistance qualities result from surface treatment with silane-based coupling agents [70–74].

Magnesium Oxide (MgO): Magnesium oxide (MgO) can improve ceramic refractoriness and electrical insulation. A key inorganic ceramic material with superior optical, thermodynamic, electrical, electronic, mechanical, and chemical properties is MgO. It can be used as a catalyst, an additive, an air purification adsorbent, a metal stabilizer, clear filler, or a surface conditioner; for a biomedical application; or to clean up toxic waste, among other things. Spinning disc reactors, chemical reactions, reverse micellar, laser ablation, and thermal decomposition are a few of the traditional processes used to create MgO nanoparticles. The sol-gel process is easier and provides high purity, greater specific surface areas, and smaller particle sizes. MgO nanocrystals were synthesized with weak bases NH_4OH and Na_2CO_3 and strong bases KOH and NaOH with a range of normalities. XRD and SEM analysis were used to characterize the production of MgO nanoparticles and identify their shape and crystal structures [75].

Barium Sulphate (BaSO$_4$): Ceramics typically utilize barium sulphate filler to increase density and radiation absorption capacity. Due to its high specific gravity, X-ray opacity, inertness, and whiteness, barium sulphate ($BaSO_4$), generally known as barite, is commonly used filler in a variety of industries. It has been applied to research crystal growth, precursor mixing, precipitation mechanisms, and morphology. Due to their distinctive structural, electrical, optical, magnetic, and chemical capabilities, nanoparticles with nanoscale dimensions are attracting growing amounts of scientific and technological interest. However, due to their high surface activity and ease of adsorption, nanoparticles can readily combine into bigger ones despite the difficulty in producing pure ultra-fine particles. Current research in nanomaterials focuses on regulating particle size, preventing aggregation, and redistributing them throughout the medium. Current research focuses on synthesis of smaller particles and inorganic additive powders with good dispersing properties in water for stable suspensions [76, 77].

Glass Fibers: Ceramics are filled with glass fibers to increase their tensile strength and hardness. The capacity of bioactive glasses (BGs) to regulate biological and chemical characteristics at the molecular level makes them useful materials for tissue engineering applications. Sol gel or melt quenching techniques can be used to create these glasses, with gel-derived BGs exhibiting enhanced bioactivity and cellular responsiveness. Controlled phase composition, structure, texture, microstructure, and surface chemistry are all possible using the sol-gel method. To enhance biological responsiveness and surface reactivity, BGs can be doped with therapeutic oxides and trace elements. For tissue engineering applications, research is concentrated on creating silicate and borate/borosilicate BGs as well as resorbable phosphate glasses. Silicate bioactive glasses (SBGs) are frequently utilized as fillers in polymer-based composites for bone tissue engineering and are the subject of extensive research for biomedical applications. Hench et al.

created the first melt-derived silicate SBG, Bioglass® 45S5, which is frequently employed as a modifier for biodegradable synthetic polymers and copolymers. In addition to their bonding abilities, SBGs also produce ionic dissolution products and osteoblastic and stem cell gene expression, stimulate angiogenesis, have anti-bacterial and anti-inflammatory characteristics, and release calcium. Chemical composition, volume fraction, size, and distribution of a bioactive phase can all affect how bioactive a composite is. Glass modifiers can enhance the bioactivity of poly (L-lactide)–based scaffolds by improving composite bioactivity [78–84].

Carbon Fibers: Carbon fibers can be used with ceramics to increase mechanical qualities while keeping the weight low. Carbon fibers offer an attractive alternative to traditional implantology materials like stainless steels and titanium alloys. They offer low thermal conductivity, easy shape formation, and resistance to corrosion and ion diffusion. Mechanical properties, such as tensile strength and resistance to UV radiation, can be further enhanced with lacquer layers and phenolic resins. However, carbon fibers have drawbacks, such as low UV radiation strength and resistance to temperatures above 200°C [85, 86].

Silicon Carbide: Silicon carbide (SiC) fillers are recognized for their strong thermal conductivity and outstanding mechanical qualities, making them ideal for high-temperature applications. With great hardness, strength, chemical and thermal stability, melting point, oxidation resistance, and erosion resistance, SiC is a versatile non-oxide ceramic. It is perfect for abrasion and cutting applications as well as high-power, high-temperature electronic components. Although the classic Acheson process, which is based on carbothermal reduction, has been commercialized, it has a lot of steps, requires a lot of energy, and uses subpar materials. Alternative techniques for producing SiC have been documented, including mechanical alloying, liquid phase sintering, sol-gel, physical vapor deposition, and chemical vapor deposition. Silicon carbide is a wide-band-gap semiconductor suitable for high temperature, power, and radiation conditions. It has high thermal conductivity and breakdown electric-field strength, with a 3.3-fold increase in 6H-SiC polytypic at 300 K. Silicon carbide, naturally found in meteorites, is human-made through the Acheson process, producing poly-crystalline SiC suitable for grinding and cutting applications [87].

Boron Nitride (BN): Because of its excellent thermal conductivity and lubricating qualities, boron nitride (BN) is useful in applications requiring heat dissipation and low friction. Robust intralayer bonding and weak van der Waals forces make BN a highly mechanically robust, electrically insulating, and thermally conducting material. It is thermally and chemically stable due to its white color and transparent surface. Its stability, however, makes it challenging to interact with other compounds. Researchers have looked into using polymers and organic particles to functionalize BN. Pre-ceramic polysilazanes have been employed as coating materials, and silane coupling agents create strong connections between organic and inorganic substances. These substances are made up of silicon atoms, groups that can be hydrolyzed, linkers, and functional groups for organs. One study looked into how to improve the affinity between the filler and matrix by coating A-BN, an aggregated binder. An epoxide group in GPTMS was employed to make a thermal conductive route. PSZ and GPTMS bridged inorganic and organic materials, preventing A-BN from losing shape and functionalizing it with epoxide groups. The resulting polymer composites were analyzed for morphology and thermal conductivity [88].

Polymer Matrix: Tissue engineering uses biodegradable polymeric materials, which can be divided into two primary categories: synthetic polymers like poly(lactic acid), poly(glycolic acid), poly(-caprolactone), and poly(3-hydroxybutyrate) and natural-based polymers like proteins or polysaccharides. Sensitive linkages are broken during biodegradation, permitting polymer erosion. Biological recognition of natural polymers exists, but the availability and concentration of the enzymes determine how quickly they degrade in vivo. Chemical composition, configurational structure, molecular weight, polydispersity, crystallinity, and material shape all control how quickly synthetic polymers degrade. For some applications, synthetic polymers provide strong mechanical strength and adaptable characteristics. However, they lack cell-recognition sequences and have hydrophobic surfaces. For biomedical applications, composites with excellent formability are created utilizing a variety of techniques, assuring the necessary characteristics and microstructures [89].

15.5 APPLICATION OF CERAMIC FILLERS, FIBERS, AND ACRYLICS

Ceramic fillers, fibers, and acrylics find wide applications in various fields like agriculture, dentistry, the juice industry, and engineering. The applications are shown in Figure 15.1. The details are discussed below.

1. **In Juice Industry:** Ceramic fillers, fibers, and acrylics are essential components in the juice industry for improving packaging materials, equipment, and containers. Ceramic fillers provide improved barrier properties, mechanical strength, and thermal stability, while fibers enhance filtration and reinforcement by removing particulates, pulp, and sediment. Acrylics, on the other hand, are used in coatings and sealants to protect packaging and maintain juice quality. These materials play a crucial role in ensuring the safety, shelf life, and overall appeal of juice products.

2. **In Dentistry:** Ceramic fillers, fibers, and acrylics are essential in dentistry for various applications. Ceramic fillers enhance dental restorative

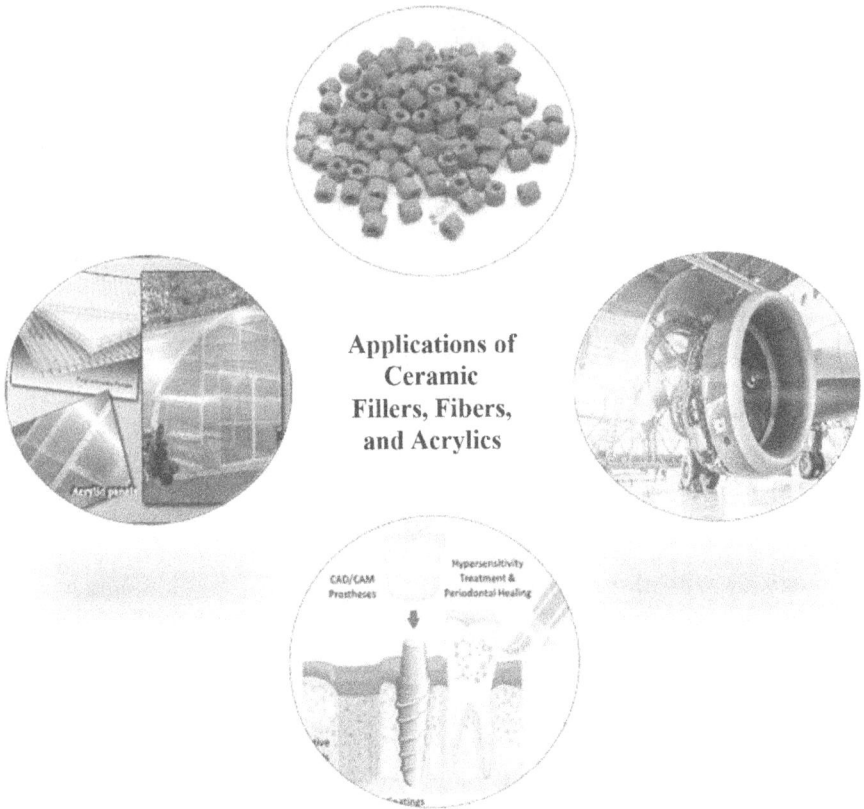

FIGURE 15.1　Applications of ceramic fillers, fibers, and acrylics in different fields.

materials, while fibers reinforce prosthetics and weakened tooth structures. They are used in tooth-colored restorations like composite fillings and ceramic crowns. Acrylics are used in various dental applications, including removable dentures, temporary restorations, orthodontic appliances, mouth guards, and repairs. These materials play a crucial role in modern dentistry, improving the functionality, aesthetics, and longevity of dental treatments [90].

3. **In Engineering:** Ceramic fillers, fibers, and acrylics are essential in enhancing material properties in various industries, including aerospace, automotive, electronics, construction, and healthcare. They enhance mechanical, thermal, and electrical properties of materials, making them suitable for various applications. Ceramic fillers can reinforce polymers, metals, and ceramics, creating composite materials with improved strength, stiffness, and wear resistance. They are also used as thermal insulators in high-temperature environments, dielectric materials in electronic components,

and coatings for corrosion resistance and high-temperature stability. Ceramic fibers possess excellent thermal stability and mechanical strength, making them suitable for high-temperature insulation, reinforcement in composites, aerospace components, and thermal protection systems. Acrylics are versatile synthetic polymers used in structural components, adhesives, coatings, and medical devices due to their biocompatibility, transparency, and ease of fabrication [91].

4. **In Agriculture:** In agriculture, ceramic fillers, fibers, and acrylics are used in various applications due to their thermal and mechanical properties. They improve the structural integrity and durability of agricultural equipment, tools, and structures, enhancing greenhouse construction, irrigation systems, storage tanks, soil moisture sensors, and crop protection. Ceramic fibers are also used in fire prevention, animal enclosures, and fertilizer encapsulation. Acrylics are versatile synthetic polymers with applications in agriculture, including mulch films, plant protection, fertilizer encapsulation, and erosion control. These applications showcase how ceramic fillers, fibers, and acrylics contribute to modern agriculture, improving productivity, efficiency, and sustainability [92].

Some of the patents related to ceramic fillers, fibers, and acrylics are listed in Table 15.1.

TABLE 15.1
Patents Related to Ceramic Fillers, Fibers, and Acrylics

S.N.	Title of Patent	Patent Number	Date of Patent	Assignee	Inventors
1.	Methods for producing metal carbide materials	10954167	March 23, 2021	Advanced Ceramic Fibers, LLC	John E. Garnier, George W. Griffith
2.	Single phase fiber reinforced ceramic matrix composites	10793478	October 6, 2020	Advanced Ceramic Fibers, LLC	John E. Garnier
3.	Boron carbide fiber reinforced articles	10208238	February 19, 2019	Advanced Ceramic Fibers, LLC	John E. Garnier, George W. Griffith
4.	Metal carbide fibers and methods for their manufacture	9803296	October 31, 2017	Advanced Ceramic Fibers, LLC	John E. Garnier
5.	Carbon fiber-reinforced silicon carbide composite and method for producing the same	JP2009227565A	December 5, 2012	Showa Denko Materials Co Ltd	Kazuya Baba, Makoto Ebihara, and Mitsuaki Unno

15.6 CONCLUSION

Composites are distinct materials with varying physical and chemical properties, essential in daily life and various applications. The fundamental function of ceramic fillers is to strengthen composite materials by more uniformly dispersing stress and enhancing strength, stiffness, and durability. Ceramic fillers are used in a variety of sectors, including aerospace, automotive, electronics, and construction. Fiber filler, also known as a fiber reinforcement or fiber additive, is a compound that incorporates fibers into various goods to improve their qualities. Fillers made from ceramic fibers are used to improve the mechanical, thermal, and electrical characteristics of ceramic composites. Ceramic composites are a type of advanced material composed of high-performance ceramic fibers embedded in a ceramic matrix. Acrylics are a versatile and popular type of paint that utilizes acrylic polymer emulsion as its primary binder. Since their inception in the mid-20th century, acrylics have gained immense popularity due to their ability to replicate various textures, offer a wide array of colors, and provide a quick-drying alternative to traditional oil paints. The method of preparation includes sol-gel technique, solid-state reaction, hydrothermal synthesis, spray pyrolysis, electrospinning, chemical vapor deposition, and spark plasma sintering. The materials used for preparation, their applications, and patents are discussed in this chapter. In the future, the in vivo models and regulatory aspects are to be considered for broader applications in science and allied fields as well as for biomedical applications.

ACKNOWLEDGMENT

The authors are grateful to the Department of Science and Technology (DST-FIST) Letter no. SR/FST/COLLEGE/2018/418, New Delhi, for providing financial assistance.

REFERENCES

1. Adzali, N.M.S., Jamaludin, S.B. & Derman, M.N. (2012). Mechanical properties, corrosion behavior and bioactivity of composite metal alloys added with ceramic for biomedical applications. Rev. Adv. Mater. Sci, 30, 262266.
2. Arvidson, K., et al., (2011). Bone regeneration and stem cells. J. Cell Mol. Med, 15, 718746.
3. Begum, K. & Islam, M.A. (2013). Natural fiber as a substitute to synthetic fiber in polymer composites: A review. Res. J. Eng. Sci, 2, 4653.
4. Marhoon, I.I. (2018). Mechanical properties of composite materials reinforced with short random glass fibers and ceramics particles. Int. J. Sci. Technol. Res, 7, 50–53.
5. Nagavally, R.R. (2017). Composite materials-history, types, fabrication techniques, advantages, and applications. Int. J. Mech. Prod. Eng, 5, 82–87.
6. Hussain, F., Hojjati, M., Okamoto, M. & Gorga, R.E. (2006). Polymer-matrix nano - composites, processing, manufacturing, and application: An overview. J. Compos. Mater, 40, 1551–1575.
7. Thirumalini, S. & Rajesh, M. (2017). Reinforcement effect on mechanical properties of bio-fiber composite. Int. J. Civil Eng. Technol, 8, 160–166.

8. Bavan, S. & Channabasappa, M.K.G. (2010). Potential use of natural fiber composite materials in India. J. Reinf. Plast. Comp, 29, 3600–3613.
9. Singh, L., Goga, G. & Rathi, M.K. (2012). Latest developments in composite materials. IOSR J. Eng, 2, 152–158.
10. Pradhan, D. & Sukla, L.B. (2017). Thin film of yttria stabilized zirconia on NiO using vacuum cold spraying process for solid oxide fuel cell. Int. J. Nano Biomater, 7, 38–47.
11. Silva, J., de Brito, J. & Veiga, R. (2008). Fine ceramics replacing cement in mortars partial replacement of cement with fine ceramics in rendering mortars. Mater Struct, 41, 1333–1344.
12. Gonzalez-Corominas, A. & Etxeberria, M. (2014). Properties of high performance concrete made with recycled fine ceramic and coarse mixed aggregates. Construction Build Mater, 68, 618–626.
13. Tamaraiselvi, T.V. & Rajeswari, S. (2004). Biological evaluation of bioceramic materials - a review. Trends Biomater. Artif. Organs, 18 (1), 9–17.
14. Emami, N., Sjodahl, M. & Söderholm, K.-J.M. (2005). How filler properties, filler fraction, sample thickness and light source affect light attenuation in particulate filled resin composites. Dent. Mater, 721–730. doi: 10.1016/j.dental.2005.01.002. PMID: 15885764.
15. Liu, Y., Jiang, X., Shi, J., Luo, Y., Tang, Y., Wu, Q. & Luo, Z. (2020). Research on the interface properties and strengthening–toughening mechanism of nanocarbon-toughened ceramic matrix composites. Nanotechnology. Rev, 9, 190–208.
16. Du, J., Zhang, H., Geng, Y., Ming, W., He, W., Ma, J., Cao, Y., Li, X. & Liu, K. (2019). A review on machining of carbon fiber reinforced ceramic matrix composites. Ceram. Int, 45, 18155–18166.
17. Diaz, O.G., Luna, G.G., Liao, Z. & Axinte, D. (2019). The new challenges of machining ceramic matrix composites (CMCs): Review of surface integrity. Int. J. Mach. Tools Manuf, 139, 24–36.
18. Arai, Y., Inoue, R., Goto, K. & Kogo, Y. (2019). Carbon fiber reinforced ultra-high temperature ceramic matrix composites: A review. Ceram. Int, 45, 14481–14489.
19. Binner, J., Porter, M., Baker, B., Zou, J., Venkatachalam, V., Diaz, V.R., D'Angio, A., Ramanujam, P., Zhang, T. & Murthy, T.S.R.C. (2019). Selection, processing, properties and applications of ultra-high temperature ceramic matrix composites, UHTCMCs—a review. Int. Mater. Rev, 65, 389–444.
20. Andraschek, N., Wanner, A.J., Ebner, C. & Riess, G. (2016). Mica/Epoxy-composites in the electrical industry: Applications, composites for insulation, and investigations on failure mechanisms for prospective optimizations. Polymers, 8, 201.
21. Pleşa, I., Noţingher, P.V., Schlögl, S., Sumereder, C. & Muhr, M. (2016). Properties of polymer composites used in high-voltage applications. Polymers, 8, 1.
22. Serrano-Aroca, Á. (2017). Latest Improvements of Acrylic-based Polymer Properties for Biomedical Applications, InTechOpen, 75–98.
23. Pratap, B., Gupta, R.K., Bhardwaj, B. & Nag, M. (2019). Resin based restorative dental materials: Characteristics and future perspectives. Jpn. Dent. Sci. Rev, 55, 126–138.
24. Ramesh, M., Narasimhan, L., Kumar, R., Srinivasan, N., Kumar, D.V. & Balaji, D. (2022). Influence of filler material on properties of fiberreinforced polymer composites: A review. Polymers, 22, 898–916.
25. Madhu, P., Sanjay, M.R., Senthamaraikannan, P., Pradeep, S., Saravanakumar, S.S. & Yogesha, B. (2018). A review on synthesis and characterization of commercially available natural fibers: Part-I. J Nat Fibers, 16, 1132–44.
26. Rohan, T., Tushar, B. & Mahesha, G.T. (2018). Review of natural fiber composites. IOP Conf Series Mater Sci Eng, 314 (1), 012020.

27. Gowda, T.Y., Sanjay, M.R., Bhat, K.S., Madhu, P., Senthamaraikannan, P. & Yogesha, B. (2018). Polymer matrix–natural fiber composites: An overview. Cogent Eng, 5 (1), 1446667.

28. Elkington, M.B., Ward, D.C., Chatzimichali, A. & Potter, K. (2015). Hand layup: Understanding the manual process. Adv. Manuf. Polymer Compos. Sci, 1, 138–151.

29. Owens, G.J., et al., (2016). Sol–gel based materials for biomedical applications. Prog. Mater. Sci, 1–79. doi:10.2147/CCIDE.S130085.

30. Alkhudhairy, F.I. (2017). The effect of curing intensity on mechanical properties of different bulk-fill composite resins. Clin. Cosmet. Invest. Dent, 1–6.

31. Klocke, F. (1997). Modern approaches for the production of ceramic components. J. Eur. Ceram. Soc, 17 (2–3), 457–465.

32. Yang, Y., Laarz, E., Kaushik, S., Mueller, E. & Sigmund, W. (2003). Forming and Drying, in Handbook of Advanced Ceramics, Elsevier Amsterdam, The Netherlands 131–183.

33. Marinescu, I.D. (Ed.). (2007). Handbook of Advanced Ceramics Machining, 1st Edition. Boca Raton, FL, United States: CRC Press.

34. Baji, A. & Mai, Y.W. (2017). Engineering Ceramic Fiber Nanostructures through Polymer-Mediated Electro-spinning. in Polymer-Engineered Nanostructures for Advanced Energy Applications, Springer: Cham, Switzerland, 3–30.

35. Esfahani, H., Jose, R. & Ramakrishna, S. (2017). Electrospun ceramic nanofiber mats today: Synthesis, properties, and applications. Materials, 10, 1238.

36. Welna, D.T., Bender, J.D., Wei, X., Sneddon, L.G. & Allcock, H.R. (2005). Preparation of boron-carbide/carbon nanofibers from a poly(norbornenyldecaborane) single-source precursor via electrostatic spinning. Adv. Mater, 17, 859–862.

37. Cowan, R.E. (1976). Slip Casting, in Treatise on Materials Science & Technology, vol. 9, pp. 153–171.

38. Edirisinghe, M.J. & Evans, J.R.G. (1986). Review: Fabrication of engineering ceramics by injection moulding. I. Materials selection. Int. J. High Technol. Ceram, 2 (1), 1–31.

39. Windsheimer, H., Travitzky, N., Hofenauer, A. & Greil, P. (2007). Laminated object manufacturing of preceramicpaper-derived Si-SiC composites. Adv. Mater, 19 (24), 4515–4519.

40. Cesarano, J., King, B.H., Denham, H.B., Cesarano, J. III & Denham, H.B. (1998). Recent Developments in Robocasting of Ceramics and Multilateral Deposition, in Proceedings of Solid Freeform Fabrication Symposium, 697–703.

41. Cawley, J.D. (1999). Solid freeform fabrication of ceramics. Curr. Opin. Solid State Mater. Sci, 4, 483–489.

42. Richerson, D.W. (2006). Modern Ceramic Engineering: Properties, Processing, and Use in Design, Third Edition. CRC Press Boca Raton, USA.

43. Yen, H.C. & Tang, H.H. (2014). Study on direct fabrication of ceramic shell mold with slurry-based ceramic laser fusion and ceramic laser sintering. Int. J. Adv. Manuf. Technol, 60 (9–12), 1009–1015.

44. Zhang, Y., He, X., Han, J. & Du, S. (1999). Ceramic green tape extrusion for laminated object manufacturing. Mater. Lett, 40 (6), 275–279.

45. Cawley, J.D., Wei, P., Liu, Z.E., Newman, W.S. & Mathewson, B.B. (1995), Al_2O^3 Ceramics Made by CAM-LEM (Computer-Aided Manufacturing Engineering Materials) Technology, in Proceedings of the Solid Freeform Fabrication Symposium, 9–16.

46. Shin, K., Koh, Y., Choi, W. & Kim, H. (2011). Production of porous poly (ε-caprolactone)/ silica hybrid membranes with patterned surface pores. Mater. Lett, 65 (12), 1903–1906.

47. Hoppe, A., Guldal, N.S. & Boccaccini, A.R. (2011). A review of the biological response to ionic dissolution products from bioactive glasses and glass-ceramics. Biomaterials, 32 (11), 2757–2774.

48. Korventausta, J., Jokinen, M., Rosling, A., Peltola, T. & Yli-Urpo, A. (2003). Calcium phosphate formation and ion dissolution rates in silica gel-PDLLA composites. Biomaterials, 24 (28), 5173–5182.

49. Chen, Q.Z., Thompson, I.D. & Boccaccini, A.R. (2006). 45S5 bioglass®-derived glass-ceramic scaffolds for bone tissue engineering. Biomaterials, 27 (11), 2414–2425.

50. Vassiliou, A.A., Bikiaris, D., Mabrouk, K.E. & Kontopoulou, M. (2021). Effect of evolved interactions in poly(butylene succinate)/fumed silica biodegradable in situ pre-pared nanocomposites on molecular weight. Material, 6, 615.

51. Veettil, S.K., Kollarigowda, R.H. & Thakur, P. (2022). Approaches to preceramic poly-mer fiber fabrication and on-demand applications. Materials, 15, 4546.

52. Alsouf, M.S., Alhazmi, M.W., Ghulman, H.A., Munshi, S.M. & Azam, S. (2016). Surface roughness and knoop indentation microhardness behavior of aluminium oxide (Al_2O_3) and polystyrene ($C_8H_8)_n$ materials. Int. J. Mech. Mechatron. Eng, 16 (6), 43–49.

53. Lee, S.W., et al., (2003). Tribological and microstructural analysis of Al_2O_3/TiO_2 nano-composites to use in the femoral head of hip replacement. Wear, 255 (7–12), 1040–1044,

54. Hu, C., Sun, J., Long, L., Wu, L., Zhou, C. & Zhang, X. (2019). Synthesis of nano zir-conium oxide and its application in dentistry. Nanotechnol Rev, 8, 396–404.

55. Feng, P., Wei, P., Li, P., Gao, C., Shuai, C. & Peng, S. (2014). Calcium silicate ceramic scaffolds toughened with hydroxyapatite whiskers for bone tissue engineering. Mater. Charact, 97, 47–56.

56. Liu, X., Chu, P.K. & Ding, C. (2004). Surface modification of titanium, titanium alloys, and related materials for biomedical applications. Mater. Sci. Eng.: R: Rep, 47 (3), 49–121.

57. Cristina, C., Alexandru, E. & Luminita, A. (2021). Synergic effect of TiO_2 filler on the mechanical properties of polymer nanocomposites. Polymers, 13, 1–24.

58. Supova, M. (2015). Substituted hydroxyapatites for biomedical applications: A review. Ceram. Int, 41 (8), 9203–9231.

59. Sadat-Shojai, M., Khorasani, M., Dinpanah-Khoshdargi, E. & Jamshidi, A. (2013). Synthesis methods for nanosized hydroxyapatite with diverse structures. Acta Biomater, 9 (8), 7591–7621.

60. Samavedi, S., Whittington, A.R. & Goldstein, A.S. (2013). Calcium phosphate ceram-ics in bone tissue engineering: A review of properties and their influence on cell behav-ior. Acta Biomater, 9 (9), 8037–8045.

61. Dorozhkin, S.V. (2012). Biphasic, triphasic and multiphasic calcium orthophosphates. Acta Biomater, 8 (3), 963–977.

62. Yubao, Z., Li, L., Aiping, Y., Xuelin, P., Xuejiang, W. & Xiang, Z. (2005). Preparation and in vitro investigation of chitosan/nano-hydroxyapatite composite used as bone sub-stitute materials. J. Mater. Sci. Mater. Med, 16 (3), 213–219.

63. Legeros, R.Z., Garrido, C.A., Lobo, S.E. & Turíbio, F.M. (2011). Biphasic calcium phos-phate bioceramics for orthopedic reconstructions: Clinical outcomes. Int. J. Biomater, 215–219. doi: 10.1155/2011/129727

64. Arinzeh, T.L., Tran, T., Mcalary, J. & Daculsi, G. (2005). A comparative study of biphasic calcium phosphate ceramics for human mesenchymal stem-cell-induced bone formation. Biomaterials, 26 (17), 3631–3638.

65. Kim, H., Koh, Y., Kong, Y., Kang, J. & Kim, H. (2004). Strontium substituted calcium phosphate biphasic ceramics obtained by a powder precipitation method. J. Mater. Sci. Mater. Med, 15 (10), 1129–1134.

66. Kim, D., Kim, T., Lee, J.D., Shin, K., Jung, J.S., Hwang, S. & Yoon, S.Y. (2013). Preparation and in vitro and in vivo performance of magnesium ion substituted biphasic calcium phosphate spherical microscaffolds as human adipose tissue-derived mesenchymal stem cell microcarriers. J. Nanomater. https://doi.org/10.1155/2013/762381

67. Shin, Y.M., Park, J., Jeong, S.I., An, S., Gwon, H., Lim, Y. & Kim, C. (2014). Promotion of human mesenchymal stem cell differentiation on bioresorbable polycaprolactone/biphasic calcium phosphate composite scaffolds for bone tissue engineering. Biotechnol. Bioprocess Eng, 19 (2), 341–349.

68. Weizhong, Y., Guangfu, Y., Dali, Z., Lijun, Y. & Linhong, C. (2007). Surface-modified biphasic calcium phosphate/poly (l-lactide) biocomposite. J. Wuhan Univ. Technol.-Mater. Sci. Ed, 24 (1), 81–86.

69. Lee, J., Lee, Y., Rim, N., Jo, S., Lim, Y. & Shin, H. (2011). Development and characterization of nanofibrous poly (lactic-co-glycolic acid)/biphasic calcium phosphate composite scaffolds for enhanced osteogenic differentiation. Macromol. Res, 19 (2), 172–179.

70. Low, N.M.P. & Beaudoin, J.J. (1994). Mechanical properties and microstructure of high alumina cement-based binders reinforced with natural wollastonite micro-fibers. Cem. Concr. Res, 24 (4), 650–660.

71. Saadaldin, S.A. & Rizkalla, A.S. (2014). Synthesis and characterization of wollastonite glass-ceramics for dental implant applications. Dent. Mater, 30 (3), 364–371.

72. Kunjalukkal, K., Padmanabhan, S., Gervaso, F., Carrozzo, M., Scalera, F., Sannino, A. & Licciulli, A. (2013). Wollastonite/hydroxyapatite scaffolds with improved mechanical, bioactive and biodegradable properties for bone tissue engineering. Ceram. Int, 39 (1), 619–627.

73. Rea, S.M., Brooks, R.A., Best, S.M., Kokubo, T. & Bonfield, W. (2004). Proliferation and differentiation of osteoblast like cells on apatite-wollastonite/polyethylene composites. Biomaterials, 25 (18), 4503–4512.

74. Chang, H. & Li, J. (2004). Fabrication and characterization of bioactive wollastonite/PHBV composite scaffolds. Biomaterials, 25 (24), 5473–5480.

75. Mehta, M., Mukhopadhyay, M., Christian, R. & Mistry, N. (2012). Synthesis and characterization of MgO nanocrystals using strong and weak bases. Powder Technol, 226, 213–221.

76. Dziadek, M., Zagrajczuk, B., Jelen, P., Olejniczak, Z. & Cholewa-Kowalska, K. (2016). Structural variations of bioactive glasses obtained by different synthesis routes. Ceram. Int, 42 (13), 14700–14709.

77. Fu, Q., Saiz, E., Rahaman, M.N. & Tomsia, A.P. (2011). Bioactive glass scaffolds for bone tissue engineering: State of the art and future perspectives. Mater. Sci. Eng. C, 31 (7), 1245–1256.

78. Kaur, G., Pandey, O.P., Singh, K., Homa, D., Scott, B. & Pickrell, G. (2014). A review of bioactive glasses: Their structure, properties, fabrication and apatite formation. J. Biomed. Mater. Res. - Part A, 102 (1), 254–274.

79. Cao, W. & Hench, L.L. (1996). Bioactive materials. Ceram. Int, 22 (6), 493–507.

80. Mami, M., Lucas-Girot, A., Oudadesse, H., Dorbez-Sridi, R., Mezahi, F. & Dietrich, E. (2008). Investigation of the surface reactivity of a sol-gel derived glass in the ternary system SiO_2-CaO-P_2O_5. Appl. Surf. Sci, 254 (22), 7386–7393.

81. Jones, J.R. (2013). Review of bioactive glass: From hench to hybrids. Acta Biomater, 9 (1), 4457–4486.

82. Sepulveda, P., Jones, J.R. & Hench, L.L. (2001). Characterization of melt-derived 45S5 and sol-gel-derived 58S bioactive glasses. J. Biomed. Mater. Res, 58 (6), 734–740.

83. Rabiee, S.M., Nazparvar, N., Azizian, M., Vashaee, D. & Tayebi, L. (2015). Effect of ion substitution on properties of bioactive glasses: A review. Ceram. Int, 41 (6), 7241–7251.

84. Pradhan, D., Panda, S. & Shukla, L.B. (2018). Recent advances in indium metallurgy: A review. Min. Proc. Ext. Met. Rev, 39, 167–180.

85. Mamunya, Y. (2011). Carbon nanotubes as conductive filler in segregated polymer composites - electrical properties. Carbon Nanotubes – Polym. Nanocompos. doi: 10.5772/18878

86. Manohar, M. & Sharma, S. (2018). A survey of the knowledge, attitude, and awareness about the principal choice of intracranial medicaments among the general dental practitioners and nonendodontic specialists. Indian J Dent. Res, 716. doi: 10.4103/ijdr.IJDR_716_16

87. Dong, C.X., Zhu, S.J., Mizuno, M. & Hashimoto, M. (2011). Fatigue behavior of HDPE composite reinforced with silane modified TiO_2. J. Mater. Sci. Technol, 27 (7), 659–667.

88. Jaehyun, W., Kwanwoo, K. & Jooheon, K. (2020). High thermal conductivity composites obtained by novel surface treatment of boron nitride. Ceram. Int, 46, 17614–17620.

89. Mazurkiewicz, S. (2010). The methods of evaluating mechanical properties of polymer matrix composites. Arch. Foundry Eng. J, 40 (3), 209–212.

90. Della Bona, A., Corazza, P.H. & Zhang, Y. (2014). Characterization of a polymer-infiltrated ceramic-network material. Dental Mater, 30 (5), 564–569.

91. Salernitano, E. & Migliaresi, C. (2003). Composite materials for biomedical applications: A review. J. Appl. Biomater. Biomech, 1, 3–18.

92. Shirazi, F.S., Mehrali, M., Oshkour, A.A., Metselaar, H.S.C., Kadri, N.A. & Abu Osman, N.A. (2014). Mechanical and physical properties of calcium silicate/alumina composite for biomedical engineering applications. J. Mech. Behav. Biomed. Mater, 30, 168–175.

16 Magnetic Lipid-Based Nanoparticles

Recent Advances in Therapeutic Applications

Raquel G. D. Andrade, Sérgio R. S. Veloso,
Elisabete M. S. Castanheira, and Ligia R. Rodrigues

16.1 INTRODUCTION

Considering the problems related to conventional drug administration, like poor solubility, nonspecific distribution, low bioavailability, rapid elimination, and intrinsic toxicity, nanomedicine has been rapidly evolving to deliver different alternative therapeutic platforms that are nowadays of countless types [1]. From inorganic nanoparticles (NPs), like magnetic NPs [2, 3], silica NPs [4, 5], and carbon nanostructures [6, 7], to the organic ones, like polymeric NPs [8], micelles [9], and liposomes [10–12], the use of individual or combined nanoparticles from different sources has been widely explored for efficient delivery systems, capable of an efficient administration, minimizing adverse side effects, as the drug can be loaded into nanocarriers and be delivered in a controlled and safe manner. Among them, magnetic NPs combined with other types of materials are commonly used for the development of multifunctional nanosystems for therapeutic applications [13–15]. They can be composed of metal (Fe, Mn, Co, Ni) oxides and ferrites (MFe_2O_4, M = Mn, Co, Ni, Zn) [16, 17]. In particular, superparamagnetic iron oxide NPs show superior magnetic properties, stability, and biocompatibility compared with other metal oxides. While magnetic NPs hold significant promise for therapeutic applications, their clinical usage faces certain limitations and challenges. One notable drawback is their tendency to aggregate and degrade when lacking a proper surface coating. To address this issue, various surface modifications have been proposed to prevent NPs' aggregation and enhance their biocompatibility. Among them, lipid-based nanoparticles (LNPs) have gained significant attention in the past years, including liposomes, solid lipid NPs (SLNs), nanostructured lipid carriers (NLCs), and nanoemulsions [18, 19]. The already known biocompatibility and biodegradability of these nanosystems, in addition to their capacity for encapsulating hydrophobic drugs, make them promising therapeutic platforms [18, 19]. In this chapter, an overview of both magnetic and lipid-based NPs is presented and the recent advances in magnetic lipid-based NPs in therapeutic applications are discussed, including drug delivery and cancer theranostics.

DOI: 10.1201/9781003454236-16

16.2   MAGNETIC NANOPARTICLES: PROPERTIES AND THERAPEUTIC APPLICATIONS

There are some exclusive characteristics that are common to all nanoparticles, namely the nanoscale size (10–100 nm), large specific surface area, and flexible surface functionalization [20]. Magnetic nanoparticles, in particular iron oxide NPs, have unique properties that arise from their transition from the bulk to the nanosize, leading to a high surface area-to-volume ratio that is responsible for the superparamagnetic behavior. Superparamagnetism is the fundamental property that allows these NPs to be applied in cellular targeting, drug delivery, magnetic hyperthermia (MH), photothermal therapy, and bioimaging [20–22]. In this regime, below a size preferably of 20 nm, nanoparticles have a fast response upon exposure to an external magnetic field or loss of magnetic moment in the absence of the magnetic field, which prevents unwanted aggregation of these nanoparticles inside the body [23]. Additionally, their strong response to the magnetic field allows a remote control that ensures their accumulation at a specific site. Despite this, naked iron oxide NPs tend to aggregate and degrade, so proper coating or surface modification is needed for enhanced physiological stability and biocompatibility [24]. Besides that, surface functionalization with targeting moieties such as small ligands, antibodies, or aptamers enables specific drug targeting through recognition of cellular receptors [25]. Notably, superparamagnetic iron oxide NPs (SPIONs) have gained significant recognition for their applications in various medical fields, which include diagnostic imaging, targeted drug delivery, cancer treatment, treatment of neurological disorders, and long-term monitoring [22, 26]. Examples of SPIONs already approved for clinical usage are Feridex® and Feraheme®, approved by the U.S. Food and Drug Administration (FDA), and Nanotherm®, approved by the European Medicines Agency (EMA) [1, 27, 28].

16.2.1   MAGNETIC NANOPARTICLES IN DRUG DELIVERY

For biological applications, nanocarriers must possess a set of suitable characteristics, including biocompatibility, facile surface functionalization, high encapsulation efficiency, long shelf-life, easy availability, and great cost-effectiveness. Besides having these characteristics, magnetic NPs further enable both passive and (magnetic) active targeting. In passive targeting, the nanoparticles can extravasate the bloodstream and infiltrate tumor cells via the enhanced permeability and retention (EPR) effect [22]. In contrast, active targeting exploits the magnetic responsiveness of nanoparticles when exposed to an external magnetic field and the functionalization with targeting ligands, such as specific antibodies or biomolecules that interact with specific receptors in target cells [29]. Additionally, the NPs can be incorporated in other nanosystems, such as polymers, hydrogels, and lipid-based systems, to further improve the pharmacokinetics, specificity, clearance, drug accumulation, and encapsulation capacity [15, 30]. Hence, the combination with other materials adds a further degree of control over drug release. For instance, magnetic nanoparticles can be functionalized with small molecules to electrostatically adsorb chemotherapeutic drugs [31], embedded in thermosensitive polymers to enable temperature-responsive release of drugs [32], conjugated with biomolecules for tumor cell targeting and drug co-delivery [33], and coated with mesoporous silica for delivery of water-insoluble drugs [34].

Hence, several magnetic NPs have been reported using different synthetic strategies and coatings to optimize their properties for drug delivery. However, attaining a suitable magnetization for efficient magnetic targeting remains a challenging task, which is essential to enable the delivery of therapeutic agents in deep tissues. Moreover, in the transition toward clinical applications, the scalability, cost considerations, and safety concerns should also be addressed.

16.2.2 Magnetic Nanoparticles in Cancer Theranostics

Theranostics consists of the integration of both imaging and therapeutic features in a single particle/agent. Within this field, magnetic nanoparticles have attracted growing research interest for the capabilities of drug delivery, magnetic hyperthermia, work as contrast agents in magnetic resonance imaging (MRI), and photothermia [35, 36]. Among the mentioned capabilities, work in the imaging technique is an important cornerstone of theranostics. In MRI, the NPs are exploited as proximity sensors that can modulate both the T_1 and the T_2 relaxation time of protons adjacent to the target-nanoparticles aggregate, thus improving the difference between normal and diseased tissues. The imaging contrast capability of ferrite nanoparticles has been applied in the detection and identification of occult lymph-node metastases [37], in the delineation of brain tumors [38], and in work as contrast agents for cardiovascular disease imaging [39]. Additionally, magnetic nanoparticles have also been studied as molecular imaging agents for the study of biological processes at the molecular and cellular levels [40, 41] or as agents for multimodal imaging systems through conjugation with other imaging agents, such as radioisotopes and fluorescent molecules [42, 43].

Magnetic hyperthermia has been widely explored as an adjuvant therapeutic strategy in theranostics. It consists of the application of a high-frequency alternating magnetic field (AMF) that leads to heat generation by the magnetic NPs through Néel-Brownian relaxation, thereby increasing the local tissue temperature (to at or above 42°C), resulting in irreversible physiological changes in tumor cells that ultimately lead to cell death [36]. Although apoptosis can only be achieved under high temperature, the higher temperature also makes cancer cells more sensitive to radiotherapy and chemotherapy. The theranostics capability of magnetic nanoparticles has been reported for different formulations and against different cancers, demonstrating the efficacy of combined drug and gene delivery against lung cancer [44], of antibody-functionalized nanoparticles for targeting of oral squamous cell carcinoma [45], and of hyperthermia therapy on osteosarcoma cells with hydroxyapatite-coated magnetic NPs [46].

Several studies have also demonstrated the suitability of magnetic NPs as photothermal agents through exposure to near infrared (NIR) laser irradiation in the first [47, 48] and second [49, 50] NIR biological windows. In this sense, magnetic NPs enable different complementary therapeutic strategies.

16.3 LIPID-BASED NANOPARTICLES

Lipid-based nanoparticles (LNPs), including liposomes, SLNs, and NLCs, are a type of nanoparticle that is primarily composed of lipids, most of them known for their amphiphilic (both hydrophobic and hydrophilic) nature, allowing them to interact with both water and lipid components. LNPs exhibit specific structural properties essential for their function and versatility in various applications and influenced by their type and composition. These NPs typically have a spherical shape with a core-shell structure. While liposomes are organized in a lipid bilayer mainly composed of phospholipids, SLNs and NLCs form micelles with a lipidic core composed of biodegradable and biocompatible lipids that are stabilized by surfactants [51, 52]. Additionally, helper lipids, like cholesterol, can be incorporated in the formulation to stabilize the lipid layer and facilitate membrane fusion [53]. Polyethylene glycol (PEG) chains can be attached to the surface of LNPs to create a PEGylated layer. This PEGylation imparts stealth properties, reducing clearance by the immune system and prolonging circulation time [53]. LNPs can be combined with other materials like polymers, inorganic nanoparticles, and proteins to produce hybrid lipid-nanoparticles displaying unique structures with enhanced properties. Recently, LNPs with incorporated magnetism have been explored as multifunctional platforms awarding the properties of both traditional LBPs and the above-mentioned properties of magnetic NPs. Magnetic LNPs have found applications in various fields, including magnetically driven drug delivery, magnetic hyperthermia, diagnostic imaging, and targeted therapies [54].

16.3.1 LIPOSOMES

Liposomes are spherical vesicles consisting of a lipid bilayer that encloses an aqueous volume. The entropically favored self-assembly is associated with the amphipathic character of lipid molecules, which also includes non-covalent interactions such as hydrogen bonding, electrostatic interactions, and van der Waals forces [55]. The size and shape can be affected by several parameters, such as the packing parameter, chemical structure, concentration, solvent, temperature, pH, and ionic strength [56, 57]. These nanosystems enable the encapsulation of both hydrophilic and hydrophobic compounds in the aqueous core and lipophilic bilayer, respectively. Thus, liposomes provide a safe means for the transport and delivery of toxic chemotherapeutic drugs [58, 59]. Regarding the fabrication methods, the liposomes are traditionally obtained by thin-film hydration, reverse phase evaporation, solvent injection, and detergent dialysis. These methods may require additional procedures, such as filtration or ultra-centrifugation to remove non-encapsulated drugs, while sonication or extrusion is employed to reduce the liposomes' size [60, 61]. The comprehensive review by Has and Sunthar is recommended for more detail on the different methodologies [62].

16.3.2 SLNs

SLNs are colloidal nanocarriers composed of a lipid matrix that is in a solid state at room and body temperatures, stabilized by a layer of surfactants (poloxamers, lecithin, Tween®, sodium lauryl sulfate). The solid lipid matrix is commonly composed of triglycerides/glycerides mixtures and even waxes [51, 52, 63]. The lipids and surfactants used in SLNs' formulations are biologically compatible, which reduces their toxicity. SLNs have gained attention as a promising alternative to traditional drug delivery systems like liposomes and emulsions. This is due to the several advantages that these nanocarriers have: (1) improved drug stability, as these nanocarriers protect encapsulated drugs from degradation; (2) enhanced bioavailability and sustained release of hydrophobic drugs; (3) a safer option than polymeric NPs, due to the absence of organic solvents in the production phase; (3) possibility of a targeted delivery, as they can be surface-modified with ligands or antibodies, facilitating targeted drug delivery to specific cells or tissues; (4) ease of preparation, which make them suitable for large-scale pharmaceutical manufacturing; and (5) versatility for a wide range of therapeutic applications, as several types of drugs can be loaded (e.g., small molecules, peptides, proteins, and nucleic acids) [18, 19, 64, 65]. Depending on the location and distribution of compounds in the SLNs, these nanosystems can show three different types of structure (Figure 16.1): in type I, (homogeneous matrix model), very lipophilic drugs are dispersed homogeneously in the lipid core; in type 2 (drug-enriched shell model), drug molecules are predominantly placed in the surfactant layer or shell of the SLN, resulting in a higher drug concentration in the shell than in the core; in type 3 (drug-enriched core model), the majority of the drug payload is located in the core region. Type 1 and type 3 can provide a controlled and prolonged drug release while type 2 is more indicated for a faster release [66]. The preparation methods influence the different structures, the most common being the high-pressure homogenization (hot and cold) but also including emulsification methods, solvent evaporation, and solvent injection. For a detailed description of the preparation methods, refer to [51] and [66]. SLNs have been investigated as drug delivery systems with many possible routes of administration such as parenteral [67], oral [68], pulmonary [69, 70], topical [71], and ocular drug delivery [72]. SLNs also play a key role in cancer treatment as nanocarriers for anti-cancer drugs such as doxorubicin [73], paclitaxel [74], and methotrexate [75], as well as delivery systems for siRNA therapeutics [76].

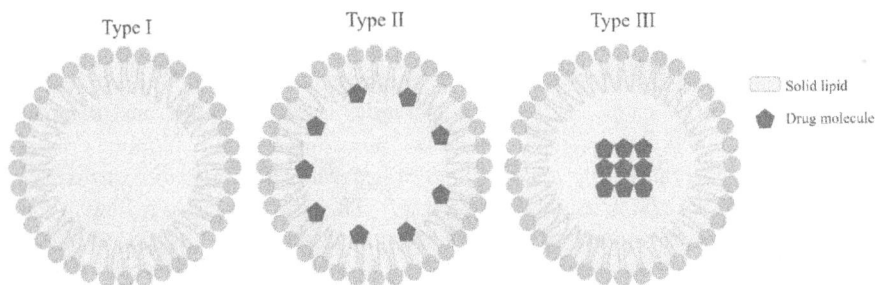

FIGURE 16.1 Schematic representation of the structures of different types of SLNs: homogeneous matrix model (type 1), drug-enriched shell model (type 2), and drug-enriched core model (type 3).

16.3.3 NLCs

NLCs are the second generation of LNPs, developed to overcome limitations associated with SLNs, namely low drug loading capacity and drug expulsion during storage. NLCs combine solid lipids with liquid lipids (fatty alcohols, medium-chain triglycerides, squalene, oleic acid) to create a more complex and disorganized lipid matrix, providing several advantages for drug encapsulation and controlled drug release [51, 63, 77]. Unlike SLNs, which have a more densely packed structure, NLCs have an amorphous and imperfect lipid matrix with spaces and imperfections in the matrix. These imperfections accommodate a higher drug payload, allowing for improved drug encapsulation, especially for poorly water-soluble drugs. Besides that, compared with SLNs, drug release during storage is lower in NLCs. This is because over time, the lipid core of SLNs becomes more densely packed due to the physicochemical transitions of lipid molecules, resulting in a more pronounced expulsion of drug molecules than in NLCs [63]. As in SLNs, there are three types of NLCs that strongly depend on the selection and ratio of the lipid mixture (Figure 16.2): type I (imperfect crystal model) contains a mixture of solid and liquid lipids with different hydrophobic chain lengths that creates a disorganized crystalline structure with voids and imperfections that improve drug incorporation; type 2 (amorphous model) consists of an amorphous lipid matrix that does not recrystallize after homogenization and cooling (hydroxyl octacosanyl, hydroxyl stearate, or isopropyl myristate), preventing drug expulsion during storage and improving its shelf-life; type 3 (multiple oil-lipid-water model) is composed of small oil compartments inside the solid lipid matrix that result from the addition of oil in a ratio that exceeds its solubility in the solid lipid; this model increases the loading capacity of compounds with higher solubility in liquid lipids than solid ones, offering a prolonged release and avoiding drug leakage [51, 63, 78]. NLCs also gained importance in several therapeutic applications and can be administered through different routes. They can be used for ocular drug delivery [79], dermal and transdermal delivery for skin disorders or in cosmetics applications [80, 81], infectious diseases [82], cardiovascular diseases [83], and gastrointestinal [84] and neurological disorders [85]. They are suitable for delivering of a wide range of pharmaceuticals including anti-cancer [86] and anti-inflammatory agents as well as antibiotics [87] and antiviral drugs [88], with improved bioavailability and controlled drug delivery [89].

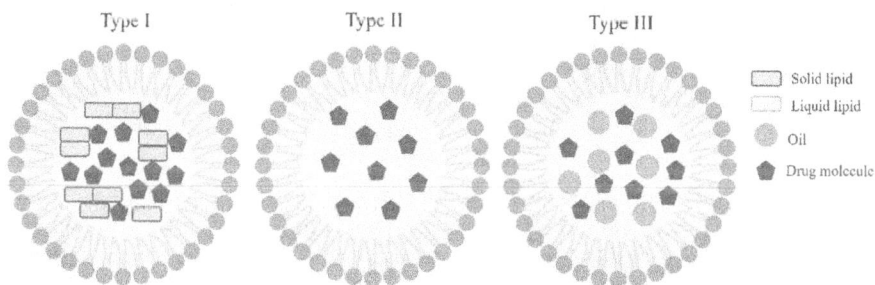

FIGURE 16.2 Schematic representation of the structures of different types of NLCs: imperfect crystal model (type I), amorphous model (type II), and multiple oil-lipid-water model (type III).

16.4 MAGNETIC LIPID-BASED NPs AND THERAPEUTIC APPLICATIONS

16.4.1 Magnetoliposomes

Despite the remarkable biological properties of conventional liposomes as delivery nanosystems, the uncontrolled and slow drug release and the passive targeting stand out as major drawbacks of their use. The need for stimuli-responsive nanocarriers and remote control of drug release led to new strategies for the development of hybrid nanoplatforms with multiple functions. Besides the introduction of pH-, redox-, light- and temperature-responsive moieties, liposomes can also respond to a magnetic stimulus when combined with magnetic nanoparticles, forming the so-called magnetoliposomes. The addition of magnetic nanoparticles provides a means for active targeting because liposomes can be remotely directed by an external magnetic field to a specific location, making possible the controlled accumulation of drugs in a target tissue [90] and performance of magnetic hyperthermia [91]. This way, magnetoliposomes can be injected in the tumor site and promote the rise of temperature up to 42–45°C, causing thermal ablation of cancer cells and higher sensitivity to drugs. Simultaneously, the use of thermosensitive liposomes enables the triggered release of the loaded drugs when the temperature is above the phase transition temperature of lipids, at which they undergo a transition from gel to liquid-crystalline state, promoting the release of the liposomal content [92].

These nanosystems can be classified into different types according to the architecture of their structure (Figure 16.3) [93]. For instance, solid magnetoliposomes (SMLs) consist of a single magnetic nanoparticle [94] or a cluster of magnetic nanoparticles immediately attached to a lipid bilayer [95, 96], while in aqueous magnetoliposomes (AMLs), the magnetic NPs are encapsulated in the aqueous core of the liposomes [97, 98]. The NPs can also be incorporated within the lipid bilayer [99] or conjugated at the surface of the outer layer of the liposome [100]. The preparation method of magnetoliposomes comprises the synthesis of magnetic nanoparticles that can be made using phospholipid vesicles as nanoreactors (in situ) or an initial synthesis with posterior combination with phospholipids (ex situ). The methods commonly used for the preparation of magnetoliposomes are similar to the ones used for liposomes and include thin-film hydration [101], ethanolic injection [98, 102], reverse-phase evaporation [97, 103], and double emulsion [104]. The magnetic NPs most used for magnetoliposome's preparation are superparamagnetic iron oxide NPs (SPIONs) due to their superparamagnetic behavior, high saturation magnetization, moderate to low toxicity, and absence of magnetization after removal of an external field [105, 106].

In fact, magnetoliposomes combine the advantages of both magnetic nanoparticles and liposomes, gathering unique physicochemical properties that enable their use in diverse biomedical applications, such as targeted drug delivery and magnetic hyperthermia [107, 108], magnetic resonance imaging [109, 110], and combined cancer therapies (chemotherapy/magnetic hyperthermia, chemo-phototherapy, chemo-photodynamic therapy, magnetic drug targeting/MRI) [111–113].

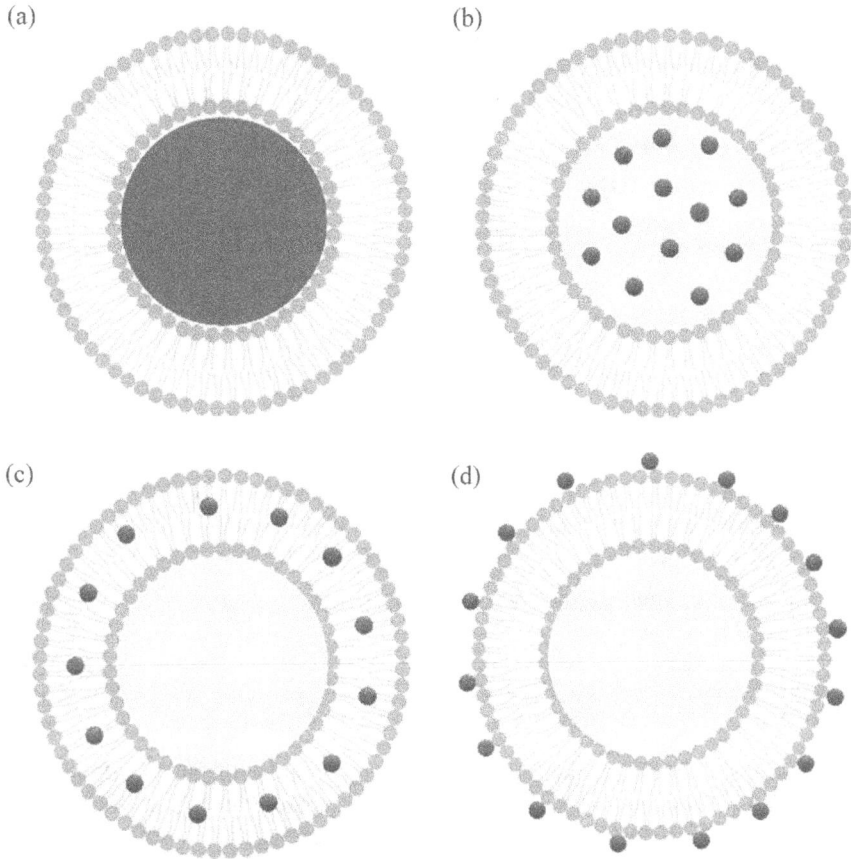

FIGURE 16.3 Schematic representation of the different types of magnetoliposomes according to their structure: (a) solid magnetoliposomes (SMLs), (b) aqueous magnetoliposomes (AMLs), (c) membrane-embedded magnetoliposomes, and (d) surface-conjugated magnetoliposomes.

These multifunctional nanoplatforms were already studied as vehicles for several anti-tumor drugs to treat different types of cancer (Table 16.1). For instance, Ribeiro et al. [114] prepared magnetoliposomes (DPPC/cholesterol 10:1) loaded with both gemcitabine and paclitaxel with encapsulation efficiencies of 57% and 68%, respectively. An improved controlled release was achieved with magnetic hyperthermia. After 72 h, gemcitabine and paclitaxel release increased from 9% and 1% to 94% and 43% after 30 min, respectively, with the application of an AMF that raised the mean temperature to 43°C in 5 minutes. This effect was also verified in human primary breast cancer cells (MGSO-3), in which a more pronounced cell viability decrease (27%) was observed for cells treated with loaded magnetoliposomes and exposed to AMF, compared to the ones treated only with loaded magnetoliposomes (60%).

TABLE 16.1

List of Cancer Therapy Strategies Using Magnetoliposomes and the Respective Therapeutic Agents and Cancer Cells Used in In Vitro and In Vivo Studies

Cancer Theranostics [119]	Therapeutic Agents	Cancer Model	Reference
Magnetic hyperthermia	siRNA-CPPs[a]	MCF-7 cells	[107]
Magnetic drug targeting	Gemcitabine Oxaliplatin	MCF-7 cells	[90]
Magnetic hyperthermia and Chemotherapy	Cisplatin	HeLa cells	[97]
Chemo-photothermal therapy and MR	Dox	MCF-7 cells and S180 cells	[103]
NIRb imaging and Magnetic hyperthermia	Paclitaxel NIR QDs	MCF-7 cells and SKOV-3 cells	[120]
Photothermal therapy and Chemotherapy	Bufalin	4T1 cells in a mouse model of lymph metastatic breast cancer	[121]

Notes: [a] siRNA-CPPs: small interfering cell-permeable peptides.
[b] NIR: near-infrared.

Thermosensitive magnetoliposomes composed of maghemite nanoparticles were proposed by Thébault et al. [115] as a theranostic nanocarrier to perform magnetic accumulation and triggered release of a vascular disrupting agent, combretastatin A4 phosphate (CA4P), by local heating using ultrasounds. This treatment, tested in CT26 murine tumors and monitored through in vivo MRI, showed a 150-fold improvement compared to the chemotherapy alone, and tumor growth inhibition was observed. Magnetoliposomes were also employed for magnetic delivery of an antigen in target lymph nodes to stimulate an immune response, demonstrating their potential use in cancer immunotherapy [116]. Despite the importance that these types of therapeutic nanoplatforms have been showing in the cancer field, it is worth noting other applications for which they have been investigated, such as for Parkinson's disease treatment [117] and as antifungal agents [118].

16.4.2 Magnetic SLNs and NLCs

Magnetic SLNs (mSLNs) and NLCs (mNLCs) represent an alternative nanocarrier that is usually formed by the incorporation of magnetic nanoparticles in the lipid matrix of SLNs and NLCs. This new type of nanoplatform combines the advantages of both entities in one nanosystem—namely the high encapsulation efficiencies of hydrophobic drugs, improved bioavailability and stability, and remote-controlled release of drugs from the lipid matrix through magnetic hyperthermia—enabling its use as an efficient multi-modal delivery nanosystem (Table 16.2). The heating to a certain range of temperatures leads to the melting of the solid lipids and consequent release of the entrapped drugs. For instance, Hsu and Su [122] developed iron oxide (γ-Fe$_2$O$_3$)–embedded SLNs, fabricated by high-pressure homogenization, that achieved a temperature rise from 37°C to 50°C in 20 min when exposed to an AMF of 60 kA/m and 25 kHz, enabling the release of about 35% of the drug molecules from the lipid nanoparticles. More recently, PEG-stabilized superparamagnetic SLNs were developed by Iacobazzi et al. [123] for magnetic targeting of sorafenib to the liver. The accumulation of the drug in the liver was explored using under-skin implanted micromagnets with two different configurations (joined and separated magnets) (Figure 16.4). The cell viability assays performed on human liver cancer cells HepG2 demonstrated an increase in the cytotoxic activity of encapsulated sorafenib as well as an enhanced accumulation of the formulations inside the cells in the presence of an external magnetic field. Results from in vivo magnetic induction measurements demonstrated that the magnetic field topography affects the magnetic targeting performance, with the separated magnets producing a higher magnetic field gradient and higher accumulation of the magnetic SLNs in the liver than the joined magnets. This study presents an alternative approach to external magnetic devices for liver magnetic delivery.

Magnetic SLNs were also investigated for cancer theranostics, combining thermo-chemotherapy and MRI, and they showed outstanding performance as T_2-weighted contrast agents [124, 125]. For example, García-Hevia et al. [125] developed a biocompatible magnetic lipid nanocomposite composed of carnauba wax and Tween 80, to carry dox and perform a synergistic combination of magnetic hyperthermia

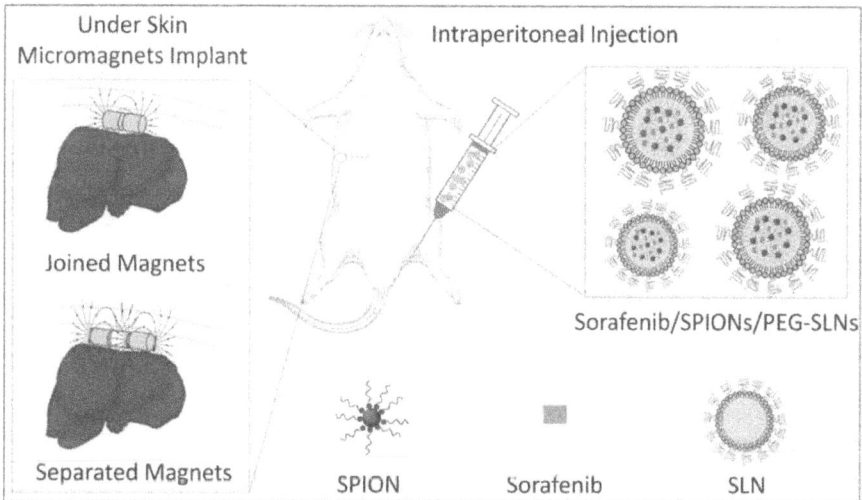

FIGURE 16.4 Schematic representation of administration of PEG-stabilized mSLNs loaded with sorafenib (sorafenib/SPIONs/PEG-SLNs) for magnetic targeting to the liver in living mice after under-skin implantation of micromagnets. (Reprinted from [126] with permission from Elsevier, 2023.)

and MRI monitoring. The nanosystems were tested both in vitro and in vivo in melanoma tumors, and the results indicated that magnetic SLNs showed enhanced cytotoxicity activity compared to the free dox. The combination of magnetic SLNs with magnetic hyperthermia for 1 h promoted a decrease of the cell viability to 40% (in vitro) and the smallest tumor growth (in vivo) compared to the dox-free and saline controls. Besides that, the researchers achieved a T_2-MRI relaxation time over 15% shorter than those of control animals injected only with saline, showing the capability of magnetic SLNs for in vivo T_2-MRI monitoring.

Like magnetic SLNs, magnetic NLCs were also employed for selective targeting and controlled release of therapeutic compounds. Rodenak-Kladniew et al. [126] designed a chitosan-coated magnetic NLC of 250 nm composed of maghemite nanoparticles and 1,8-cineole, an oily bioactive compound with pharmacological properties that also acts in this nanosystem as liquid lipid, incorporated into a matrix of myristyl myristate. The cytotoxicity of the formulation was tested in hepatocellular carcinoma (HepG2) and lung adenocarcinoma (A549) cells, as well as in normal WI-38 cells, and it showed significant viability inhibition in the cancer cells and no cytotoxicity for normal ones. NLCs co-loaded with SPIONs and methotrexate (MTX) were prepared by Ong et al. [127] to deliver this drug and perform magnetic hyperthermia in breast cancer cells. These nanosystems showed good colloidal stability, bio- and hemocompatibility, and a time-dependent cytotoxic effect against MDA-MB-231 breast cancer cells. Moreover, the authors demonstrated a localized temperature increase at cellular level that resulted in apoptotic cell death.

TABLE 16.2

List of Treatment Strategies Using Magnetic SLNs and NLCs and the Respective Target Cells Used in In Vitro and In Vivo Studies

Type of Nanocarrier + (Solid/Liquid Lipids)	Treatment	Target Cells/Organ	Reference
mSLN (carnauba wax)	Thermo-chemotherapy using the drug oncoA[a] coupled with MRI	Human lung carcinoma cell line (A549 cell line)	[128]
mSLNs (glycerol monostearate)	Thermo-chemotherapy using the drug temozolomide	Glioblastoma cancer model (U-87 cell line) and brain-endothelial cell model	[129]
mSLNs (stearic acid, tripalmitin)	Chemotherapy using the drug dox	MCF-7 breast cancer cell line	[130]
mNLC (solid lipid–stearic acid; liquid lipid–oleic acid)	Drug: ascorbyl palmitate	-	[131]
mNLC (solid lipid–poloxamer 188; liquid lipid – oleic acid)	Ligand targeting using A54 polypeptide for MRI	Hepatic cancer model (Bel-7402 cell line)	[132]

Note: [a] OncoA: oncocalyxone A.

16.5 CONCLUSIONS

Magnetic materials gained a significant role in the field of nanomedicine due to their unique properties and versatile applications. These materials, typically composed of iron oxide NPs, have magnetic properties that can be harnessed for various purposes in the diagnosis and treatment of diseases, providing a platform for targeted drug delivery, imaging, chemotherapy, and diagnostics. This chapter addressed magnetic LNPs and highlighted their importance for improved drug stability and control over drug delivery and release. The combination of lipid and magnetic nanoparticles, forming magnetoliposomes and magnetic SLNs and NLCs, enables more precise and efficient treatments while reducing the side effects associated with conventional therapies. These nanosystems demonstrated a promising in vitro and in vivo anti-tumor activity through magnetic hyperthermia and controlled drug delivery. Moreover, they merge treatment with diagnosis (theranostics) into a single platform, enabling real-time monitoring of the treatment's progress through imaging and offering a more personalized and efficient approach to healthcare. Despite the promising application in cancer theranostics, more investigation in the treatment of other diseases should be considered.

ACKNOWLEDGMENTS

Portuguese Foundation for Science and Technology (FCT) under the scope of the strategic funding of CF-UM-UP (UIDB/04650/2020), CEB (UIDB/04469/2020) and CBMA (UIDB/04050/2020). R.G.D. Andrade and S.R.S. Veloso acknowledge FCT for PhD grants 2020.05781.BD and SFRH/BD/144017/2019, respectively.

REFERENCES

1. Zhang, C.; Yan, L.; Wang, X.; Zhu, S.; Chen, C.; Gu, Z.; Zhao, Y. Progress, Challenges, and Future of Nanomedicine. *Nano Today*, 2020, *35*, 1–13. doi:10.1016/j.nantod.2020.101008.
2. Makridis, A.; Topouridou, K.; Tziomaki, M.; Sakellari, D.; Simeonidis, K.; Angelakeris, M.; Yavropoulou, M.P.; Yovos, J.G.; Kalogirou, O. In Vitro Application of Mn-Ferrite Nanoparticles as Novel Magnetic Hyperthermia Agents. *J Mater Chem B*, 2014, *2*, 8390–8398. doi:10.1039/c4tb01017e.
3. Meidanchi, A.; Motamed, A. Preparation, Characterization and in Vitro Evaluation of Magnesium Ferrite Superparamagnetic Nanoparticles as a Novel Radiosensitizer of Breast Cancer Cells. *Ceram Int*, 2020, *46*, 17577–17583. doi:10.1016/j.ceramint.2020.04.057.
4. Selvarajan, V.; Obuobi, S.; Ee, P.L.R. Silica Nanoparticles—A Versatile Tool for the Treatment of Bacterial Infections. *Front Chem*, 2020, *8*, 1–16. doi:10.3389/fchem.2020.00602.
5. Ahmed, H.; Gomte, S.S.; Prathyusha, E.; Agrawal, M.; Alexander, A. Biomedical Applications of Mesoporous Silica Nanoparticles as a Drug Delivery Carrier. *J Drug Deliv Sci Technol*, 2022, *76*, 103729. doi:10.1016/j.jddst.2022.103729.
6. Norizan, M.N.; Moklis, M.H.; Ngah Demon, S.Z.; Halim, N.A.; Samsuri, A.; Mohamad, I.S.; Knight, V.F.; Abdullah, N. Carbon Nanotubes: Functionalisation and Their Application in Chemical Sensors. *RSC Adv*, 2020, *10*, 43704–43732. doi:10.1039/d0ra09438b.

7. Yadav, D.; Amini, F.; Ehrmann, A. Recent Advances in Carbon Nanofibers and Their Applications – A Review. *Eur Polym J*, 2020, *138*, 109963. doi:10.1016/j.eurpolymj.2020.109963.

8. Kumari, A.; Yadav, S.K.; Yadav, S.C. Biodegradable Polymeric Nanoparticles Based Drug Delivery Systems. *Colloids Surf B Biointerfaces*, 2010, *75*, 1–18. doi:10.1016/j.colsurfb.2009.09.001.

9. Ghosh, B.; Biswas, S. Polymeric Micelles in Cancer Therapy: State of the Art. *J Controlled Release*, 2021, *332*, 127–147. doi:10.1016/j.jconrel.2021.02.016.

10. Jaafari, M.R.; Hatamipour, M.; Alavizadeh, S.H.; Abbasi, A.; Saberi, Z.; Rafati, S.; Taslimi, Y.; Mohammadi, A.M.; Khamesipour, A. Development of a Topical Liposomal Formulation of Amphotericin B for the Treatment of Cutaneous Leishmaniasis. *Int J Parasitol Drugs Drug Resist*, 2019, *11*, 156–165. doi:10.1016/j.ijpddr.2019.09.004.

11. Hong, S.S.; Oh, K.T.; Choi, H.G.; Lim, S.J. Liposomal Formulations for Nose-to-Brain Delivery: Recent Advances and Future Perspectives. *Pharmaceutics*, 2019, *11*, 1–18. doi:10.3390/pharmaceutics11100540.

12. Gkionis, L.; Campbell, R.A.; Aojula, H.; Harris, L.K.; Tirella, A. Manufacturing Drug Co-Loaded Liposomal Formulations Targeting Breast Cancer: Influence of Preparative Method on Liposomes Characteristics and in Vitro Toxicity. *Int J Pharm*, 2020, *590*, 119926. doi:10.1016/j.ijpharm.2020.119926.

13. Fathy, M.M.; Fahmy, H.M.; Balah, A.M.M.; Mohamed, F.F.; Elshemey, W.M. Magnetic Nanoparticles-Loaded Liposomes as a Novel Treatment Agent for Iron Deficiency Anemia: In Vivo Study. *Life Sci*, 2019, *234*, 116787. doi:10.1016/j.lfs.2019.116787.

14. Dorjsuren, B.; Chaurasiya, B.; Ye, Z.; Liu, Y.; Li, W.; Wang, C.; Shi, D.; Evans, C.E.; Webster, T.J.; Shen, Y. Cetuximab-Coated Thermo-Sensitive Liposomes Loaded With Magnetic Nanoparticles and Doxorubicin for Targeted EGFR-Expressing Breast Cancer Combined Therapy. *Int J Nanomedicine*, 2020, *15*, 8201–8215. doi:10.2147/IJN. S261671.

15. Veloso, S.R.S.; Andrade, R.G.D.; Castanheira, E.M.S. Review on the Advancements of Magnetic Gels: Towards Multifunctional Magnetic Liposome-Hydrogel Composites for Biomedical Applications. *Adv Colloid Interface Sci*, 2021, *288*. doi:10.1016/j.cis.2020.102351.

16. Shubayev, V.I.; Pisanic, T.R.; Jin, S. Magnetic Nanoparticles for Theragnostics. *Adv Drug Deliv Rev*, 2009, *61*, 467–477. doi:10.1016/J.ADDR.2009.03.007.

17. Akbarzadeh, A.; Samiei, M.; Davaran, S. Magnetic Nanoparticles: Preparation, Physical Properties, and Applications in Biomedicine. *Nanoscale Res Lett*, 2012, *7*. doi:10.1186/1556-276X-7-144.

18. Kumar, R. Lipid-Based Nanoparticles for Drug-Delivery Systems. in *Nanocarriers for Drug Delivery: Nanoscience and Nanotechnology in Drug Delivery* 2019, 249–284. doi:10.1016/B978-0-12-814033-8.00008-4.

19. García-Pinel, B.; Porras-Alcalá, C.; Ortega-Rodríguez, A.; Sarabia, F.; Prados, J.; Melguizo, C.; López-Romero, J.M. Lipid-Based Nanoparticles: Application and Recent Advances in Cancer Treatment. *Nanomaterials*, 2019, *9*, 638. doi:10.3390/NANO9040638.

20. Palanisamy, S.; Wang, Y.M. Superparamagnetic Iron Oxide Nanoparticulate System: Synthesis, Targeting, Drug Delivery and Therapy in Cancer. *Dalton Trans*, 2019, *48*, 9490–9515. doi:10.1039/c9dt00459a.

21. Kritika, N.; Roy, I. Therapeutic Applications of Magnetic Nanoparticles: Recent Advances. *Mater Adv*, 2022, *3*, 7425–7444. doi:10.1039/d2ma00444e.

22. Materón, E.M.; Miyazaki, C.M.; Carr, O.; Joshi, N.; Picciani, P.H.S.; Dalmaschio, C.J.; Davis, F.; Shimizu, F.M. Magnetic Nanoparticles in Biomedical Applications: A Review. *Appl Surf Sci Adv*, 2021, *6*. doi:10.1016/j.apsadv.2021.100163.

23. Kolhatkar, A.G.; Jamison, A.C.; Litvinov, D.; Willson, R.C.; Lee, T.R. Tuning the Magnetic Properties of Nanoparticles. *Int J Mol Sci*, 2013, *14*, 15977–16009.
24. Aslam, H.; Shukrullah, S.; Naz, M.Y.; Fatima, H.; Hussain, H.; Ullah, S.; Assiri, M.A. Current and Future Perspectives of Multifunctional Magnetic Nanoparticles Based Controlled Drug Delivery Systems. *J Drug Deliv Sci Technol*, 2022, *67*, 102946. doi:10.1016/j.jddst.2021.102946.
25. Tran, H.V.; Ngo, N.M.; Medhi, R.; Srinoi, P.; Liu, T.; Rittikulsittichai, S.; Lee, T.R. Multifunctional Iron Oxide Magnetic Nanoparticles for Biomedical Applications: A Review. *Materials*, 2022, *15*. doi:10.3390/ma15020503.
26. Mittal, A.; Roy, I.; Gandhi, S. Magnetic Nanoparticles: An Overview for Biomedical Applications. *Magnetochemistry*, 2022, *8*. doi:10.3390/magnetochemistry8090107.
27. Rodríguez, F.; Caruana, P.; De la Fuente, N.; Español, P.; Gámez, M.; Balart, J.; Llurba, E.; Rovira, R.; Ruiz, R.; Martín-Lorente, C.; et al. Nano-Based Approved Pharmaceuticals for Cancer Treatment: Present and Future Challenges. *Biomolecules*, 2022, *12*, 1–27. doi:10.3390/biom12060784.
28. Nowak-Jary, J.; Machnicka, B. Pharmacokinetics of Magnetic Iron Oxide Nanoparticles for Medical Applications. *J Nanobiotechnol*, 2022, *20*. https://doi.org/10.1186/s12951-022-01510-w
29. Zhao, S.; Yu, X.; Qian, Y.; Chen, W.; Shen, J. Multifunctional Magnetic Iron Oxide Nanoparticles: An Advanced Platform for Cancer Theranostics. *Theranostics*, 2020, *10*, 6278–6309. doi:10.7150/THNO.42564.
30. Pusta, A.; Tertis, M.; Crăciunescu, I.; Turcu, R.; Mirel, S.; Cristea, C. Recent Advances in the Development of Drug Delivery Applications of Magnetic Nanomaterials. *Pharmaceutics*, 2023, *15*, 1872. doi:10.3390/PHARMACEUTICS15071872.
31. Oh, Y.; Moorthy, M.S.; Manivasagan, P.; Bharathiraja, S.; Oh, J. Magnetic Hyperthermia and pH-Responsive Effective Drug Delivery to the Sub-Cellular Level of Human Breast Cancer Cells by Modified $CoFe_2O_4$ Nanoparticles. *Biochimie*, 2017, *133*, 7–19. doi:10.1016/J.BIOCHI.2016.11.012.
32. Najafipour, A.; Gharieh, A.; Fassihi, A.; Sadeghi-Aliabadi, H.; Mahdavian, A.R. MTX-Loaded Dual Thermoresponsive and pH-Responsive Magnetic Hydrogel Nanocomposite Particles for Combined Controlled Drug Delivery and Hyperthermia Therapy of Cancer. *Mol Pharm*, 2021, *18*, 275–284. doi:10.1021/ACS.MOLPHARMACEUT.0C00910/ASSET/IMAGES/MEDIUM/MP0C00910_0013.GIF.
33. Lerra, L.; Farfalla, A.; Sanz, B.; Cirillo, G.; Vittorio, O.; Voli, F.; Grand, M.L.; Curcio, M.; Nicoletta, F.P.; Dubrovska, A.; et al. Graphene Oxide Functional Nanohybrids with Magnetic Nanoparticles for Improved Vectorization of Doxorubicin to Neuroblastoma Cells. *Pharmaceutics*, 2019, *11*, 3. doi:10.3390/PHARMACEUTICS11010003.
34. Otalvaro, I.O.; Álvarez, T.R.; Gurovic, M.S.V.; Lassalle, V.; Agotegaray, M.; Avena, M.; Brigante, M. Magnetic Mesoporous Silica Nanoparticles for Drug Delivery Systems: Synthesis, Characterization and Application as Norfloxacin Carrier. *J Pharm Sci*, 2022, *111*, 2879–2887. doi:10.1016/J.XPHS.2022.05.024.
35. Coene, A.; Leliaert, J. Magnetic Nanoparticles in Theranostic Applications. *J Appl Phys*, 2022, *131*. doi:10.1063/5.0085202.
36. Li, X.; Li, W.; Wang, M.; Liao, Z. Magnetic Nanoparticles for Cancer Theranostics: Advances and Prospects. *J Controlled Release*, 2021, *335*, 437–448. doi:10.1016/J.JCONREL.2021.05.042.
37. Yan, Y.; Liu, Y.; Li, T.; Liang, Q.; Thakur, A.; Zhang, K.; Liu, W.; Xu, Z.; Xu, Y. Functional Roles of Magnetic Nanoparticles for the Identification of Metastatic Lymph Nodes in Cancer Patients. *J Nanobiotechnol*, 2023, *21*, 1–13. doi:10.1186/S12951-023-02100-0.

38. Israel, L.L.; Galstyan, A.; Holler, E.; Ljubimova, J.Y. Magnetic Iron Oxide Nanoparticles for Imaging, Targeting and Treatment of Primary and Metastatic Tumors of the Brain. *J Control Release*, 2020, *320*, 45–62. doi:10.1016/J.JCONREL.2020.01.009.

39. Cicha, I.; Alexiou, C. Cardiovascular Applications of Magnetic Particles. *J Magn Magn Mater*, 2021, *518*, 167428. doi:10.1016/J.JMMM.2020.167428.

40. Kim, J.; Lee, N.; Hyeon, T. Recent Development of Nanoparticles for Molecular Imaging. *Philos Trans R Soc A*, 2017, *375*. doi:10.1098/RSTA.2017.0022.

41. Kolosnjaj-Tabi, J.; Wilhelm, C.; Clément, O.; Gazeau, F. Cell Labeling with Magnetic Nanoparticles: Opportunity for Magnetic Cell Imaging and Cell Manipulation. *J Nanobiotechnology*, 2013, *11 Suppl 1*, 1–19. doi: 10.1186/1477-3155-11-S1-S7.

42. Thomas, G.; Boudon, J.; Maurizi, L.; Moreau, M.; Walker, P.; Severin, I.; Oudot, A.; Goze, C.; Poty, S.; Vrigneaud, J.M.; et al. Innovative Magnetic Nanoparticles for PET/MRI Bimodal Imaging. *ACS Omega*, 2019, *4*, 2637–2648. doi: 10.1021/acsomega.8b03283.

43. Shin, T.H.; Choi, Y.; Kim, S.; Cheon, J. Recent Advances in Magnetic Nanoparticle-Based Multi-Modal Imaging. *Chem Soc Rev*, 2015, *44*, 4501–4516. doi:10.1039/C4CS00345D.

44. Wang, C.; Ravi, S.; Garapati, U.S.; Das, M.; Howell, M.; Mallela, J.; Alwarappan, S.; Mohapatra, S.S.; Mohapatra, S. Multifunctional Chitosan Magnetic-Graphene (CMG) Nanoparticles: A Theranostic Platform for Tumor-Targeted Co-Delivery of Drugs, Genes and MRI Contrast Agents. *J Mater Chem B*, 2013, *1*, 4396–4405. doi:10.1039/C3TB20452A.

45. Legge, C.J.; Colley, H.E.; Lawson, M.A.; Rawlings, A.E. Targeted Magnetic Nanoparticle Hyperthermia for the Treatment of Oral Cancer. *J Oral Pathol Med*, 2019, *48*, 803–809. doi:10.1111/JOP.12921.

46. Mondal, S.; Manivasagan, P.; Bharathiraja, S.; Moorthy, M.S.; Nguyen, V.T.; Kim, H.H.; Nam, S.Y.; Lee, K.D.; Oh, J. Hydroxyapatite Coated Iron Oxide Nanoparticles: A Promising Nanomaterial for Magnetic Hyperthermia Cancer Treatment. *Nanomaterials*, 2017, *7*, 426. doi:10.3390/NANO7120426.

47. Shen, S.; Kong, F.; Guo, X.; Wu, L.; Shen, H.; Xie, M.; Wang, X.; Jin, Y.; Ge, Y. CMCTS Stabilized Fe3O4 Particles with Extremely Low Toxicity as Highly Efficient Near-Infrared Photothermal Agents for in Vivo Tumor Ablation. *Nanoscale*, 2013, *5*, 8056–8066. doi:10.1039/C3NR01447A.

48. Veloso, S.R.S.; Marta, E.S.; Rodrigues, P.V.; Moura, C.; Amorim, C.O.; Amaral, V.S.; Correa-Duarte, M.A.; Castanheira, E.M.S. Chitosan/Alginate Nanogels Containing Multicore Magnetic Nanoparticles for Delivery of Doxorubicin. *Pharmaceutics*, 2023, *15*, 2194. doi: 10.3390/pharmaceutics15092194.

49. Cabana, S.; Curcio, A.; Michel, A.; Wilhelm, C.; Abou-Hassan, A. Iron Oxide Mediated Photothermal Therapy in the Second Biological Window: A Comparative Study between Magnetite/Maghemite Nanospheres and Nanoflowers. *Nanomaterials*, 2020, *10*, 1548. doi:10.3390/NANO10081548.

50. Busquets, M.A.; Fernández-Pradas, J.M.; Serra, P.; Estelrich, J. Superparamagnetic Nanoparticles With Efficient Near-Infrared Photothermal Effect at the Second Biological Window. *Molecules*, 2020, *25*, 5315. doi:10.3390/MOLECULES25225315.

51. Viegas, C.; Patrício, A.B.; Prata, J.M.; Nadhman, A.; Chintamaneni, P.K.; Fonte, P. Solid Lipid Nanoparticles vs. Nanostructured Lipid Carriers: A Comparative Review. *Pharmaceutics*, 2023, *15*, 1593. doi:10.3390/PHARMACEUTICS15061593.

52. Samimi, S.; Maghsoudnia, N.; Eftekhari, R.B.; Dorkoosh, F. Lipid-Based Nanoparticles for Drug Delivery Systems. in *Characterization and Biology of Nanomaterials for Drug Delivery: Nanoscience and Nanotechnology in Drug Delivery* 2019, 47–76. doi:10.1016/B978-0-12-814031-4.00003-9.

53. Nogueira, E.; Gomes, A.C.; Preto, A.; Cavaco-Paulo, A. Design of Liposomal Formulations for Cell Targeting. *Colloids Surf B Biointerfaces*, 2015, *136*, 514–526. doi:10.1016/J.COLSURFB.2015.09.034.

54. Luiz, M.T.; Dutra, J.A.P.; Viegas, J.S.R.; de Araújo, J.T.C.; Tavares Junior, A.G.; Chorilli, M. Hybrid Magnetic Lipid-Based Nanoparticles for Cancer Therapy. *Pharmaceutics*, 2023, *15*, 751. doi:10.3390/PHARMACEUTICS15030751.

55. Hanshaw, R.G.; Stahelin, R.V.; Smith, B.D. Noncovalent Keystone Interactions Controlling Biomembrane Structure. *Chemistry*, 2008, *14*, 1690–1697. doi:10.1002/CHEM.200701589.

56. Nsairat, H.; Khater, D.; Sayed, U.; Odeh, F.; Al Bawab, A.; Alshaer, W. Liposomes: Structure, Composition, Types, and Clinical Applications. *Heliyon*, 2022, *8*, e09394. doi:10.1016/J.HELIYON.2022.E09394.

57. Lombardo, D.; Calandra, P.; Pasqua, L.; Magazù, S. Self-Assembly of Organic Nanomaterials and Biomaterials: The Bottom-Up Approach for Functional Nanostructures Formation and Advanced Applications. *Materials*, 2020, *13*, 1048. doi:10.3390/MA13051048.

58. Glassman, P.M.; Muzykantov, V.R. Pharmacokinetic and Pharmacodynamic Properties of Drug Delivery Systems. *J Pharmacol Exp Ther*, 2019, *370*, 570–580. doi:10.1124/JPET.119.257113.

59. Olusanya, T.O.B.; Ahmad, R.R.H.; Ibegbu, D.M.; Smith, J.R.; Elkordy, A.A. Liposomal Drug Delivery Systems and Anticancer Drugs. *Molecules*, 2018, *23*, 907. doi:10.3390/MOLECULES23040907.

60. Huang, Z.; Li, X.; Zhang, T.; Song, Y.; She, Z.; Li, J.; Deng, Y. Progress Involving New Techniques for Liposome Preparation. *Asian J Pharm Sci*, 2014, *9*, 176–182. doi:10.1016/J.AJPS.2014.06.001.

61. Trucillo, P.; Campardelli, R.; Reverchon, E. Liposomes: From Bangham to Supercritical Fluids. *Processes*, 2020, *8*, 1022. doi:10.3390/PR8091022.

62. Has, C.; Sunthar, P. A Comprehensive Review on Recent Preparation Techniques of Liposomes. *J Liposome Res*, 2020, *30*, 336–365. doi:10.1080/08982104.2019.1668010.

63. Viegas, C.; Seck, F.; Fonte, P. An Insight on Lipid Nanoparticles for Therapeutic Proteins Delivery. *J Drug Deliv Sci Technol*, 2022, *77*, 103839. doi:10.1016/J.JDDST.2022.103839.

64. Obeid, M.A.; Tate, R.J.; Mullen, A.B.; Ferro, V.A. Lipid-Based Nanoparticles for Cancer Treatment. in *Lipid Nanocarriers for Drug Targeting* 2018, 313–359, doi:10.1016/B978-0-12-813687-4.00008-6.

65. Naseri, N.; Valizadeh, H.; Zakeri-Milani, P. Solid Lipid Nanoparticles and Nanostructured Lipid Carriers: Structure, Preparation and Application. *Adv Pharm Bull*, 2015, *5*, 305–313. doi:10.15171/apb.2015.043.

66. Borges, A.; de Freitas, V.; Mateus, N.; Fernandes, I.; Oliveira, J. Solid Lipid Nanoparticles as Carriers of Natural Phenolic Compounds. *Antioxidants*, 2020, *9*, 998. doi:10.3390/ANTIOX9100998.

67. Galvão, J.G.; Santos, R.L.; Silva, A.R.S.T.; Santos, J.S.; Costa, A.M.B.; Chandasana, H.; Andrade-Neto, V.V.; Torres-Santos, E.C.; Lira, A.A.M.; Dolabella, S.; et al. Carvacrol Loaded Nanostructured Lipid Carriers as a Promising Parenteral Formulation for Leishmaniasis Treatment. *Eur J Pharm Sci*, 2020, *150*. doi:10.1016/J.EJPS.2020.105335.

68. Basha, S.K.; Dhandayuthabani, R.; Muzammil, M.S.; Kumari, V.S. Solid Lipid Nanoparticles for Oral Drug Delivery. *Mater Today Proc*, 2021, *36*, 313–324. doi:10.1016/J.MATPR.2020.04.109.

69. Bi, R.; Shao, W.; Wang, Q.; Zhang, N. Solid Lipid Nanoparticles as Insulin Inhalation Carriers for Enhanced Pulmonary Delivery. *J Biomed Nanotechnol*, 2009, *5*, 84–92. doi:10.1166/JBN.2009.036.

70. Hu, L.D.; Jia, Y.; Ding, W. Preparation and Characterization of Solid Lipid Nanoparticles Loaded With Epirubicin for Pulmonary Delivery. *Pharmazie*, 2010, *65*, 585–587. doi:10.1691/PH.2010.0023.

71. Rigon, R.B.; Fachinetti, N.; Severino, P.; Santana, M.H.A.; Chorilli, M. Skin Delivery and in Vitro Biological Evaluation of Trans-Resveratrol-Loaded Solid Lipid Nanoparticles for Skin Disorder Therapies. *Molecules*, 2016, *21*, 116. doi:10.3390/MOLECULES21010116.

72. Li, J.; Guo, X.; Liu, Z.; Okeke, C.I.; Li, N.; Zhao, H.; Aggrey, M.O.; Pan, W.; Wu, T. Preparation and Evaluation of Charged Solid Lipid Nanoparticles of Tetrandrine for Ocular Drug Delivery System: Pharmacokinetics, Cytotoxicity and Cellular Uptake Studies. *Drug Dev Ind Pharm*, 2014, *40*, 980–987. doi:10.3109/03639045.2013.795582.

73. Valdivia, L.; García-Hevia, L.; Bañobre-López, M.; Gallo, J.; Valiente, R.; Fanarraga, M.L. Solid Lipid Particles for Lung Metastasis Treatment. *Pharmaceutics*, 2021, *13*, 93. doi:10.3390/PHARMACEUTICS13010093.

74. Xu, W.; Bae, E.J.; Lee, M.K. Enhanced Anticancer Activity and Intracellular Uptake of Paclitaxel-Containing Solid Lipid Nanoparticles in Multidrug-Resistant Breast Cancer Cells. *Int J Nanomedicine*, 2018, *13*, 7549–7563. doi: 10.2147/IJN.S182621.

75. Muntoni, E.; Martina, K.; Marini, E.; Giorgis, M.; Lazzarato, L.; Salaroglio, I.C.; Riganti, C.; Lanotte, M.; Battaglia, L. Methotrexate-Loaded Solid Lipid Nanoparticles: Protein Functionalization to Improve Brain Biodistribution. *Pharmaceutics*, 2019, *11*, 65. doi:10.3390/PHARMACEUTICS11020065.

76. Wang, J.L.; Hanafy, M.S.; Xu, H.; Leal, J.; Zhai, Y.; Ghosh, D.; Williams, R.O.; David Charles Smyth, H.; Cui, Z. Aerosolizable SiRNA-Encapsulated Solid Lipid Nanoparticles Prepared by Thin-Film Freeze-Drying for Potential Pulmonary Delivery. *Int J Pharm*, 2021, *596*, 120215. doi:10.1016/J.IJPHARM.2021.120215.

77. Subramaniam, B.; Siddik, Z.H.; Nagoor, N.H. Optimization of Nanostructured Lipid Carriers: Understanding the Types, Designs, and Parameters in the Process of Formulations. *J Nanopart Res*, 2020, *22*, 1–29. doi: 10.1007/s11051-020-04848-0.

78. Chauhan, I.; Yasir, M.; Verma, M.; Singh, A.P. Nanostructured Lipid Carriers: A Groundbreaking Approach for Transdermal Drug Delivery. *Adv Pharm Bull*, 2020, *10*, 150. doi:10.34172/APB.2020.021.

79. Chakole, C.M.; Chauhan, M.K. Research Progress of Nanostructured Lipid Carriers in Ocular Drug Delivery. *Drug Deliv Lett*, 2021, *11*, 203–219. doi: 10.3390/ph16030474.

80. Waghule, T.; Rapalli, V.K.; Gorantla, S.; Saha, R.N.; Dubey, S.K.; Puri, A.; Singhvi, G. Nanostructured Lipid Carriers as Potential Drug Delivery Systems for Skin Disorders. *Curr Pharm Des*, 2020, *26*, 4569–4579. doi:10.2174/1381612826666200614175236.

81. Iqbal, B.; Ali, J.; Ganguli, M.; Mishra, S.; Baboota, S. Silymarin-Loaded Nanostructured Lipid Carrier Gel for the Treatment of Skin Cancer. *Nanomedicine*, 2019, *14*, 1077–1093. doi:10.2217/NNM-2018-0235.

82. Nogueira, N.C.; de Sá, L.L.F.; de Carvalho, A.L.M. Nanostructured Lipid Carriers as a Novel Strategy for Topical Antifungal Therapy. *AAPS PharmSciTech*, 2022, *23*, 1–10. doi: 10.1208/s12249-021-02181-w.

83. Alam, T.; Ansari, M.A.; Baboota, S.; Ali, J. Nanostructured Lipid Carriers of Isradipine for Effective Management of Hypertension and Isoproterenol Induced Myocardial Infarction. *Drug Deliv Transl Res*, 2022, *12*, 577–588. doi: 10.1007/s13346-021-00958-x.

84. Rouco, H.; Diaz-Rodriguez, P.; Gaspar, D.P.; Gonçalves, L.M.D.; Cuerva, M.; Remuñán-López, C.; Almeida, A.J.; Landin, M. Rifabutin-Loaded Nanostructured Lipid Carriers as a Tool in Oral Anti-Mycobacterial Treatment of Crohn's Disease. *Nanomaterials*, 2020, *10*, 2138. doi:10.3390/NANO10112138.

85. Jnaidi, R.; Almeida, A.J.; Gonçalves, L.M. Solid Lipid Nanoparticles and Nanostructured Lipid Carriers as Smart Drug Delivery Systems in the Treatment of Glioblastoma Multiforme. *Pharmaceutics*, 2020, *12*, 860. doi:10.3390/PHARMACEUTICS12090860.

86. Haider, M.; Abdin, S.M.; Kamal, L.; Orive, G. Nanostructured Lipid Carriers for Delivery of Chemotherapeutics: A Review. *Pharmaceutics*, 2020, *12*, 288. doi:10.3390/PHARMACEUTICS12030288.

87. Bazán Henostroza, M.A.; Diniz Tavares, G.; Nishitani Yukuyama, M.; De Souza, A.; José Barbosa, E.; Carlos Avino, V.; dos Santos Neto, E.; Rebello Lourenço, F.; Löbenberg, R.; Araci Bou-Chacra, N. Antibiotic-Loaded Lipid-Based Nanocarrier: A Promising Strategy to Overcome Bacterial Infection. *Int J Pharm*, 2022, *621*, 121782. doi:10.1016/J.IJPHARM.2022.121782.

88. Pindiprolu, S.K.S.S.; Kumar, C.S.P.; Kumar Golla, V.S.; Likitha, P.; K, S.C.; Esub Basha, S.K.; Ramachandra, R.K. Pulmonary Delivery of Nanostructured Lipid Carriers for Effective Repurposing of Salinomycin as an Antiviral Agent. *Med Hypotheses*, 2020, *143*, 109858. doi:10.1016/J.MEHY.2020.109858.

89. Beloqui, A.; Solinís, M.; Rodríguez-Gascón, A.; Almeida, A.J.; Préat, V. Nanostructured Lipid Carriers: Promising Drug Delivery Systems for Future Clinics. *Nanomedicine*, 2016, *12*, 143–161. doi:10.1016/J.NANO.2015.09.004.

90. Ye, H.; Tong, J.; Liu, J.; Lin, W.; Zhang, C.; Chen, K.; Zhao, J.; Zhu, W.; Ye, H.; Tong, J.; et al. Combination of Gemcitabine-Containing Magnetoliposome and Oxaliplatin-Containing Magnetoliposome in Breast Cancer Treatment: A Possible Mechanism with Potential for Clinical Application. *Oncotarget*, 2016, *7*, 43762–43778. doi:10.18632/ONCOTARGET.9671.

91. T.S, A.; Shalumon, K.T.; Chen, J.-P. Applications of Magnetic Liposomes in Cancer Therapies. *Curr Pharm Des*, 2019, *25*, 1490–1504. doi:10.2174/1389203720666190521114936.

92. Gogoi, M.; Jaiswal, M.K.; Sarma, H.D.; Bahadur, D.; Banerjee, R. Biocompatibility and Therapeutic Evaluation of Magnetic Liposomes Designed for Self-Controlled Cancer Hyperthermia and Chemotherapy. *Integr Biol*, 2017, *9*, 555–565. doi:10.1039/C6IB00234J.

93. Veloso, S.R.S.; Andrade, R.G.D.; Castanheira, E.M.S. Magnetoliposomes: Recent Advances in the Field of Controlled Drug Delivery. *Expert Opin Drug Deliv*, 2021, *18*, 1323–1334. doi:10.1080/17425247.2021.1915983.

94. Sangregorio, C.; Wiemann, J.K.; O'Connor, C.J.; Rosenzweig, Z. A New Method for the Synthesis of Magnetoliposomes. *J Appl Phys*, 1999, *85*, 5699–5701. doi:10.1063/1.370256.

95. Cardoso, B.D.; Rodrigues, A.R.O.; Bañobre-López, M.; Almeida, B.G.; Amorim, C.O.; Amaral, V.S.; Coutinho, P.J.G.; Castanheira, E.M.S. Magnetoliposomes Based on Shape Anisotropic Calcium/Magnesium Ferrite Nanoparticles as Nanocarriers for Doxorubicin. *Pharmaceutics*, 2021, *13*, 1248. doi: 10.3390/pharmaceutics13081248.

96. Rio, I.S.R.; Rodrigues, A.R.O.; Rodrigues, A.P.; Almeida, B.G.; Pires, A.; Pereira, A.M.; Araújo, J.P.; Castanheira, E.M.S.; Coutinho, P.J.G. Development of Novel Magnetoliposomes Containing Nickel Ferrite Nanoparticles Covered with Gold for Applications in Thermotherapy. *Materials (Basel)*, 2020, *13*. doi:10.3390/MA13040815.

97. Toro-Cordova, A.; Flores-Cruz, M.; Santoyo-Salazar, J.; Carrillo-Nava, E.; Jurado, R.; Figueroa-Rodriguez, P.A.; Lopez-Sanchez, P.; Medina, L.A.; Garcia-Lopez, P. Liposomes Loaded With Cisplatin and Magnetic Nanoparticles: Physicochemical Characterization, Pharmacokinetics, and In-Vitro Efficacy. *Molecules*, 2018, *23*, 2272. doi:10.3390/MOLECULES23092272.

98. Rodrigues, A.R.O.; Mendes, P.M.F.; Silva, P.M.L.; Machado, V.A.; Almeida, B.G.; Araújo, J.P.; Queiroz, M.J.R.P.; Castanheira, E.M.S.; Coutinho, P.J.G. Solid and Aqueous Magnetoliposomes as Nanocarriers for a New Potential Drug Active Against Breast Cancer. *Colloids Surf B Biointerfaces*, 2017, *158*, 460–468. doi:10.1016/J.COLSURFB.2017.07.015.

99. Salvatore, A.; Montis, C.; Berti, D.; Baglioni, P. Multifunctional Magnetoliposomes for Sequential Controlled Release. *ACS Nano*, 2016, *10*, 7749–7760. doi: 10.1021/acsnano.6b03194.

100. Floris, A.; Ardu, A.; Musinu, A.; Piccaluga, G.; Fadda, A.M.; Sinico, C.; Cannas, C. SPION@liposomes Hybrid Nanoarchitectures with High Density SPION Association. *Soft Matter*, 2011, *7*, 6239–6247. doi: 10.1186/s11671-017-2119-4.

101. Hardiansyah, A.; Yang, M.C.; Liu, T.Y.; Kuo, C.Y.; Huang, L.Y.; Chan, T.Y. Hydrophobic Drug-Loaded PEGylated Magnetic Liposomes for Drug-Controlled Release. *Nanoscale Res Lett*, 2017, *12*, 1–11. doi: 10.1186/s11671-017-2119-4.

102. Ribeiro, B.C.; Alvarez, C.A.R.; Alves, B.C.; Rodrigues, J.M.; Queiroz, M.J.R.P.; Almeida, B.G.; Pires, A.; Pereira, A.M.; Araújo, J.P.; Coutinho, P.J.G.; et al. Development of Thermo- and PH-Sensitive Liposomal Magnetic Carriers for New Potential Antitumor Thienopyridine Derivatives. *Materials*, 2022, *15*, 1737. doi:10.3390/MA15051737.

103. Shen, S.; Huang, D.; Cao, J.; Chen, Y.; Zhang, X.; Guo, S.; Ma, W.; Qi, X.; Ge, Y.; Wu, L. Magnetic Liposomes for Light-Sensitive Drug Delivery and Combined Photothermal–Chemotherapy of Tumors. *J Mater Chem B*, 2019, *7*, 1096–1106. doi:10.1039/C8TB02684J.

104. Pradhan, P.; Giri, J.; Banerjee, R.; Bellare, J.; Bahadur, D. Preparation and Characterization of Manganese Ferrite-Based Magnetic Liposomes for Hyperthermia Treatment of Cancer. *J Magn Magn Mater*, 2007, *311*, 208–215. doi:10.1016/J.JMMM.2006.10.1179.

105. Andrade, R.G.D.; Veloso, S.R.S.; Castanheira, E.M.S. Shape Anisotropic Iron Oxide-Based Magnetic Nanoparticles: Synthesis and Biomedical Applications. *Int J Mol Sci*, 2020, *21*, 2455. doi:10.3390/IJMS21072455.

106. Du, B.; Han, S.; Li, H.; Zhao, F.; Su, X.; Cao, X.; Zhang, Z. Multi-Functional Liposomes Showing Radiofrequency-Triggered Release and Magnetic Resonance Imaging for Tumor Multi-Mechanism Therapy. *Nanoscale*, 2015, *7*, 5411–5426. doi:10.1039/C4NR04257C.

107. Yang, Y.; Xie, X.; Xu, X.; Xia, X.; Wang, H.; Li, L.; Dong, W.; Ma, P.; Yang, Y.; Liu, Y.; et al. Thermal and Magnetic Dual-Responsive Liposomes with a Cell-Penetrating Peptide-SiRNA Conjugate for Enhanced and Targeted Cancer Therapy. *Colloids Surf B Biointerfaces*, 2016, *146*, 607–615. doi:10.1016/J.COLSURFB.2016.07.002.

108. Ferreira, R.V.; Martins, T.M.D.M.; Goes, A.M.; Fabris, J.D.; Cavalcante, L.C.D.; Outon, L.E.F.; Domingues, R.Z. Thermosensitive Gemcitabine-Magnetoliposomes for Combined Hyperthermia and Chemotherapy. *Nanotechnology*, 2016, *27*, 085105. doi:10.1088/0957-4484/27/8/085105.

109. Martínez-González, R.; Estelrich, J.; Busquets, M.A.; Sivakov, V.; Salifoglou, A. Liposomes Loaded With Hydrophobic Iron Oxide Nanoparticles: Suitable T2 Contrast Agents for MRI. *Int J Mol Sci*, 2016, *17*, 1209. doi:10.3390/IJMS17081209.

110. Kostevšek, N.; Cheung, C.C.L.; Serša, I.; Kreft, M.E.; Monaco, I.; Franchini, M.C.; Vidmar, J.; Al-Jamal, W.T. Magneto-Liposomes as MRI Contrast Agents: A Systematic Study of Different Liposomal Formulations. *Nanomaterials*, 2020, *10*, 889. doi:10.3390/NANO10050889.

111. Skupin-Mrugalska, P.; Sobotta, L.; Warowicka, A.; Wereszczynska, B.; Zalewski, T.; Gierlich, P.; Jarek, M.; Nowaczyk, G.; Kempka, M.; Gapinski, J.; et al. Theranostic Liposomes as a Bimodal Carrier for Magnetic Resonance Imaging Contrast Agent and Photosensitizer. *J Inorg Biochem*, 2018, *180*, 1–14. doi:10.1016/J.JINORGBIO.2017.11.025.

112. Anilkumar, T.S.; Lu, Y.J.; Chen, J.P. Optimization of the Preparation of Magnetic Liposomes for the Combined Use of Magnetic Hyperthermia and Photothermia in Dual Magneto-Photothermal Cancer Therapy. *Int J Mol Sci*, 2020, *21*, 5187. doi: 10.3390/ijms21155187.

113. An, Y.; Yang, R.; Wang, X.; Han, Y.; Jia, G.; Hu, C.; Zhang, Z.; Liu, D.; Tang, Q. Facile Assembly of Thermosensitive Liposomes for Active Targeting Imaging and Synergetic Chemo-/Magnetic Hyperthermia Therapy. *Front Bioeng Biotechnol*, 2021, *9*, 691091. doi: 10.3389/fbioe.2021.691091.

114. Ribeiro, R.F.L.; Ferreira, R.V.; Pedersoli, D.C.; Paiva, P.R.P.; Cunha, P.S.; Goes, A.M.; Domingues, R.Z. Cytotoxic Effect of Thermosensitive Magnetoliposomes Loaded With Gemcitabine and Paclitaxel on Human Primary Breast Cancer Cells (MGSO-3 Line). *J Nanopart Res*, 2020, *22*, 1–16. doi: 10.15761/ICST.1000128.

115. Thébault, C.J.; Ramniceanu, G.; Boumati, S.; Michel, A.; Seguin, J.; Larrat, B.; Mignet, N.; Ménager, C.; Doan, B.T. Theranostic MRI Liposomes for Magnetic Targeting and Ultrasound Triggered Release of the Antivascular CA4P. *J Controlled Release*, 2020, *322*, 137–148. doi:10.1016/J.JCONREL.2020.03.003.

116. Sheng, J.; Liu, Y.; Ding, H.; Wu, L.; Liu, L.; Si, G.; Shen, Y.; Yang, F.; Gu, N. Magnetic Delivery of Antigen-Loaded Magnetic Liposomes for Active Lymph Node Targeting and Enhanced Anti-Tumor Immunity. *Adv Healthc Mater*, 2023, 2301232. doi:10.1002/ADHM.202301232.

117. Cifuentes, J.; Cifuentes-Almanza, S.; Ruiz Puentes, P.; Quezada, V.; González Barrios, A.F.; Calderón-Peláez, M.A.; Velandia-Romero, M.L.; Rafat, M.; Muñoz-Camargo, C.; Albarracín, S.L.; et al. Multifunctional Magnetoliposomes as Drug Delivery Vehicles for the Potential Treatment of Parkinson's Disease. *Front Bioeng Biotechnol*, 2023, *11*, 1181842. doi: 10.3389/fbioe.2023.1181842.

118. Pereira, M.; Rodrigues, A.R.O.; Amaral, L.; Côrte-Real, M.; Santos-Pereira, C.; Castanheira, E.M.S. Bovine Lactoferrin-Loaded Plasmonic Magnetoliposomes for Antifungal Therapeutic Applications. *Pharmaceutics*, 2023, *15*, 2162. doi:10.3390/PHARMACEUTICS15082162.

119. Lu, Y.J.; Chuang, E.Y.; Cheng, Y.H.; Anilkumar, T.S.; Chen, H.A.; Chen, J.P. Thermosensitive Magnetic Liposomes for Alternating Magnetic Field-Inducible Drug Delivery in Dual Targeted Brain Tumor Chemotherapy. *Chem Eng J*, 2019, *373*, 720–733. doi:10.1016/J.CEJ.2019.05.055.

120. Deng, Y.; Huang, H.; Chen, M.; Chen, G.; Zou, W.; Zhao, Y.; Zhao, Q. Comprehensive Effects of Near-Infrared Multifunctional Liposomes on Cancer Cells. *Molecules*, 2020, *25*, 1098. doi:10.3390/MOLECULES25051098.

121. Hu, W.; Qi, Q.; Hu, H.; Wang, C.; Zhang, Q.; Zhang, Z.; Zhao, Y.; Yu, X.; Guo, M.; Du, S.; et al. Fe_3O_4 Liposome for Photothermal/Chemo-Synergistic Inhibition of Metastatic Breast Tumor. *Colloids Surf A Physicochem Eng Asp*, 2022, *634*, 127921. doi:10.1016/J.COLSURFA.2021.127921.

122. Hsu, M.H.; Su, Y.C. Iron-Oxide Embedded Solid Lipid Nanoparticles for Magnetically Controlled Heating and Drug Delivery. *Biomed Microdevices*, 2008, *10*, 785–793. doi: 10.1007/s10544-008-9192-5.

123. Iacobazzi, R.M.; Vischio, F.; Arduino, I.; Canepa, F.; Laquintana, V.; Notarnicola, M.; Scavo, M.P.; Bianco, G.; Fanizza, E.; Lopedota, A.A.; et al. Magnetic Implants in Vivo Guiding Sorafenib Liver Delivery by Superparamagnetic Solid Lipid Nanoparticles. *J Colloid Interface Sci*, 2022, *608*, 239–254, doi:10.1016/J.JCIS.2021.09.174.

124. Vieira Rocha, C.; Costa Da Silva, M.; Banõbre-López, M.; Gallo, J. (Para)Magnetic Hybrid Nanocomposites for Dual MRI Detection and Treatment of Solid Tumours. *Chem Commun*, 2020, *56*, 8695–8698. doi:10.1039/D0CC03020A.

125. García-Hevia, L.; Casafont, Í; Oliveira, J.; Terán, N.; Fanarraga, M.L.; Gallo, J.; Bañobre-López, M. Magnetic Lipid Nanovehicles Synergize the Controlled Thermal Release of Chemotherapeutics with Magnetic Ablation While Enabling Non-Invasive Monitoring by MRI for Melanoma Theranostics. *Bioact Mater*, 2022, *8*, 153–164. doi:10.1016/J.BIOACTMAT.2021.06.009.

126. Rodenak-Kladniew, B.; Noacco, N.; Pérez de Berti, I.; Stewart, S.J.; Cabrera, A.F.; Alvarez, V.A.; García de Bravo, M.; Durán, N.; Castro, G.R.; Islan, G.A. Design of Magnetic Hybrid Nanostructured Lipid Carriers Containing 1,8-Cineole as Delivery Systems for Anticancer Drugs: Physicochemical and Cytotoxic Studies. *Colloids Surf B Biointerfaces*, 2021, *202*, 111710. doi:10.1016/J.COLSURFB.2021.111710.

127. Ong, Y.S.; Bañobre-López, M.; Costa Lima, S.A.; Reis, S. A Multifunctional Nanomedicine Platform for Co-Delivery of Methotrexate and Mild Hyperthermia towards Breast Cancer Therapy. *Mater Sci Eng: C*, 2020, *116*, 111255. doi:10.1016/ J.MSEC.2020.111255.

128. de Moura, C.L.; Gallo, J.; García-Hevia, L.; Pessoa, O.D.L.; Ricardo, N.M.P.S.; Bañobre-López, M. Magnetic Hybrid Wax Nanocomposites as Externally Controlled Theranostic Vehicles: High MRI Enhancement and Synergistic Magnetically Assisted Thermo/Chemo Therapy. *Chemistry*, 2020, *26*, 4531–4538. doi:10.1002/ CHEM.201904709.

129. Tapeinos, C.; Marino, A.; Battaglini, M.; Migliorin, S.; Brescia, R.; Scarpellini, A.; De Julián Fernández, C.; Prato, M.; Drago, F.; Ciofani, G. Stimuli-Responsive Lipid-Based Magnetic Nanovectors Increase Apoptosis in Glioblastoma Cells through Synergic Intracellular Hyperthermia and Chemotherapy. *Nanoscale*, 2018, *11*, 72–88. doi:10.1039/C8NR05520C.

130. Soltani, A.; Pakravan, P. Preparation and Characterization of Magnetic Solid Lipid Nanoparticles as a Targeted Drug Delivery System for Doxorubicin. *Adv Pharm Bull*, 2023, *13*, 301. doi:10.34172/APB.2023.033.

131. Kalaycioglu, G.D. Preparation of Magnetic Nanoparticle Integrated Nanostructured Lipid Carriers for Controlled Delivery of Ascorbyl Palmitate. *MethodsX*, 2020, *7*, 101147. doi:10.1016/J.MEX.2020.101147.

132. Lu, C.Y.; Ji, J.S.; Zhu, X.L.; Tang, P.F.; Zhang, Q.; Zhang, N.N.; Wang, Z.H.; Wang, X.J.; Chen, W.Q.; Hu, J.B.; et al. T2-Weighted Magnetic Resonance Imaging of Hepatic Tumor Guided by SPIO-Loaded Nanostructured Lipid Carriers and Ferritin Reporter Genes. *ACS Appl Mater Interfaces*, 2017, *9*, 35548–35561. doi: 10.1021/acsami.7b09879.

Index

A

abolition, 254
abound, 49
ABRICATION, 6
ABS, 228
absence, 16, 23–24, 42, 208, 211, 227, 250, 253, 315, 318, 320
absorbance, 90–92, 94
absorbent, 165
absorber, 55
absorbers, 91
absorption, 1, 6, 11, 15, 48, 90–92, 94–95, 117, 195, 197, 244, 247, 265–266
absorptivity, 92
abundant, 39, 165, 169, 245
acacia, 143–144
acceptor, 124, 175
accessories, 97
accordance, 208, 214
accordions, 17
accumulate, 243, 261
accuracy, 2, 93, 107, 147, 192, 266, 270, 274
accurately, 6, 57, 118–119, 130, 148, 150, 249
acetaminophen, 13
acetate, 16, 162
acetyl, 211
acetylacetonate, 113
achievement, 232
acidified, 253
acidifying, 189, 253
acohol, 161
Acquisition, 95, 97
acquisition, 95, 97
Acrylamide, 144
acrylamide, 144
acrylic, 169, 171, 211, 242
Actuation, 249
Acute, 151
adaptability, 1, 7, 86, 212–213, 216, 222, 233
adaptable, 6–7, 91, 140, 205, 221, 227
adaptive, 3, 214, 230, 250
additive, 76, 227–229, 232–233, 249
additives, 3, 60, 240, 243
adeno, 115
adenocarcinoma, 274, 324
adherence, 11
adipocytes, 146
adjustability, 250
adjuvant, 316
administered, 150, 319

administering, 117, 271
adoptive, 118
adsorb, 2, 165–166, 168, 246, 278, 315
adsorbed, 162, 165, 211
adsorbents, 160, 163, 166, 169, 175, 179
adsorption, 14, 16, 27, 47, 113, 132, 160–171, 173–179, 254, 260, 266
advancement, 4–6, 14, 18, 39, 46, 210, 228, 230, 232, 258–259, 271
advantage, 2, 5, 31, 51, 57, 111, 116, 118, 131, 153, 160, 268
advantageously, 147
Advantages, 8, 70, 72–73, 253, 261
aerogel, 168
aeronautics, 228
aerosol, 114
aesthetic, 167, 243
afflicted, 267
AFM, 61, 89–90, 134
aforementioned, 10, 86, 213, 229, 234, 258–259, 266, 269
Ag, 176
agar, 131
agarose, 15, 147, 155
agglomerate, 56, 81, 264
aggregation, 5, 9, 29, 107, 110, 162–163, 194, 261–262, 264–265, 314–315
agitated, 146
agnatically, 247
Agriculture, 155, 240–241, 243, 245, 247, 249, 251, 253
agriculture, 48, 242
ailment, 268
aircraft, 2–3, 244–245
airplanes, 250
alanine, 30
Albumin, 109
albumin, 109, 259, 272, 274
alcohol, 13, 109, 112–113, 128, 139, 154, 163, 168, 192, 230, 264
alcohols, 91, 319
aldehyde, 109
aldehydes, 91–92
Alg, 149
Alginate, 149
alginate, 141, 149, 151, 154–155, 164, 175–176, 178, 199, 221, 230, 265
Alginates, 265
algorithm, 94–95
Ali, 139, 154
Aliabadi, 155

For Product Safety Concerns and Information please contact our EU
representative GPSR@taylorandfrancis.com
Taylor & Francis Verlag GmbH, Kaufingerstraße 24, 80331 München, Germany

www.ingramcontent.com/pod-product-compliance
Lightning Source LLC
Chambersburg PA
CBHW060758220326
41598CB00022B/2478